Continuum Models and Discrete Systems Volume 2

Continuum Models and Discrete Systems Volume 2

Edited by

G A Maugin
Université Pierre et
Marie Curie, Paris

Longman Scientific & Technical

Copublished in the United States with
John Wiley & Sons, Inc., New York

Longman Scientific & Technical,
Longman Group UK Ltd,
Longman House, Burnt Mill, Harlow,
Essex CM20 2JE, England
and Associated Companies throughout the world.

Copublished in the United States with
John Wiley & Sons, Inc., 605 Third Avenue, New York, NY 10158

First published 1991

British Library Cataloguing in Publication Data
Continuum models and discrete systems.
Vol. 2
1. Continuous media. Mechanics. Partial differential equations
I. Maugin, Gerard A, 1944- II. Series
531

ISBN 0-582-05591-1

Library of Congress Cataloguing-in-Publication Data
(Revised for vol. 2)
Continuum models and discrete systems.
(Interaction of mechanics and mathematics series)
1. Continuum mechanics. I. Maugin, G. A. (Gérard A.), 1944– II. Series.
QA808.C67 1990 531 89–13143
ISBN 0-470-21592-5 (v. 1: Wiley)
ISBN 0-470-21593-3 (v. 2: Wiley)

Printed and bound in Great Britain at the Bath Press, Avon

Foreword

The International Society for the Interaction of Mechanics and Mathematics (ISIMM) was founded in 1977. Its purpose is to promote cooperative research involving the fields of mechanics and pure mathematics.

Its Executive Committee decided that, from time to time, scholarly works relevant to the Society's interests should, by invitation, by published under its auspices. The present volume is one in this series which, it is hoped, will help to advance the objective of the Society.

The Editorial Board

Contents

CONTENTS

PART 2 DEFORMABLE SOLIDS WITH MICROSTRUCTURE

CONTENTS

CONTENTS

Editor's preface

This volume of *Continuum Models and Discrete Systems* contains the texts of most of the short contributions presented at the Sixth International Symposium on Continuum Models and Discrete Systems (CMDS6) which was held on the campus of the University of Bourgogne, Dijon, France, from June 25 to June 29, 1989, and gathered about a hundred registered participants.

As with earlier symposia in this series, which began in Jodlowy Dwor, Poland, in June 1975, the purpose of the Symposium was to bring together internationally renowned researchers concerned with various aspects of continuum modelling of discrete systems, to discuss new topics of common interest, and to disseminate information about recent developments. During the four days of the Symposium twenty-two invited lectures and forty-three contributions were presented.

Four broad themes had been selected for the Symposium: (i) complex fluids, (ii) deformable solids with microstructure, (iii) microstructure, thermodynamics and geometry, and (iv) non-linear excitations and coherent structures. This division is directly reflected in the structure of the present volume. It steered the whole symposium, but it also covers many of the cases where the microstructure related to the discrete system has a prevailing influence and manifests itself in a sensible manner on the macroscale. Clearly, semi-dilute and concentrated fluid suspensions, granular structures, deformable solids which are perfect or defective at the crystalline or poly-crystalline level, composites of various types presenting an ordered or random structure, etc., enter directly into the subject matter of the Symposium. The wealth of effects thus covered in material science and modern phenomenological physics is extremely large. Simultaneously, this interest has fostered the development of new appropriate techniques in applied mathematics such as cell automata theory, asymptotic and stochastic homogenization techniques, techniques in non-linear dispersive wave theory, thermodynamically admissible geometrical approaches (i.e., Hamiltonian structures, gauge theories, Lie pseudo-groups), coherent structures due to scale effects and the combination or competition of these with other effects yielding, for instance, the propagation of non-linear waves of permanent form and the striking time evolution of some patterns. The contributions collected in this volume were selected by the Scientific Committee in order to illustrate recent developments in a most representative and balanced manner.

The Scientific Committee of the Symposium consisted of

M. Beran	Israel
D. Brulin (past Chairman)	Sweden

EDITOR'S PREFACE

E.G.D. Cohen	U.S.A.
G. Duvaut	France
G. Herrmann	U.S.A.
E. Kröner (past Chairman)	F.R.G.
I.A. Kunin	U.S.A.
G.A. Maugin (Chairman)	France
A.J.M. Spencer (past Chairman)	U.K.
K. Markov	Bulgaria
H. Zorski (past Chairman)	Poland

The local organization was handled with great efficiency and unselfish devotion by Michel Peyrard and Michel Remoissenet, both Professors of Physics at the University of Bourgogne. It is a pleasure to record my thanks to them and also to the members of the Scientific Committee for their advice and encouragement; to the President of the University of Bourgogne, Professor Bertrand, for hosting the Symposium; to our sponsors (Centre National de la Recherche Scientifique, Département des Sciences Physiques pour l'Ingénieur, Paris; the French Ministry of Defense through the Direction des Recherches, Etudes et Techniques, Paris: Electricité de France through its Direction des Etudes et Recherches, Clamart; Institut Français du Pétrole, Rueil-Malmaison; the Office of Naval Research of the U.S. Navy, and the European Research Office of the U.S. Army), to all Chairmen of sessions, to Longman Scientific and Technical for the production of these proceedings and, last but not least, to all participants and to contributors for the preparation of their manuscripts.

The Symposium CMDS6 was placed under the auspices of the *International Society for the Interaction of Mechanics and Mathematics*. These proceedings are published in the ISIMM series. Both the Symposium CMDS6 and its proceedings are dedicated to Professor Ekkehart Kröner, Stuttgart, F.R.G., acting President of ISIMM, one of the founding fathers of the series of CMDS Symposia and a pioneer in homogenization of random media and in the geometrical theory of crystalline defects. Without Ekkehart, we would *not* have met in Dijon in June 1989.

G.A. Maugin
September 1989

List of contributors

Al Assa'ad A.A.	Groupe Phénomènes d'Interface, Ecole Nationale Supérieure des Techniques Avancées, Centre de l'Yvette, Chemin de la Hunière, F-91120 Palaiseau Cédex, France
Anthony K.-H.	Fachbreich 6/Naturwissenschaft. I Physik, Universität Gesanthochschule, Warburgstrasse 100/Post 1621, D-4790 Paderborn, F.R.G.
Askar A.	Department of Mathematics, Bogazici Universitesi, Bebek, Istanbul, Turkey
Banach Z.	I.P.P.T.-P.A.N., Swietokrzyska 21, PL-00-049 Warsaw, Poland
Berveiller M.	L.P.M.M., Faculté des Sciences, Université de Metz, Ile du Saulcy, F-57045 Metz Cédex, France
Bishop A.R.	Theoretical Division and Center for Nonlienr Studies, Los Alamos National Laboratory, Los Alamos, New Mexico 87545, U.S.A.
Bonomi E.	G.A.S.O.V. Group, Ecole Polytechnique Fédérale, CH-1015 Lausanne, Switzerland
Braun O.	Institute of Physics, Ukrainian Acad. of Science, Prospect Nauki 46, 252028 Kiev, U.S.S.R.
Bretheau Th.	L.P.M.T.M., Université de Paris-Nord, Avenue J.B. Clément, F-93430 Villetaneuse, France
Brieger L.M.	Department of Mathematics, Ecole Polytechnique Fédérale, CH-1015 Lausanne, Switzerland
Canova G.	L.P.M.M., Faculté des Sciences, Université de Metz, Ile du Saulcy, F-57045 Metz Cédex, France

LIST OF CONTRIBUTORS

Darrozes J.-S.	Groupe Phénomènes d'Interface, Ecole Nationale Supérieure des Techniques Avancées, Centre de l'Yvette, Chemin de la Hunière, F-91120 Palaiseau, France
Dauxois T.	Physique Non Linéaire: Ondes et Structures Cohérentes, Faculté des Sciences, 6 Boulevard Gabriel, F-21000 Dijon, France
Davini C.	Istituto di meccanica teorica ed applicata, UniversitàUdine, Viale Ungheria, I-33100, Udine, Italy
Dounaev I.M.	Department of Strength of Materials, Polytechnic Institute of Krasnodar, Moskovskaya 2, 350042 Krasnodar, U.S.S.R.
Doremus P.	Institut de Mécanique de Grenoble, B.P. 53X, F-38041 Grenoble Cédex, France
Ellinghaus R.	Institut für Theoretische Physik, Technische Universität Berlin, Hardenbergstrasse 36, D-1000 Berlin 12, F.R.G.
Favier D.	L.G.P.M.M., Ecole National Supérieure de Physique de Grenoble, I.N.P.G., B.P. 46, F-38402 Saint-Martin d'Hères Cédex, France
Fähnle M.	Institut für Physik, Max Planck Institute für Metallforschung, Heisenbergstrasse 1, D-7000 Stuttgart 80, F.R.G.
Furthmuller J.	Institut für Physik, Max Planck Institute für Metallforschung, Heisenbergstrasse 1, D-7000 Stuttgart 80, F.R.G.
Gouin H.	Département de Mathématiques, Faculté des Sciences et Techniques, Av. Escadrille Normandie-Niemen, F-13397 Marseille Cédex 13, France
Guelin P.	Institut de Mécanique de Grenoble, B.P. 53X, F-38041 Grenoble Cédex, France

LIST OF CONTRIBUTORS

Hervé E.	L.P.M.T.M., Université de Paris Nord, Avenue J.B. Clément, F-93430 Villetaneuse, France
Huet C.	Ecole Polytechnique Fédérale, Département des Matériaux, Chemin de Bellerive 32, CH-1015 Lausanne, Switzerland
Kivshar Yu.	Institute of Physics, Ukrainian Acad. of Science, Prospect Nauki 46, 252028 Kiev, U.S.S.R.
Kotowski R.	I.P.P.T.-P.A.N., Swietokrzyska 21, PL-00-049 Warsaw, Poland
Kubin L.P.	Laboratoire d'Etude des Microstructures CNRS / ONERA, B.P. 72, F-92322 Châtillon Cédex, France
Leblond J.-B.	Laboratoire de Modélisation en Mécanique, Université Pierre-et-Marie Curie, Tour 66, 4 place Jussieu, F-75252 Paris Cédex 05, France
Lhuillier D.	Laboratoire de Modélisation en Mécanique, Université Pierre-et-Marie Curie, Tour 66, 4 place Jussieu, F-75252 Paris Cédex, France
Limat L.	L.H.M.P., Ecole Supérieure de Physique et de Chimie Industrielles, 10 rue Vauquelin, F-75231 Paris Cédex 05, France
Litewka A	Technical University of Poznan, Institute of Building Structures, ul. Piotrowo 5, PL-60-965 Poznan, Poland
Louchet F.	Laboratoire de Thermodynamique et de Physico-Chimie Métallurgique, I.N.P.G., B.P. 75, F-38402 Saint-Martin d'Hères Cédex, France
Maradudin A.A.	Department of Physics and Institute for Surface and Interface Science, University of California, Irvine, CA 92717, U.S.A.
Markov K.	University of Sofia, Faculty of Mathematics and Informatics, 5 Boulevard A. Ivanov, 1126 Sofia, Bulgaria

LIST OF CONTRIBUTORS

Mayer A.	Department of Physics, University of California, Irvine, CA 92717, U.S.A.
Mazilu P.	Institut für Umformteknik, TH-Darmstadt, Petersen Strasse 30, D-6100 Darmstadt, F.R.G.
Melehy M.A.	Elect. and Syst. Eng. Dept. - U217, University of Connecticut, 260 Glenbrook Road, Storrs CT 06269-3157, U.S.A.
Mićunović M	Faculty of Mechanical Engineering, University Svetozar Markovic, Sestre Janjic 6, YU-34000 Kragujevac, Yugoslavia
Morro A.	Facoltà di Engegneria, Università di Genova, Viale Causa 13, I-16145 Genova, Italy
Murdoch A.I.	Department of Mathematics, University of Strathclyde, Livingstone Tower, 26 Richmond Street, Glasgow G1 1XH, U.K.
Muschik W.	Institut für Theoretische Physik, Technische Universität Berlin, Hardenbergstrasse 36, D-1000 Berlin 12, F.R.G.
Navi P.	Ecole Nationale des Ponts et Chaussées, CERAM, 1 avenue Montaigne, F-93167 Noisy-le-Grand Cédex, France
Nowacki J.P.	I.P.P.T.-P.A.N., Swietokryska 21, PL-00-049 Warsaw, Poland
Ostoja-Starzewski M	School of Aeronautics and Astronautics, Purdue University, West Lafayette, Indiana 47907, U.S.A.
Papenfuss C	Institut für Theoretische Physik, Technische Universität Berlin, Hardenbergstrasse 36, D-1000 Berlin 12, F.R.G.
Parker D.F.	Department of Theoretical Mechanics, The University of Nottingham, University Park, Nottingham NG7 2RD, U.K.

LIST OF CONTRIBUTORS

Parry G.	School of Mathematics, University of Bath, Claverton Down, Bath BA2 7AY, U.K.
Parton V.Z.	Moscow Institute of Chemical Machines, Department of Mathematics, Karl Marx Street 21/4, Moscow B-99 107884, U.S.S.R.
Pawellek R.	Institut für Physik, Max Planck Institute für Metallforschung, Heisenbergstrasse 1, D-7000 Stuttgart 80, F.R.G.
Pegon P	Joint Research Center C.C.E., I-21020 Ispra (Varese), Italy
Perrin G.	Laboratoire de Modélisation en Mécanique, Université Pierre-et-Marie Curie, Tour 66, 4 place Jussieu, F-75252 Paris Cédex 05, France
Peyrard M.	Laboratoire O.R.C. EFR.M.I.P.C.-Faculté des Sciences, Université de Bourgogne, 6 Boulevard Gabriel, F-21100 Dijon, France
Piau J.-M.	Institut de Mécanique de Grenoble, B.P. 53X, F-38041 Grenoble Cédex, France
Planat M.	L.P.M.O.-C.N.R.S., 32 avenue de l'Observatoire, F-2500 Besançon, France
Roelfstra P.	Ecole Polytechnique Fédérale, Département des Matériaux, Chemin de Bellerive 32, CH-1015 Lausanne, Switzerland
Sabar H.	L.P.M.M. Fac. des Sciences, Université de Metz, Ile du Saulcy, F-57045 Metz Cédex, France
Straughan B	Department of Mathematics, University of Glasgow, Glasgow G12 8QW, U.K.
Telega J.J.	I.P.P.T.-P.A.N., Swietokryska 21, PL-00-049 Warsaw, Poland
Tourabi A.	Institut de Mécanique de Grenoble, B.P. 53X, F-38041 Grenoble Cédex, France

LIST OF CONTRIBUTORS

Trzesowski A.	I.P.P.T.-P.A.N., Swietokrzyska 21, PL-00-049 Warsaw, Poland
Wack B.	Institut de Mécanique de Grenoble, B.P. 53X, F-38041 Grenoble Cédex, France
Zaoui A.	L.P.M.T.M., Université de Paris-Nord, Avenue J.B. Clément, F-93430 Villetaneuse, France.
Zelenskaya I.	Institute of Physics, Ukrainian Acad. of Science, Prospect Nauki 46, 252028 Kiev, U.S.S.R.
Zorski H.	I.P.P.T.-P.A.N., Department of Fluid Mechanics, Swietokrzyska 21, PL-00-049 Warsaw, Poland

PART 1 : COMPLEX FLUIDS

Z. BANACH

On the analogue of Grad's moment procedure for a one-dimensional gas of quasi-particles

1 Introduction

In the moment method of Grad's type [10, 11] applied to the Boltzmann equation, one considers only such molecular densities f as can be expanded in a series of three-dimensional Hermite polynomials [12]. The expansion coefficients, which are nothing else than the Hermite moments of f, should be considered to be independent gas-state variables satisfying the differential equations of transfer of first order both in the time t and in the position x. Of course, for general problems and collision integrals the practical advantage of the method is based on a somewhat *ad hoc* truncation procedure.

Given the model Boltzmann-Peierls equation [13], the primary object of this note is to extend the range of validity of Grad's moment procedure to the case of a one-dimensional gas composed of quasi-particles, more to indicate the universal character of the problems involved [1-6] than aiming at any measure of completeness [7, 8].

2 The model Boltzmann-Peierls equation

For the present purpose, the "non-equilibrium occupation probability" f, aside from its dependence upon the wave number k, $k \in \mathbb{R} := (-\infty, +\infty)$, is a function of the place x and the time t satisfying the Boltzmann-Peierls equation of the form [13]

$$\partial_t f + \nabla_k \Omega \circ \nabla_x f - \nabla_x \Omega \circ \nabla_k f = J(f), \tag{2.1}$$

where, as usual, J stands for the collision operator. Here $\hbar\, \Omega(k, x, t)$ represents the energy of a single quasi-particle in the mode k, where $2\pi\hbar$ denotes Planck's constant.

Hereinafter we simply set $\Omega(k, x, t) = \frac{1}{2} ck^2$, c being a "constitutive" constant (stated once and for all) independent of k, x, or t, and in addition we suppose that $J(f)$ is given by

$$J(f) = -\nu(f - f_0), \quad \nu > 0, \tag{2.2a}$$

where

$$f_0(\omega) := (e^{\omega} - 1)^{-1}, \quad \omega := \frac{\hbar\,\Omega}{k_B T} = \frac{1}{2} z^2 , \tag{2.2b}$$

$$z := \Theta\, k, \quad \Theta := (\hbar c / k_B T)^{1/2}, \tag{2.2c}$$

$$\frac{\hbar}{2\pi}\int dk\, \Omega f_0 = e := \frac{\hbar}{2\pi}\int dk\, \Omega f , \tag{2.2d}$$

$$\nu := \hat{\nu}(\Theta) . \tag{2.2e}$$

Since f_0 gives the same principal moment (the energy density e per unit length) as f, we may call f_0 and T the local Bose-Einstein density that corresponds to the distribution function f and the local (kinetic theory or absolute) temperature associated with f, respectively. Of course, the quantity k_B occurring in (2.2c) should be regarded as Boltzmann's constant and the exact form $\hat{\nu}(\Theta)$ of the reciprocal of the effective (k-independent) relaxation time ν^{-1} depends on the specific scattering model considered [13].

Within the framework set up here, the Boltzmann-Peierls equation becomes

$$\partial_t f + ck \circ \nabla f = -\nu(f - f_0), \quad \nabla := \nabla_x . \tag{2.3}$$

Let

$$W(z) := \frac{\omega^2 e^{\omega}}{(e^{\omega} - 1)^2} , \tag{2.4a}$$

$$\langle \psi \rangle := \int dz\, W(z)\psi(z), \quad \alpha_n := (\langle z^{2n} \rangle)^{-1/2} , \tag{2.4b}$$

so that

$$\langle z^{2n} \rangle = 2^{n+\frac{1}{2}}\,\Gamma\,(n + \frac{5}{2})\; \zeta\,(n + \frac{3}{2}) ; \tag{2.4c}$$

the symbols Γ and ζ refer to the Gamma function and the Riemann function, respectively. By means of (2.2d) and (2.4b) we can express the energy density e in terms of the temperature T; thus

$$e = \frac{\hbar c}{3\pi\alpha_0^2 \Theta^3} = \frac{\hbar c}{3\pi\alpha_0^2} \left(\frac{k_B T}{\hbar c}\right)^{3/2} . \tag{2.5}$$

Although our approach [7, 8] seems to be of potential interest regardless of its application to concrete macroscopic systems, the usefulness of (2.2)-(2.5) arises from every endeavour to study the collective behaviour of the aggregate of long-wavelength magnon excitations in a continuum.

3 The Tchebychef representation of the distribution function f

A. *A slightly modified lemma of Dijkstra and van Leeuwen*

The equations of transfer for Θ and the "non-equilibrium (Tchebychef) moments" b_n (n = 1,2,...) of f are some of the possible tools through which the Boltzmann-Peierls equation can be exploited. Of course, any attempt to work with the kinetic equation via an infinite hierarchy of equations of transfer immediately faces the problem of determining the distribution function f from the gas-state variables (Θ, b_n).

In order to overcome (to some extent) the above difficulty, we begin with the following definitions:

$$< \psi_1 | \psi_2 > := < \psi_1 \psi_2 >, \quad \| \psi \| := (< \psi | \psi >)^{1/2} , \tag{3.1a}$$

$$\mathbb{H} : = \{ \psi : \| \psi \| < \infty \} . \tag{3.1b}$$

Elementary inspection shows that $\mathbb{H}$ is the (real Hilbert) space of functions $\psi(z)$, $z \in \mathbb{R}$, which are square-integrable on $\mathbb{R}$ with weight $W(z)$. Since $W: \mathbb{R} \Rightarrow \mathbb{R}^+ := [0,\infty)$ is a Lebesgue measurable function with the property that for certain positive constants $\mathbb{M}$ and $\mathbb{C}$ one arrives at the inequality $W(z) \leq \mathbb{M} \exp(-\mathbb{C}z^2)$, the collection of polynomials defined on $\mathbb{R}$ is a dense subset of $\mathbb{H}$. (Concerning more details of general interest, see the original lemma of Dijkstra and van Leeuwen formulated on p. 468 in Ref.[9].)

We construct a sequence $\{B_n ; n = 0,1,...\}$ of the so-called Tchebychef polynomials $B_n(z)$ as follows. Orthogonalizing with respect to the inner product in $\mathbb{H}$ the set $\{z^n; n = 0,1,...\}$ of non-negative powers of z, $z \in \mathbb{R}$, we obtain a system $\{B_n; n = 0,1,...\}$ of polynomials uniquely determined by the two conditions: (i) $B_n(z)$ is a polynomial of precise degree n in which the coefficient τ_n of z^n is positive; (ii) the system $\{B_n; n = 0,1,...\}$ is orthonormal, i.e., $< B_n | B_m > = \delta^n_m$, where δ^n_m denotes the Kronecker delta. From the *slightly modified* lemma of Dijkstra and van Leeuwen [9] we then conclude that a family of Tchebychef polynomials $B_n(z)$ forms a basis in $\mathbb{H}$.

Definition. $\hat{f} := (e^{\omega} - 1)\,(f - f_0)\,[\omega\, e^{\omega} f_0\,]^{-1}$.

Now, if $\hat{f}$ belongs to $\mathbb{H}$, then the density f has a unique expansion

$$f = f_0\left[\,1 + \frac{\omega\, e^{\omega}}{e^{\omega} - 1} \sum_{n=0}^{\infty} b_n(x, t) \circ B_n(z)\,\right] \tag{3.2}$$

and this expansion converges in the *mean* to the function f. By virtue of the orthogonality properties of B_n, we find that the Tchebychef moments (the expansion coefficients of $\hat{f}$), denoted by b_n, are directly derivable from $f - f_0$:

$$b_n = \int \mathrm{d}z\ \omega\,(f - f_0)\, B_n(z) \qquad (\Rightarrow b_0 = 0)\,. \tag{3.3}$$

Looking at the definitions of the variables of physical import, we see that the heat flux q as given by

$$q := \frac{\hbar}{2\pi} \int \mathrm{d}k\, \Omega\, (\nabla_k\, \Omega)\, f \tag{3.4}$$

can easily be calculated from b_1; indeed, it is possible to show that

$$q = \frac{\hbar\, c^2}{2\pi\alpha_1\, \Theta^4}\, b_1\,. \tag{3.5}$$

B. Elementary properties of the Tchebychef polynomials B_n

In this section we list *without proof* basic properties of $B_n(z)$. No effort at completeness has been made [7, 8].

(1) The Tchebychef polynomials B_{n-1}, B_n , and B_{n+1} obey the relation

$$zB_n(z) = \tau_n(\tau_{n+1})^{-1}\ B_{n+1}(z) + \tau_{n-1}(\tau_n)^{-1}\, B_{n-1}(z)\,, \tag{3.6}$$

where $\tau_n, \tau_n > 0$, is the highest coefficient of $B_n(z)$, as we know.

(2) The Tchebychef polynomials satisfy inequalities of the form

$$|B_n(z)| \leq \sqrt{n}\,\exp(\frac{1}{2}\omega),\quad n \geq 1 \; . \tag{3.7}$$

(3) There is a positive constant $\mathbb{B}$ not depending upon n, such that, for $n \geq 1$,

$$\tau_n(\tau_{n+1})^{-1} = \langle zB_n\, B_{n+1} \rangle \leq \mathbb{B}\, n^2 \, . \tag{3.8}$$

If we set

$$S^{\gamma}(x, t) := \sum_{n=1}^{\infty} n^{\gamma}\, |b_n(x, t)|\,,\quad \gamma \geq 0\,, \tag{3.9}$$

we arrive at the following

Theorem. Let $S^{1/2}(x, t) < \infty$. Then the series appearing on the right-hand side of (3.2) converges both pointwise absolutely for each z and in the sense of the norm $\| \circ \|$ in $\mathbb{H}$. (Cf., Ref. [5] and the literature quoted there.)

4 Equations of transfer

Substituting (3.2) into (2.3) yields the differential system of equations of transfer for Θ and b_n; in the result we obtain

$$\alpha_0^{-2}\, \partial_t \Theta + c\alpha_1^{-1}\left(\frac{2}{\Theta} b_1\, \nabla\Theta - \frac{1}{2}\,\nabla b_1\right) = 0\,, \tag{4.1a}$$

$$\partial_t b_n = \Theta^{-1}\left(3b_n + \sum_{p=0}^{n} Z_p^n b_p\right) \partial_t \Theta + c\Theta^{-2}\left(4\beta_n + \sum_{p=0}^{n} Z_p^n \beta_p\right) \nabla\Theta$$

$$- c\Theta^{-1}\, \nabla\beta_n + c\Theta^{-2}\, \alpha_0\, \alpha_1^{-1}\, \delta_1^n\, \nabla\Theta - \nu b_n\,, \tag{4.1b}$$

where

$$\beta_n := \tau_n(\tau_{n+1})^{-1}\, b_{n+1} + \tau_{n-1}(\tau_n)^{-1}\, b_{n-1}\,, \tag{4.2a}$$

$$Z_n^n := n,\quad Z_p^n := 2\langle FB_nB_p \rangle,\quad n > p\,, \tag{4.2b}$$

$$F(\omega) := \omega(e^{\omega} + 1)\,(e^{\omega} - 1)^{-1}\,. \tag{4.2c}$$

In deriving (4.1b), we have made use of (3.6) and other properties of $B_n(z)$.

In order to get a determined system of field equations, we must decide to "approximate" the expansion (3.2) by a finite sum of Tchebychef polynomials, say, up to B_r, $r \geq 1$.

5 Motivation

In taking stock of the connection between kinetics and thermodynamics, one of the most important features of (2.4a) is the unique way the specific choice $W(z)$ affects the approximate dependence of both the entropy density

$$h := k_B(2\pi)^{-1}\int dk\,[\,(1+f)\ln(1+f) - f\ln f\,] \tag{5.1a}$$

and the entropy flux

$$\Phi := k_B(2\pi)^{-1}\int dk\,(\nabla_k\Omega)\,[\,(1+f)\ln(1+f) - f\ln f\,] \tag{5.1b}$$

upon the Tchebychef moments b_n. In the neighbourhood of the state of local equilibrium, if instead of the logarithm $\ln(1+X)$ $(|X|<1)$ we use the first two terms in its Taylor expansion $X - \frac{1}{2}X^2 + \dots$, we obtain for h and Φ

$$h - h_0 \cong -k_B(4\pi\Theta)^{-1}\sum_{n=1}^{\infty} b_n \circ b_n, \quad h_0 := 2\frac{\partial e}{\partial T}, \tag{5.2a}$$

$$\Phi - \frac{1}{T}q \cong -k_B c(4\pi\Theta^2)^{-1}\sum_{n=1}^{\infty} b_n \circ \beta_n\,, \tag{5.2b}$$

and the series appearing in (5.2) converge, because we suppose that $S^1(x,t) < \infty$. [In this context, see (4.2a) and (3.8).]

Thus the method of defining the weight (2.4a) and of introducing the Tchebychef polynomials $B_n(z)$ culminates in, and leads to, the approximate formula (5.2a) for $h - h_0$ with no cross terms coupling the different Tchebychef moments b_n (*the effect of diagonalization*). Due to this fact, we may regard (3.2) as a strict quasi-particle analogue of Grad's expansion of the Boltzmann molecular density f in terms of Hermite polynomials and call the Tchebychef moments b_n analogues of Grad's Hermite coefficients (see also Sections VII and VIII in Ref. [7]).

References

[1] Banach, Z., On the balance laws of extended thermodynamics for non-ideal gases, *Physica* **129A**, 95 (1984).

[2] Banach, Z., Extended thermodynamics of fluids versus the revised Enskog equation, *Physica* **145A**, 105 (1987)

[3] Banach, Z., The maximum entropy principle and the moment truncation procedure, in *Proc. 5th International Symposium on Continuum Models of Discrete Systems*, A.J.M. Spencer, ed. (A.A. Balkema, Rotterdam, 1987), p.111.

[4] Banach, Z., On the fundamentals of extended thermodynamics of a one-dimensional rarefied gas, *J. Stat. Phys.* **48**, 813 (1987).

[5] Banach, Z., On the mathematical structure of Eu's modified moment method, *Physica* **159A**, 343 (1989).

[6] Banach, Z. and Piekarski, S., Irreducible tensor description. I. A classical gas, *J. Math. Phys.* **30**, 1816 (1989).

[7] Banach, Z. and Piekarski, S., Irreducible tensor description. II. A quasiparticle gas, *J. Math. Phys.* (to appear in the July 1989 issue).

[8] Banach, Z. and Piekarski, S., Irreducible tensor description. III. Thermodynamics of a low-temperature phonon gas, *J. Math. Phys.* **30**, 1826 (1989).

[9] Dijkstra, J.J. and van Leeuwen, W.A., Mathematical aspects of relativistic kinetic theory, *Physica* **90A**, 450 (1978).

[10] Grad, H., On the kinetic theory of rarefied gases, *Commun. Pure Appl. Math.* **2**, 331 (1949).

[11] Grad, H., Principles of the kinetic theory of gases, in *Handbuch der Physik*, Vol. XII, S. Flugge, ed. (Springer-Verlag, Berlin, 1958), p. 205.

[12] Grad, H., Note on N-dimensional Hermite polynomials, *Commun. Pure Appl. Math.* **2**, 325 (1949).

[13] Gurevich, V.L., *Kinetics of Phonon Systems* (Nauka, Moskva, 1980, in Russian).

Banach Z.
I.P.P.T.-P.A.N.,
Swietokrzyska 21
PL-00-049 Warsaw
POLAND

L.M. BRIEGER AND E. BONOMI

A stochastic cellular automaton model of non-linear diffusion and diffusion with reaction

1 Introduction

The construction of a computer simulation of given physical phenomena has begun traditionally with a mathematical model consisting of differential equations defined in the continuum. This is followed by the definition of an algorithm, or numerical model, to discretize and numerically solve the equations. Finally, the numerical model is implemented on the computer, yielding the computer simulation and the numerical results which are studied to demonstrate, predict or clarify, for example, the physical phemonena in question. Thus, the final computer model and its results are significantly removed from the original model of the physics: the numerical model approximates the solution of the mathematical model and, machine round-off in floating point calculations means that the computer implementation itself only approximates the numerical model. (In fact, a principal concern in the field of numerical analysis is the quality of these approximations, i.e., assuring that the final numerical solutions generated by the computer are faithful to the true solution of the original mathematical model.)

Cellular automata offer an alternative to this approach to modelling [6]. In a cellular automaton model, space is discretized by a lattice whose grid points (sites) are permitted a finite number of values (states). The state of each site evolves step by step in the model. This evolution is governed by a set of local microscopic laws (rules) which, if chosen properly, produce in the model the macroscopic behaviour to be simulated. The rules can be implemented as logical functions performed between bits representing the states of the sites, in which case the machine implementation entails absolutely no approximation of the model but is exact. The results are then the direct image of the model and demonstrate the behaviour of the model itself and not of the model additionally influenced by approximative numerical methods. Nowadays, modern computing capacity and special machine architecture render cellular automata practical for large numerical simulations. The intrinsic granularity of cellular automata algorithms leads naturally to their implementation on vector and parallel machines. Machine architecture tailored to the large number of site-by-site calculations of cellular automata dynamics, executing them extremely efficiently, has given rise to one class of dedicated machines[7].

In our study, we have taken advantage of the well-established relation between the random walk of a Brownian particle and the diffusion equation [5] to develop a stochastic cellular automaton model for the simulation of diffusion, linear and non-linear. We extend this diffusion with reaction in a simulation of the carbonation of concrete, a corrosive process which occurs in the presence of drying [2].

2 The automaton model

The model of Brownian motion provided by the random walk of a particle on a square lattice is defined by a simple local rule: displace the particle with equal probability (1/4) in any one of the four directions on the lattice. This defines a probability distribution, $P(r, t_n)$, discrete in time and space, which is just the probability that position $r = x_{ij}$ is occupied by the Brownian particle at time t_n. Δt indicates the time interval between consecutive steps of the particle and Δx the grid discretization. $P(r, t_n)$ evolves diffusively on the lattice [5], and as Δx and Δt approach zero, with the ratio $D\ \Delta x^2/4\Delta t$ constant, $P(r, t)$ solves the following continuous diffusion equation (Fick's law of diffusion) in which D is the (constant) diffusion coefficient and ∇^2 indicates the Laplacian:

$$\frac{\partial P(r,t)}{\partial t} = \mathrm{D}\, \nabla^2 P(r,t)\ . \tag{1}$$

Consider the master equations for this model. The process is Markovian and the probability that the particle occupies site r at time t_{n+1} is a function of the neighbouring probabilities at time t_n :

$$P_r^{n+1} = \sum_{q \in \mathcal{N}(r)} P_q^n W_{qr}^n\ . \tag{2}$$

$\mathcal{N}(r)$ is the five-site von Neumann neighbourhood of r (r and its nearest neighbours on the grid), and W_{qr}^n denotes the conditional probability of moving from site q to site r at time t_n, given the particle at site q. $W_{qr}^n = 0$ whenever q and r are not in the same neighbourhood, and the following conservation condition is respected:

$$\sum_{r \in \mathcal{N}(s)} W_{sr}^n = 1\ . \tag{3}$$

Using (3), we can rewrite (2) and obtain the following description for the evolution of P :

$$P_r^{n+1} - P_r^n = \sum_{\substack{q \in \mathcal{N}(r) \\ q \neq r}} P_q^n W_{qr}^n - P_r^n \sum_{\substack{q \in \mathcal{N}(r) \\ q \neq r}} W_{rq}^n\ . \tag{4}$$

In the random walk model for one particle, expression (4) takes the form

$$P_r^{n+1} - P_r^n = \frac{1}{4}\left[\sum_{\substack{q \in \mathcal{N}(r) \\ q \neq r}} P_q^n - 4P_r^n\right] \tag{5}$$

which is the finite difference approximation of equation (1), with

$$D = \frac{\Delta x^2}{4\,\Delta t}\,.$$

Thus the random walk model reproduces the finite difference approximation of diffusion equation (1) on the grid. As the discretization goes to zero, the finite difference solution converges to the solution of the continuous equation.

In our model we adapt the stochastic rule to a population of particles, respecting an exclusion principle which allows at most one particle per site. We consider two implementations of these dynamics for a population of particles: asynchronous, in which an event consists of moving a single particle, and synchronous, in which the entire population of particles is updated simultaneously.

For the asynchronous case, with N sites in the automaton configuration, one site at a time is chosen at random from among the N sites. If Δt is the time interval for one sweep of the configuration, that is, of N events, then it can be shown that the master equations for this model yield exactly the finite difference form (5) of the diffusion equation. Thus the particle population diffuses on the lattice, and the model reproduces the solution of the discrete approximation of Fick's equation (1).

In the synchronous model, conflicts between particles vying for a free site must be resolved. This induces interactions between particles, resulting in master equations which differ from the diffusion equation (1) by a correlation term.

To illustrate the disparity between Fick's diffusion and this synchronous model, we consider in detail the one-dimensional case in which particles move left or right with probability 1/2, respecting the exclusion principle. The one-dimensional neighbourhood of x_j is simply x_{j-1}, x_j, x_{j+1}, and the master equations for the model give

$$P_j^{n+1} = P_j^n = \frac{1}{2}\left[\sum_{\substack{i \in \mathcal{N}(r) \\ i \neq j}} P_i^n - 2P_j^n\right] + E_j^n\,, \tag{6}$$

where

$$E_j^n = \frac{1}{8}\left\{ P_{j-2}^n(1 - P_{j-1}^n)P_j^n - 2P_{j-1}^n\,(1 - P_j^n)P_{j-1}^n + P_j^n\,(1 - P_{j-1}^n)\,P_{j-2}^n \right\}.$$

This is again the one-dimensional finite difference form (5) of the diffusion equation, altered by the term E_j^n,. which, for uniform populations (P constant), is zero. The effect of this extra term on the simulation is illustrated in an example of one-dimensional diffusion at equilibrium, Figure 1. The diffusion equation

$$\frac{\partial u}{\partial t} = \frac{1}{2}\frac{\partial^2 u}{\partial x^2} \quad \text{with} \quad \begin{cases} u(0,t) = 0 \\ u(1,t) = 1 \end{cases} \tag{7}$$

has equilibrium solution $u(x, t) = x$ on the unit interval. The steady state of the model appears in Figure 1: the smooth curve is the fixed point of the numerical iteration defined by (6), with $P(x, 0) = x$, and the "experimental" points are the measurements of the average particle distribution, as a function of x, over 100 simultaneous automaton experiments, at timestep 20,000. Initial conditions for the automaton respect the straight-line equilibrium solution of (7). Equilibrium for the synchronous system is not this straight-line solution, nor does it approach the straight line as the discretization is made increasingly fine.

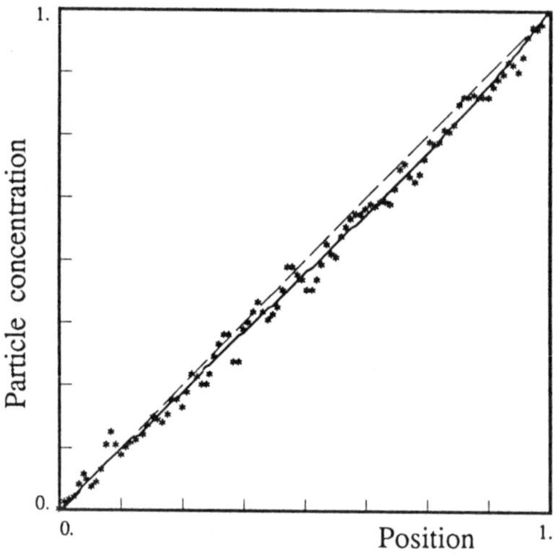

Figure 1. Law of Fourier. The dashed line is the steady-state solution of eq. (7); the solid curve is the exact concentration at equilibrium for the synchronous automaton; the $*$'s indicate the automaton concentration measurements at timestep 20,000.

The asynchronous dynamics reproduce more closely Fick's equation of diffusion and serve as the reference in our examinations of the master equations for the models. However, it is the synchronous model which holds the interest as a massively parallel algorithm, and it is on the synchronous model that we have concentrated our practical efforts. The figures in this article are all "snapshots" of the synchronous implementation (50×100 sites).

3 Non-linear diffusion

Equation (1) in Section 2 is an example of Fick's law of diffusion in which the coefficient D is constant. Non-uniform diffusion, as in an inhomogeneous medium, can also be modelled by Fick's equation

$$\frac{\partial u}{\partial t} = \operatorname{div}(\mathcal{D} \cdot \mathbf{grad}\ u) \tag{8}$$

in which the coefficient $\mathcal{D}$ is a positive function of position or of the solution itself. Fick's law remains the physical model of the phenomenon, with the effects of the inhomogeneities described by an effective diffusion coefficient which captures the phenomenology of the system and governs the model accordingly. For example, fluid diffusion in a porous medium can be modelled with Fick's equation if the coefficient $\mathcal{D}$ in (8) adequately reflects the combined phenomena which influence the fluid transport, including pure diffusion and capillary transport in a complex pore structure [4], for example. The coefficient can be supplied either from theoretical considerations or from experimental observations, and when it is a function of the solution u, eq. (8) is non-linear. Drying in concrete modelled as non-linear diffusion, studied in [1] and applied in [3], represents an application of such an approach.

Our automaton model exemplifies this modelling approach: we have constructed a model which simulates non-linear diffusion. Assuming the diffusion coefficient given, we use it to influence the displacement probabilities in the automaton in such a way as to reproduce the effect of the coefficient and the behaviour of eq. (8).

The finite difference form of eq. (8), using central differences for the second derivative, can be written as follows:

$$u_r^{n+1} = u_r^n = \frac{\Delta t}{\Delta x^2} \Big(\sum_{\substack{q \in \mathcal{N}(r) \\ q \neq r}} u_q^n \, \mathcal{D}_{qr}^n - u_r^n \sum_{\substack{q \in \mathcal{N}(r) \\ q \neq r}} \mathcal{D}_{qr}^n \Big), \tag{9}$$

where $\mathcal{D}_{qr}^n$ represents the diffusion coefficient $\mathcal{D}$ evaluated between sites q and r at time n. In the automaton implementation, if a particle chooses one of the four possible directions with probability $(1/4\ D)$ where D is a given function which we evaluate locally, then the master equations for the asynchronous model give just the finite difference approximation (9) of eq. (8) in which

$$\mathcal{D} = \frac{\Delta x^2}{4 \Delta t} k\, D, \quad 0 \le d \le 1$$

and where k is a scaling constant. Thus with the right choice of diffusion coefficient D implemented in the model, the automaton reproduces the solution of (8) for a given arbitrary (normalized) $\mathcal{D}$.

Figure 2 shows a comparison between synchronous automaton results and a finite

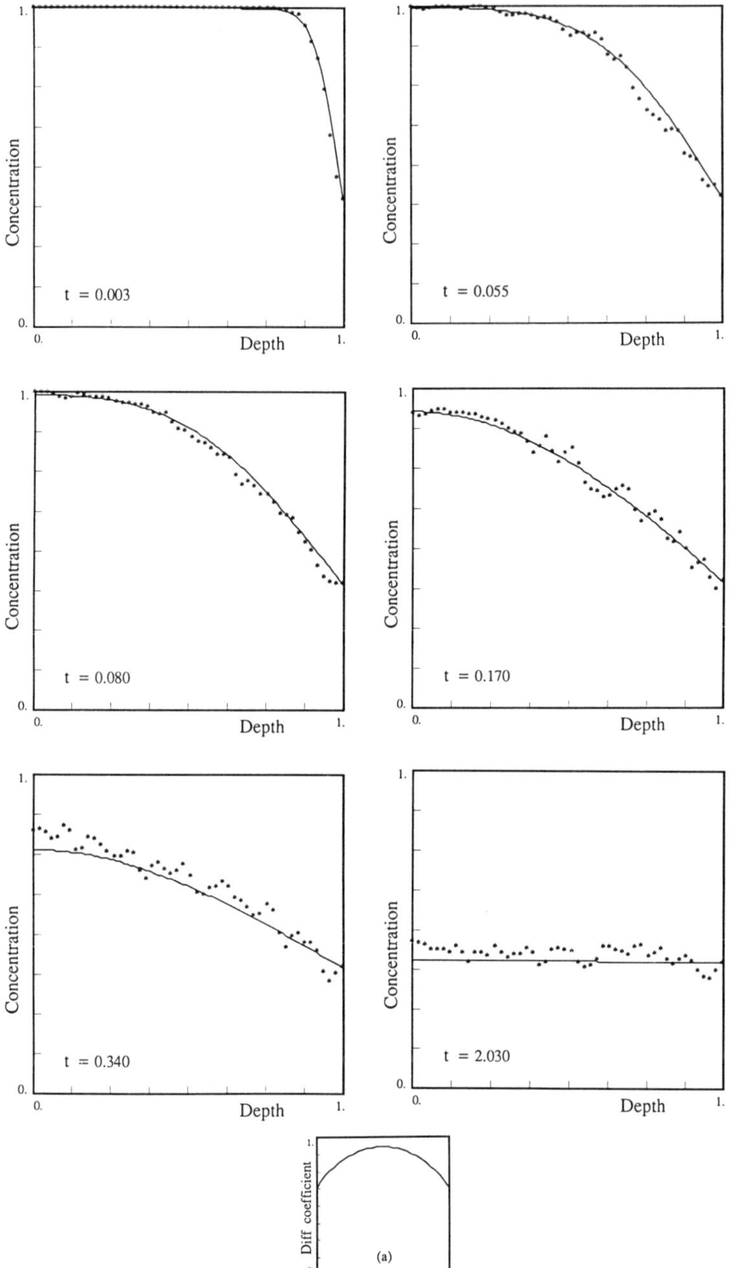

Figure 2. Non-linear diffusion for the synchronous model. (a) The diffusion coefficient $\mathcal{D}$ versus the concentration u. The smooth curves show the finite element solution of eq. (8) in one dimension at the indicated times; the experimental points are the "snapshot" automaton concentrations, at the same moment of simulated time.

element solution for non-linear diffusion equation (8) in one dimension, with the coefficient $\mathcal{D}$ a function of u as shown. The two-dimensional automaton configuratio implemented with a fixed boundary condition along $x = 1$ and zero flux conditions along the other borders, is given a one-dimensional representation by plotting the column-by-column average concentration as a function of x. The smooth curves show the finite element solution of the one-dimensional equation at the indicated times, and the experimental points are the "snapshot" automaton concentrations, measured at the same moment of simulated time.

Such a comparison as that of Figure 2 necessitates that we introduce units into the heretofore dimensionless automaton. A comparison between the automaton master equations and eq. (9), leads us to deduce the following definition of real-world time in terms of automaton timesteps:

$$\Delta t_w = \frac{1}{4k}\left(\frac{\Delta x_w}{\Delta x_a}\right)^2 \Delta t_a , \tag{10}$$

where k is the scaling constant, Δx_w and Δt_w represent the real (world) discretizations in space and time and Δx_a and Δt_a represent the corresponding automaton discretizations.

4 Diffusion with reaction: carbonation

Having calibrated the cellular automaton model with respect to simple diffusion, we consider now the application of the model to a more complex problem: the simulation of a diffusion-driven reaction in a homogeneous medium, concrete [3].

The carbonation of concrete is a corrosion process which is provoked by the drying of the concrete and its exposure to air which contains carbon dioxide, CO_2, entering the concrete pores which have been emptied by drying, reacts with calcium hydroxide, $Ca(OH)_2$, present in the concrete as a hydration product. The reaction forms calcium carbonate, $CaCO_3$, and liberates water in a series of reactions usually represented simply as $CO_2 + Ca(OH)_2 \rightarrow CaCO_3 + H_2O$. This transformation of $Ca(OH)_2$ into $CaCO_3$, once complete, lowers the pH of pore water below that which protects interior steel reinforcements from corrosion. The "carbonation zone", in which all the $Ca(OH)_2$ has been transformed, progresses toward the interior reinforcements with the inward diffusion of the CO_2, threatening the integrity of the reinforcements once it reaches them. However, this progress depends also on pore water, which can block the influx of the CO_2; in water-saturated conditions the carbonation front does not advance at all. This carbonation reaction normally proceeds at the rate of millimeters per decade, depending on such factors as the porosity of the concrete, its $Ca(OH)_2$ content, atmospheric relative humidity and exposure to wetting-drying cycles, cracking, surface treatment, etc. To date, there is no adequate mathematical model which furnishes a reliable macroscopic description of the various interacting factors and their influence on the carbonation environment.

The ultimate goal behind a model of carbonation in concrete is to characterize the inward progress of the reaction zone under a variety of conditions. With a reliable model, existing structures can be assessed for remaining service life as influenced by carbonation, and codes can be established to guarantee the desired durability, under specific environmental conditions, for new structures.

We begin here the development of an automaton model of carbonation, based on the stochastic model of diffusion presented in the preceding sections. The adaptation of the original model to include several species and a chemical reciton is relatively simple. Of considerable importance is the fact that there is no accompanying numerical method which must be redefined and implemented order to solve the revised model. The job of constructing the simulation is thus considerably easier than the numerical solution of a standard mathematical model based on partial differential equations. In the following we give an illustration of the cellular automaton as a simulation tool rather than a systematic study of carbonation.

In the automaton simulation, CO_2 and H_2O are modelled as two populations of diffusing particles obeying the exclusion principle (at most one particle per site). The movement of each population on the automaton lattice is governed by a diffusion coefficient furnished by experimental observations and representative of the porous medium. $Ca(OH)_2$ and $CaCO_3$ are supposed held in the pore walls and hence are represented as stationary populations of fixed particles in the background. The carbonation reaction occurs whenever the random walk of the CO_2 produces a collision with a $Ca(OH)_2$ particle, provoking the spontaneous transformation of these two species. Boundary conditions representative of atmospheric relative humidity and CO_2 content are imposed at the right border, and the other borders are closed (zero flux conditions). Figure 3 shows the simulated behaviour of the carbonation reaction in a mortar specimen and the corresponding one-dimensional distribution curves. The sample is initially saturated with water and dries in an atmosphere of CO_2. We see the equilibration of the water content over time and the progress of the reaction front. The model clearly depicts three distinct zones: the "carbonation zone" (the zone in which the $Ca(OH)_2$ has been completely consumed and the pH diminished), the reaction zone (the width of the reaction front) and the uncontaminated region.

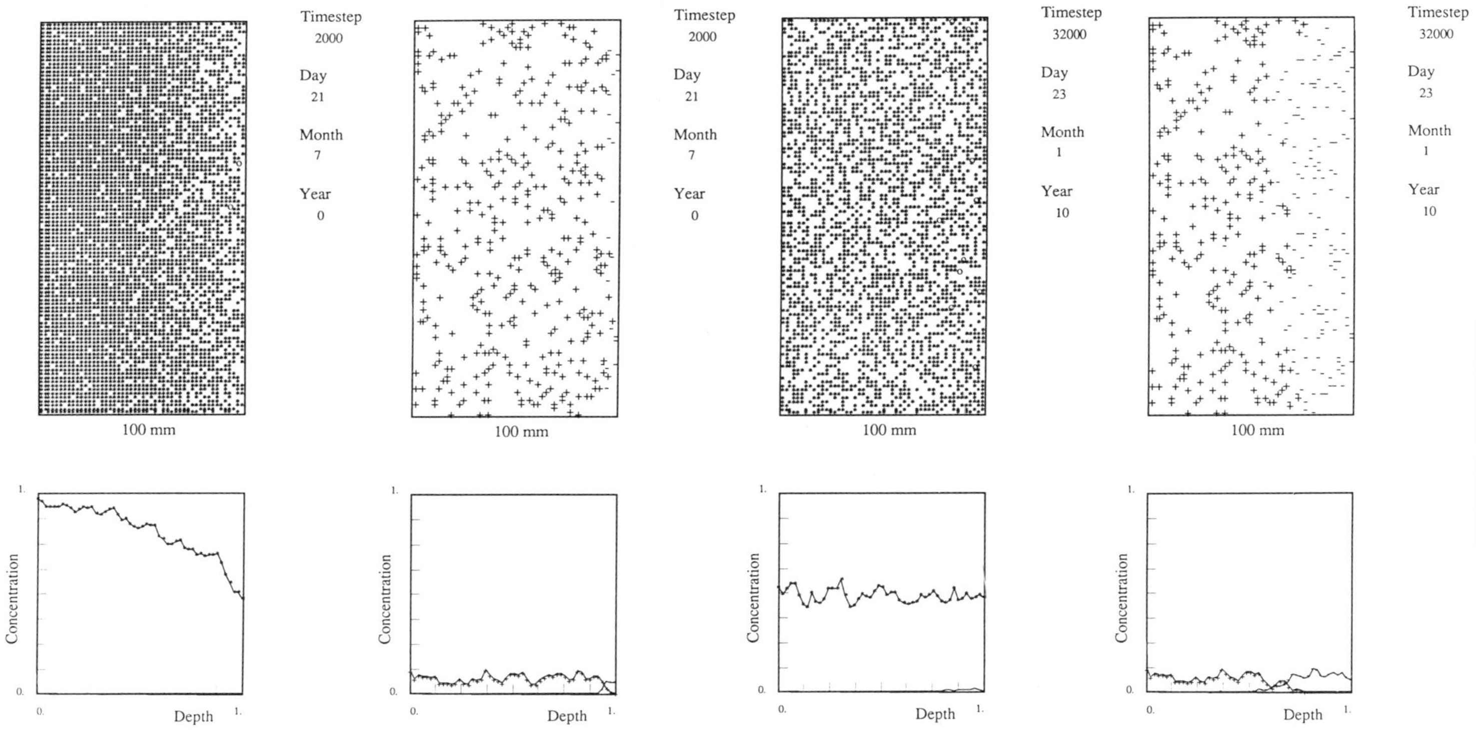

Figure 3. Carbonation of concrete. On the left in the two examples of the simulation are shown the two diffusing species, water (∗) and CO_2(o); on the right are the two static species, $Ca(OH)_2$(+) and $CaCO_3$(–), indicative of the passage of the reaction front.

5 Conclusion

In the present paper we have investigated a probabilistic cellular automaton model. One purpose is the introduction of a modelling tool which facilitates the simulation of a complex physical process; an application illustrated here is a reaction-diffusion process in an inhomogeneous medium (the carbonation of concrete). The other purpose is the investigation of the automaton model as a parallel algorithm, considered here for solving the diffusion equation by simulation. The strategy involved is to use simple microscopic probabilistic dynamics to reproduce deterministic macroscopic behaviour. The realism of the implemented rules as reflections of microscopic physical phenomena is not the main goal; what is critical is that the cooperative effect of the local rules correctly reproduce the macroscopic physical observables.

A cellular automaton implementation is the direct image of a physical model, without the overhead of numerical schemes which approximate mathematical solutions. The state of a site is represented in bits and the dynamics of the system are governed by a set of bit operations. In the model described in this article, the operations which cause the evolution of the sites are chosen by stochastic rules from among the set of possible bit operations.

As is evident from the figures in this article, fluctuations are present in the simulations. Increasing the number of sites in a configuration and enlarging the neighbourhood for averaging will diminish the fluctuations but will also require more computational effort. Since the cellular automaton algorithms, with their naturally fine granularity, can be implemented on parallel machines, this requirement may not be prohibitive. Indeed, on dedicated machines, computational time might actually be drastically reduced.

These simulations were carried out using the CRAY-2 with Silicon Graphics workstations at EPF, Lausanne.

References

[1] Bazant Z.P. and Najjar L.J., Drying of concrete as a non-linear diffusion problem, *Cement and Concrete Research*, **1**, 461-473 (1971).

[2] Brieger L.M. and Bonomi E., A cellular automaton model of diffusion and reaction in a porous medium: concrete, presented at the *Symposium on Fractal Aspects of Materials: Disordered Systems*, Fall MRS meeting, Boston, 1988.

[3] Brieger L.M. and Wittmann F.H., Numerical simulation of carbonation of concrete, *Proc. Int. Colloq. on Materials Science and Restoration*, Esslingen, 635-638, 1986.

[4] Dullien F.A.L., *Porous Media - Fluid Transport and Pore Structure*, Academic Press, New York, 1979.

[5] Feller W., *An Introduction to Probability Theory and its Applications*,

Wiley, New York, 1964.

[6] Lallemand P., Models and Simulations of Lattice Gases, these CMDS6 Proceedings. See also the references furnished in this contribution.

[7] Toffoli T. and Margolus N., *Cellular Automata Machines*, MIT Press, Cambridge, Mass. 1987.

Brieger L.M.
Department of Mathematics
Ecole Polytechnique Fédérale
CH-1015 Lausanne
SWITZERLAND

Bonomi E.
G.A.S.O.V. Group
Ecole Polytechnique de Fédérale,
CH-1015 Lausanne
SWITZERLAND

P. DOREMUS AND J.M. PIAU

Constitutive equation of a yield stress fluid based on the network theory

1 Introduction

Certain materials possess the property of flowing like fluids whenever they are used, although they can resist in such a way so that they do not seem to flow at all for years under the effect of reduced finite amplitude stresses applied by measuring apparatus or else by gravity or capillarity. These materials will be referred to as yield stress fluids rather than plastic materials which is the terminology used in solid mechanics. Thus it concerns a physical and experimental concept of yield. Generally speaking, these materials are obviously used as fluids and not as solids, and we have no a priori reason for selecting a plastic behaviour scheme.

Until now, the behaviour of yield stress fluids in experiments which subject them to small deformations, as well as corresponding theoretical modelling, have been described in a limited number of publications [17, 2, 21, 22]. More generally, various approaches to the constitutive behaviour for formulating viscoplastic materials in these experiments are possible according to whether or not a mathematical relationship defining a yield criterion is introduced explicitly or not (this is thus referred to as the mathematical concept of yield).

As in the case of the majority of current models, the Bingham model and its further three-dimensional [12, 18] developments, the mathematical yield criterion, whenever it appears explicitly is often developed using the stress level. In this case the behaviour below the mathematical yield is separated (rigid, elastic, etc.) from that above the yield (viscous, viscoelastic, etc.) [13, 23]. The constitutive equation is thus written with the help of two relations which can give rise to certain mathematical peculiarities while the problem is being solved if the stress tensor value is to be determined on both sides of the mathematical yield. If we have an idea beforehand of the behaviour of the fluid to be modelled, the freedom of choice of two types of behaviour in the fluid and solid zone can, however, be a worthwhile advantage.

Other approaches do exist which tend to seek a mathematical representation of experimental behaviour without using mathematical yield criteria [15, 6].

Whatever the success of the purely phenomenological laws, it is beneficial to complete them with an understanding of physical mechanisms, and in this work we have chosen a structural description of the flowing media. It enables us to propose a mechanism which leads to apparent yields like those which occur in rheometric measurements of yield stress fluids. Our aim is to model yield stress fluids for which the solvent can be schematized by a network model. This model has already been improved by several authors and it provides interesting results. A non-affine motion of the structure was applied [9, 19] and analytical results for simple flows were obtained [7]. The case of a non-Gaussian network has also been studied [5]. The model of the entangled polymer network [8, 10, 24] seems to enable at least part of the physics of the yield stress fluid to be represented and will be used as a basis for our procedure.

Many yield stress fluids are produced from at least two components. When the filler is mixed with a polymer fluid, the polymer-filler junctions appear in addition to the polymer-polymer entanglements. These newly created junctions arising from the absorption of polymer by the filler are the basis of the mechanism causing yield stresses. Thus the aim of the present article is to adapt the network model in such a way as to take the polymer-filler junctions into account [3, 4].

2 Double network model and segment statistics of the macroscopic kinematics

In the model presented by Yamamoto, the segment kinetics are given by the equation

$$\frac{\partial f}{\partial t} + \operatorname{div}(\dot{\mathbf{R}} f) = G(\mathbf{R}, N) - H(\mathbf{R}, N) f \tag{1}$$

with f the probability density such that $f\,dR^3$ represents the number of elastic segments of the network per unit volume, made up of N elements whose non-dimensional length at time t is between R and $R + dR$. The non-dimensional vector $\mathbf{R}$ is defined as the ratio of the length of the elastic segment r at the stretched length Nl (Figure 1). G and Hf represents respectively the rate of creation and destruction of the segments with N elements.

Figure 1

We consider J as a kinematic variable that characterizes a flow and such that it is zero when the material is at rest. (Afterwards it will take the value $J = 2 \operatorname{tr} D^2$ where D is the rate of deformation tensor).

A. Description of the polymer-filler junctions

We shall deduce a simple mechanism which accounts for the generation of cross-linking from the analysis of an experiment where a yield stress is present: the relaxation of a material after cessation of flow.

In this stage $(J = 0)$ stresses of the polymer relax. The energy due to Brownian motion is sufficient to modify the cross-linking distribution in such a way that the final state is that of isotropic stress. For the case of a material with a yield stress, the Brownian motion energy of the macromolecules linked to the filler is of a lower order of magnitude than that associated with the polymer-filler absorption. Thus, during a relaxation experiment the structure of the polymer-filler network (hereafter denoted as a filled network), is capable of sustaining a state of non-isotropic stress which is identified with the yield stress.

It is possible to take into account considerations relative to the bond mechanism by imposing kinetic functions of creation and destruction of the filled network segment which cancel out when the material is at rest $(J = 0)$.

If f', G' and H' are the functions which describe the kinetics of the filled network, then: $G'_{J=0} = 0$, $H'_{J=0} = 0$ from which it follows:

$$\left[\frac{\partial f'}{\partial t} + \operatorname{div}\,(\dot{\mathbf{R}} f')\right]_{J=0} = 0\,. \tag{2}$$

Here, we shall use $\mathbf{R}$ as the argument of the kinetics G' and H', a common practice in the theory of networks, while J is necessary as the argument for the generation of yield stresses.

The presence of a flow field strains the links between the filler and the polymer. These links are weaker than the covalent links of a polymer chain and break when material is flowing.

In this configuration a polymer molecule forms links with the filler or with other molecules. At a given instant, a polymer element can belong to the liquid phase and in another to the filled network. A macromolecular segment is in the filled network if it is between two bonds having different elements of the filler. In any other case this segment belongs to the liquid phase. We shall come back later to the description of these two networks.

B. *Description of the liquid phase*

First let us look at the liquid phase. It is possible to describe the structure of this phase using kinetics such as the one proposed by Fuller and Leal [7]. As usual, the polymer constituting this phase is trapped in the filled network. Therefore one should expect to find differences in polymer behaviour when it is in the filled network and when it is not. For example, the characteristic relaxation time of the liquid phase of a material with a yield stress can be more important than that of the isolated polymer (diffusion of macromolecules impeded by the structure of the filled network). For this reason the argument J will be also used for the liquid phase. One can write:

$$G = G(R, J)\,, \tag{3}$$

$$H = H(R, J). \tag{4}$$

In this way the action of the filled network on the liquid phase is incorporated in the functions G and H. In so doing the analytic calculations are possible. However this procedure excludes the determination of the liquid phase properties from those of the pure polymer.

It is well known that the introduction of a kinematic parameter such as J in a rheological law that aims to portray the behaviour of a polymer is perhaps not advisable [1]. These criticisms bear generally over the domain of linear viscoelasticity subject to weak deformations. Some remarks can be made concerning this point. However, for the modelling of the material in this work, these criticisms can only be made for the liquid phase as the filled network does not have a behaviour comparable to that of a material without yield stress. As has already been mentioned, the liquid phase evolves within another network of completely different behaviour. Therefore, it is not evident that this phase has a linear viscoelastic regime. Lastly, the formulation that we propose here allows one, by including a dependence on a parameter J, to have as a particular case a linear viscoelastic regime with small deformations, according to the value of certain parameters in the segment kinetics.

C. Description of the networks

Let us come back to the description of the network of the material. A structural segment of the liquid phase corresponding to a polymer segment laying between two bonds (polymer–polymer or polymer–filler) is denoted by R_1 (Figure 2). This definition implies that the interbond molecular weight dominates the behaviour of the liquid phase. A structural segment of the filled network corresponding to a segment of the polymer chain laying between two bonds (polymer–filler) is denoted by R_c (Figure 2). The length R_c is a determinant of the behaviour of the filled network for the following reasons. It is the ratio of the mean interfiller particles distance to the length of a macromolecule which permits us to determine the possible existence of a structure such as the one we are dealing with. While relaxation is taking place, it is also the prevailing tension on the segment which controls the order of magnitude of the yield stress. Thus, the behaviour of the segment R'_1 (Figure 2) belonging to the filled network and laying between two polymer–polymer bonds, appears to us to be less determinant than that of R_c. Moreover, it is reasonable to think that the polymer–polymer entanglements, less resistant than the polymer–filler bonds, will not perturb too much the average tension of segment R_c containing R'_1.

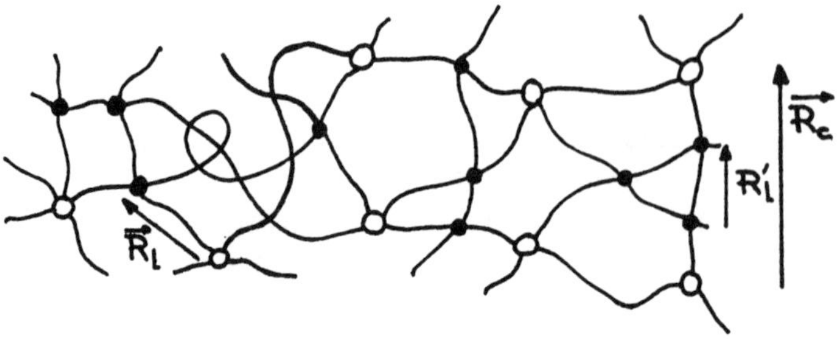

Figure 2

D. Expression of the stress tensor

The creation and destruction functions already introduced [16] can be generalized to include a dependence on J. For the liquid phase it is possible to suggest

$$G = g_0\, g_1(J)\left(\frac{3N}{2\pi}\right)^{3/2} \exp\left(-\frac{3N}{2}\, g_2(J)\mathbf{R}^2\right), \tag{5}$$

$$H = H_1(J)\,. \tag{6}$$

In a similar manner, for the filler, we can write

$$G' = g'_0\, g'_1(J)\left(\frac{3N}{2\pi}\right)^{3/2} \exp\left(-\frac{3N}{2}\, g'_2(J)\mathbf{R}^2\right), \tag{7}$$

$$H' = h'_1(J)\,. \tag{8}$$

The microstructure motion, for the liquid phase, is given by eq. (9) and for the filler by eq. (10)

$$\dot{\mathbf{R}} = (L - \xi D)\mathbf{R}\,, \tag{9}$$

$$\dot{\mathbf{R}} = (L - \xi' D)\mathbf{R}\,, \tag{10}$$

where L is the velocity gradient, D the rate of deformation tensor and ξ, ξ' the structural slip constant parameters introduced by Gordon and Schowalter. The entanglement kinetics for the liquid phase will be given by the relation

$$\frac{\partial f}{\partial t} + \operatorname{div}(\dot{\mathbf{R}} f) = G - H f \tag{11}$$

and for the filler by the equation

$$\frac{\partial f'}{\partial t} + \operatorname{div}(\dot{\mathbf{R}} f') = G' - H' f' . \tag{12}$$

We assume that the free energy of the material is equal to the sum of the free energies of both networks. The stress tensor for such kinetic associations is described for Gaussian networks [11]

$$\mathbb{T} = \sum_N \int_{R^3} 3NkT\,(\mathbf{R} \otimes \mathbf{R})\,[\,(1 - \xi) f + (1 - \xi') f'\,]\,\mathrm{d}^3R . \tag{13}$$

In this work we use a simplified formulation, without any summation on the values of N which is supposed to be a constant,

$$\mathbb{T} = \int_{R^3} 3NkT\,(\mathbf{R} \otimes \mathbf{R})\,[\,(1 - \xi) f + (1 - \xi') f'\,]\,\mathrm{d}^3R . \tag{14}$$

3 Description of networks in an initial state

Shear flows and two-dimensional extension flows are among the simple two-dimensional homogeneous flows used most often. The response of the material is generally obtained by a simple succession of such kinematics. For example after a state of rest, a shift in the deformation rate is applied which successively leads to shear stress during start up flow and then during steady flow and finally to relaxation from steady flow onwards. The initial state must be known accurately for each transient state, i.e., start-up flow or relaxation. Generally speaking, the most commonly chosen initial state in the case of start-up flow, is the isotropic state usually obtained after a sufficiently long rest period. Almost immediately after putting the liquid into the rheometer, a phase during which the flow is complicated, the sample to be tested must remain motionless long enough to ensure that the stresses caused by placing it have time to relax.

In the case of a yield stress fluid however, the relaxation phase cannot lead to a homogeneous and isotropic state of the stresses. The preparatory phase can be analysed in the following way, using the equations laid down for the model.

The analysis of the polymer network of the liquid phase given by the function f is simple and demonstrates how to reach the isotropic and homogeneous state after relaxation. In the case when relaxation is carried out just after the placing phase, the solution of eq. (1) is given by [7]

$$f = f_{\mathrm{i}} \exp\left[-H_{J=0} t\right] + \frac{G_{J=0}}{H_{J=0}}\left(1 - \exp\left[- H_{j=0} t\right]\right) \tag{15}$$

with f_{i} equal to the value of f describing the state of the liquid just prior to relaxation. This equation shows that, after a long enough time, the initial state of the liquid, f_{i}, is no longer significant. It tends towards a state determined by the value:

$$f_0 = \frac{G_{J=0}}{H_{J=0}} \tag{16}$$

which will be taken as the initial state in the results presented for the start-up tests concerning shear flow and elongational flow. It must be remembered that f_0, the value obtained, implies that the function H cannot be cancelled out when $J = 0$. Similarly in the case of the network modelling the filler and described by f', the choice $H'_{J=0} = 0$ gives for the relaxation equation

$$f' = f'_{\mathrm{i}} + G'_{J=0} t \,. \tag{17}$$

To satisfy eq. (2), $G'_{j=0} = 0$ has been taken and so f' and the isotropic part of the stress tensor do not increase indefinitely with time t. Consequently, this results in the conservation of the state in which the network - which models the filler - was found: $f'_0 = f'_{\mathrm{i}}$. The relaxation phase which follows a given flow, does not allow a filled network (and consequently the entire material) to reach a state of non-stress. The result is the experimental observation of flow yields. The analysis of this filled network means that several choices of the f' function revealing the initial state of the stresses, can be of practical interest. More specifically, the material functions will depend on the hypothesis concerning the initial state. In the rest of this article, in order to calculate the fluid response, we will choose an initial value of the distribution

$$f_0 = \frac{g_0\, g_1(J=0)}{h_1(J=0)} \left(\frac{3N}{2\pi}\right)^{3/2} \exp\left[- \frac{3N}{2}\, g_2\,(J=0)\, R^2\right]. \tag{18}$$

For the filled network, we can choose an initial isotropic state of stress associated with probability density

$$f'_0 = \frac{g'_0}{h'_0} \left(\frac{3N}{2\pi} \right)^{3/2} \exp\left[-\frac{3N}{2} R^2 \right]. \tag{19}$$

Other expressions for f'_0 will also lead to isotropic states of stresses. Expression (19) for f'_0 corresponds to an isotropic state obtained during oscillatory shear flow with decreasing amplitude. The value of f'_0 is an isotropic state depends on the deformation history initially imposed. For example, a biaxial extension test followed by a uniaxial extension permits us to find a different value of f'_0 to that given by (19).

This apparent multiplicity of isotropic initial states complicates the analysis of the transient experimental tests which are generally started up with non-deviatoric stress tensor or with non-tangential or normal stress.

4 Conclusion

We have studied a model of a yield stress fluid based on the network theory. An explicit formulation of kinetic laws of molecular segments has been given [3, 4], and analytical results are presented for rheometric flows: shear, extension and inverse flows, in steady or non-steady states.

Four of the principal results are:

- Apparent yield stress and yield stress relaxation are predicted and found to be identical.
- When stresses are below the yield value, the model has an inelastic behaviour. It must be noted that purely elastic behaviour on a given deformation area implies that, for both the structure of the filled network and of the liquid phase, no junction is modified during deformation occurring in this area.
- The model has no discrete memory. The initial state is forgotten, when steady flow is reached in the case of the liquid phase and the filled network.
- The flow history memory is frozen during a relaxation. The freeze of the filled network memory at rest is due to the fact that junctions hold on.

The laws obtained, as well as the general nature of the results, due to the physics of the yield used, are of considerable help in clarifying experimental procedures related to the rheometry of yield stress fluids. With the present model, stresses within a yield stress fluid for different experiments are readily calculated as the sum of the liquid phase stresses plus the filled network stresses. By adjusting the model parameters, it is possible to reproduce rheometric results classically found in the literature [17, 20]. For example, one can predict the shear stress behaviour of a yield stress fluid at low velocity gradient and in relaxation, shown in Figures 3 and 4).

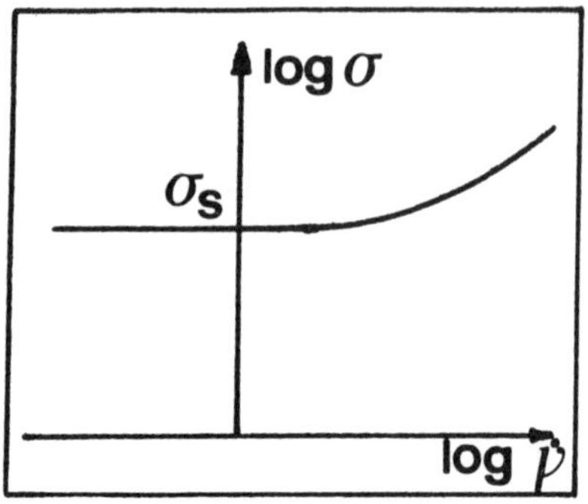

Figure 3. Shear stress evolution for a yield stress fluid as a function of shear rate at low $\dot{\gamma}$.

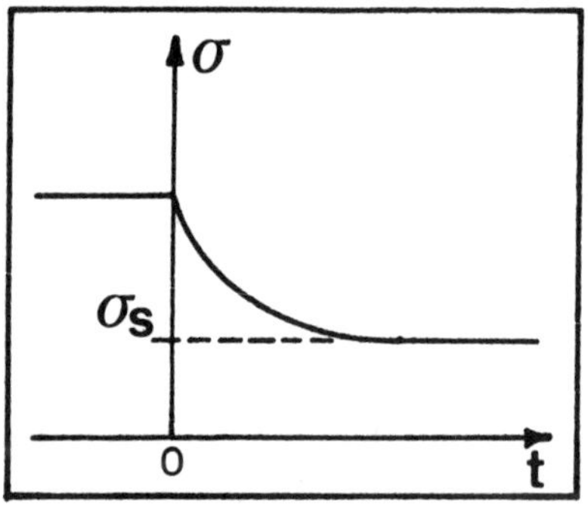

Figure 4. Shear stress relaxation for a yield stress fluid as a function of time.

References

[1] Astarita, G., and Marruci, G., *Principles of Non-Newtonian Fluid Mechanics*, McGraw-Hill, London, 1974.

[2] Cheng, D.C.-H., *Rheol. Acta.*, 25 (1986), 542.

[3] Doremus, P., Thèse de Doctorat d'Etat àl'INP et l'UJF de Grenoble, defended 7/7/89.

[4] Doremus, P., and Piau, J.M., Comptes Rendus du 22e Colloque Annuel du Groupe Français de Rhéologie, (1987), 113.

[5] Ekong, E.A., and Jayarahan, K., *Jour. of Pheol.*, **28**, (1) (1984), 45.

[6] Favier, D., Guélin, P., and Pégon, P., Comptes Rendus du 20e Colloque Annuel du Groupe Français de Rhéologie, Ed. D. Bourgoin, D. Geiger, (1985) 53.

[7] Fuller, G.G., and Leal, L.G., *Jour. of Poly. Sci.*, **19** (1981), 5.

[8] Giesekus, H., *Viscoelasticity and Rheology*, Ed. Lodge A.S., Renardy M., Nohel, J.A., Acad. Press, London (1985), 157.

[9] Gordon, J.R., and Schowalter, W.R., *Trans. Soc. Rheology*, **16** (1972), 79.

[10] Green, M.S., and Tobolsky, A.V., *Jour. of Chem. Phys.*, **14**, 2 (1946), 80.

[11] Grmela, M., and Carreau, P.J., *Pheol. Acta*, **21** (1982), 1-14.

[12] Hohenemser, K., and Prager, W., *ZAMM*, **12** (1932), 216.

[13] Hutton, J.F., *Rheol. Acta*, **14** (1975), 979.

[14] Janeschitz-Kriegl, H., *Polymer Melt Rheology and Flow Birefringence*, Springer-Verlag, Berlin, 1983.

[15] Krempl, E., *Mechanics of Engineering Materials*, 1984, Wiley, Chapter 19.

[16] Lodge, A.S., *Trans. Faraday Soc.*, **52**, 120 (1956), 7.

[17] Magnin, A., and Piau, J.M., *Jour. of Non-Newtonian Fluid Mech.*, **23** (1987), 91.

[18] Oldroyd, J.G., *Proc. Camb. Phil. Soc.* **43** (1947), 100.

[19] Phan Thien, N., and Tanner, R.I., *Jour. of Non-Newtonian Fluid Mech.*, **2** (1977), 353.

[20] Sacchettini, M., Magnin, A., Piau, J.M., and Pierrard, J.M., *Jour. of Theoretical and Applied Mech.*, 1985, 165-199.

[21] Tanaka, H., and White, J.L., *Polym. Eng. Sci.*, **20** (1980), 949.

[22] Vinogradov, G.V., Malkin, A.Y., Plotnikova, E.P., Sabsai, O.Y., and Nikolayeva, N.E., *Int. J. Polym. Mat.*, **21** (1972).

[23] White, J.L., *Rheol. Acta*, **20** (1981), 381-389.

[24] Yamamoto, M., *Jour. of the Phys. Soc. of Japan*, **11**, 4 (1955), 413-421.

Doremus P.
Institut de Mécanique de Grenoble
B.P. 53X
F-38041 Grenoble Cédex
FRANCE

Piau J.-M.
Institut de Mécanique de Grenoble
B.P. 53X
F-38041 Grenoble Cédex
FRANCE

I.M. DOUNAEV

A continuum model of microheterogeneous elastomers

1 Introduction

Elastomers (e.g., rubber) are microheterogeneous materials composed of a matrix of macromolecular chains and of micro-inclusions having different structures, connected to the matrix (blocks and fibres of rigid polymers or metals, particles of filling materials, domains with high density of chemical cross bonds). The matrix of flexible macromolecular chains which have a weak reciprocal action and are in a chaotic motion (Brownian motion) determines the capacity of elastomers to attain large elastic strains which are practically reversible. The viscoelastic microinclusions have a better regulated structure. During the process of strains due to the action of forces transmitted by the matrix, the equilibrium is attained more slowly and is followed by a dissipation of energy. Further we will propose another model in the macroscopic theory of thermoviscoelasticity of elastomers, based on the statistic and thermodynamic analysis of this structure and on the following assumptions. The matrix is a statistical system of macromolecular chains whose average specific free energy W_1 is determined by [2–4] as well as in the network theory of Treloar–James–Guth. The microinclusions form a statistical system whose essential features are an average linear dimension (50–5000 Å) and an average free energy. Every microinclusion is a viscoelastic continuum whose properties are known and whose free average energy W_2 is determined as a functional for the thermoviscoelastic body. The matrix and the microinclusions, within a macroparticle, have the same temperature. The displacement of the joining points of the chain ends at the surface of microinclusions and is equal to the displacement of the microinclusion in this point, and the vector of the force of interaction is also continuous in these points. These conditions are fulfilled in the mean, for the microinclusions in a macroparticle. The total free energy of the macroparticle is the sum of the energies of the matrix and the microinclusions. The volume of the macroparticle depends only on the temperature.

2 The free specific energy of microheterogeneous elastomers

We begin by writing the thermodynamic laws of irreversible processes, which can be applied to the macroparticle and the subsystems

$$dU - \delta Q + \delta A, \quad T dS = \delta Q + \Lambda dt, \quad dW + S dT = \delta A - \Lambda dt, \tag{1}$$

where U is the internal specific energy, S is the specific entropy, T is the temperature, $W = U - TS$ is the specific free energy, Λ is the dissipative function, and Q is the quantity of heat. According to the accepted working hypothesis, we write the specific free energy in the homogeneous state of stress and strain of the unit cube as

$$W = W_1 [\lambda(t), \varphi(t), T(t)] + W_2 [\varphi(\tau), T(t)] + p [J - F(T)], \tag{2}$$

where $\lambda_i(t)$ are the principal dilatations of the macroparticle, $\varphi_i(t)$ are the principal dilatations of the microinclusions, p is a scalar function of time, $F(T)$ is the function determining the dependence on temperature of the volume of the macroparticle, and $J = \lambda_1 \lambda_2 \lambda_3$, $0 \le \tau \le t$. Here and further on, the expressions having one or several repeated indices imply simple summation. The specific free energy of the elastic matrix is equal to [2–4]

$$W_1 = f_1(T) - kT(t) [< \ln Q^* > - < \ln Q_0^* >], \; Q_0^* = Q^* |_{t=0}, \tag{3}$$

where Q^* is the statistical integral (configurative sum) of the elastic matrix, k is Boltzmann's constant, and the square brackets mean the mathematic expectations of the stochastic variables and their respective functions. We shall calculate the statistical integral (the configurative sum) Q^*. Let $(p, q) = 1,2,3, \ldots , N^0$ the number of the points determining the position of the centres of mass of the microinclusions, p_τ, q_ν the numbers of the points which determine the position of the ends of all the chains, "stuck" to the microinclusions whose centres of mass are in the points p, q ($\tau = 1,2,3, \ldots$ $\nu = 1,2,3, \ldots$) while the number of "sticking" points on p and q (the end of chains) may be different and two or several segments of chains may be fixed in one point. Then, the statistical integral of the elastic matrix is calculated using the same assumptions as in the network theory for elastomers having an homogeneous structure [2–4] (Treloar–James–Guth) and it is equal to

$$Q^* = A^* \exp\left[- \frac{3}{2b^2} \sum_{p_\tau > q_\nu} \sum Z^{-1}_{p_\tau q_\nu} \left(x^i_{0p_\tau} - x^i_{0p_\nu} \right) \left(x^i_{0p_\tau} - x^i_{0p_\nu} \right) \right]. \tag{4}$$

Here $Z_{p_\tau q_\nu}$ is the number of segments in the chain, b is the effective length of the segment, $x^i_{0p_\tau}$, $x^i_{0q_\nu}$ are the average and the most probable values of the coordinates of the ends of the chains which transform the configurative sum into the absolute maximum [2,3]. In the expression (4) and further on, the upper and lower indices (repeated ones) and subindices τ, ν indicate the simple summation which spreads formally and on the two points not connected directly by the chains, if for them $Z_{p_\tau q_\nu} = \infty$. Let the centres of mass average coordinates of microinclusions be transformed affinely proportionally to the homogeneous strain of the unit cube $\lambda_i(t)$ and the average values of principal dilatations of microinclusions be $\varphi_i(t)$. Then the coor–

dinates $x^i_{0p_\tau}$, $x^i_{0q_\nu}$, and the configurative sum (4) of the isotropic network of centres of mass of the microinclusions after elementary transformations will be equal to (in (5) the summation should not be performed over i) :

$$x^i_{0p_\tau} = \lambda_i X^i_{0p} + \varphi_i d^i_{0p_\tau} \qquad x^i_{0q_\nu} = \lambda_i X^i_{0p} + \varphi_i d^i_{0q_\nu}, \tag{5}$$

$$Q^* = A_0^* J^G \exp\left[- (K \lambda_i \lambda_i + 2D \lambda_i \varphi_i + B \varphi_i \varphi_i) \right], \tag{6}$$

$$K = \frac{1}{b^2} \sum_{p_\tau > q_\nu}\sum Z^{-1}_{p_\tau q_\nu}\left(X^i_{0p} - X^i_{0q}\right)\left(X^i_{0p} - X^i_{0q}\right),$$

$$D = \frac{1}{b^2} \sum_{p_\tau > q_\nu}\sum Z^{-1}_{p_\tau q_\nu}\left(X^i_{0p} - X^i_{0q}\right)\left(d^i_{0p_\tau} - d^i_{0q_\nu}\right), \tag{7}$$

$$B = \frac{1}{b^2} \sum_{p_\tau > q_\nu}\sum Z^{-1}_{p_\tau q_\nu}\left(d^i_{0p_\tau} - d^i_{0q_\nu}\right)\left(d^i_{0p_\tau} - d^i_{0q_\nu}\right),$$

$$X^i_{0p_\tau} = X^i_{0p} + \left(X^i_{0p_\tau} - X^i_{0p}\right) = X^i_{0p} + d^i_{0p_\tau},\ d^i_{0p_\tau} = X^i_{0p_\tau} - X^i_{0p}, \tag{8}$$

$$X^i_{0q_\nu} = X^i_{0q} + \left(X^i_{0q_\nu} - X^i_{0q}\right) = X^i_{0q} + d^i_{0q_\nu},\ d^i_{0q_\nu} = X^i_{0q_\nu} - X^i_{0q}, \tag{9}$$

where d^i_{0p}, $d^i_{0q_\nu}$ are the projections on the axis X^i of the vectors $\mathbf{d}_{0p_\tau}$, $\mathbf{d}_{0q_\nu}$ which are connecting the microinclusions centres of mass (p, q) with the points $p_\tau\ q_\nu$, situated on their surfaces, $(X^i_{0p} - X^i_{0q})$ are the projections on the axis X^i of the vector $\mathbf{r}^{(0)}_{pq}$, which connects the centres of mass of a pair of microinclusions. Substitution of (6), (7) into (3), leads to

$$W_1 = f_1(T) + \frac{k\,T}{2}\Big[\langle K \rangle (\lambda_i \lambda_i - 3) + 2\langle D \rangle (\lambda_i \varphi_i - 3)$$
$$+ \langle B \rangle (\varphi_i \varphi_i - 3) - 2\langle G \rangle \ln J \Big]. \tag{10}$$

Let us compute the coefficients $\langle K \rangle, \langle D \rangle, \langle B \rangle, \langle G \rangle$. Let us introduce the notations (not to sum by τ, ν)

$$\left(r^{(0)}_{p_\tau q_\nu}\right)^2 = \left(X^i_{0p_\tau} - X^i_{0q_\nu}\right)\left(X^i_{0p_\tau} - X^i_{0q_\nu}\right),\ \left(X^i_{0p_\tau} - X^i_{0q_\nu}\right) = \ell^i_{p_\tau q_\nu} \left| \mathbf{r}^{(0)}_{p_\tau q_\nu} \right|$$

$$R^{(0)}_{p_\tau q_\nu} = (Z_{p_\tau q_\nu} b)^{-1} \mid \mathbf{r}^{(0)}_{p_\tau q_\nu} \mid, \quad c_{p_\tau q_\nu} = \mid d^{(0)}_{p_\tau q_\nu} \mid / \mid \mathbf{r}^{(0)}_{p_\tau q_\nu} \mid$$

$$(d^{\,i}_{p_\tau} - d^{\,i}_{0q_\nu}) = m^{i}_{p_\tau q_\nu} \mid \mathbf{d}^{(0)}_{p_\tau q_\nu} \mid, \ \cos(\gamma_{p_\tau q_\nu}) = \ell^{\,i}_{p_\tau q_\nu} m^{i}_{p_\tau q_\nu},$$

$$\ell^{i}_{p_\tau q_\nu} \ell^{i}_{p_\tau q_\nu} = 1, \ m^{i}_{p_\tau q_\nu} m^{i}_{p_\tau q_\nu} = 1, \ \mathbf{d}^{(0)}_{p_\tau q_\nu} = \mathbf{d}_{0p_\tau} + \mathbf{d}_{0q_\nu}, \tag{11}$$

where $\mathbf{r}^{(0)}_{p_\tau q_\nu}$ is the vector connecting the ends of the chains situated between the micro-inclusions (p, q) in the points τ and ν in the state of non-strain, $R^{(0)}_{p_\tau q_\nu}$ is the average relative length of the chain, which connects the microinclusions (p, q) in the points τ, ν, in the state of non-strain, and $\ell^{i}_{p_\tau q_\nu}$ are the director cosines of the vector $\mathbf{r}^{(0)}_{p_\tau q_\nu}$. Substituting expressions (8), (9), (11) into the relation (7), we have

$$K = \sum_{p_\tau > q_\nu}\sum Z_{p_\tau q_\nu} (R^{(0)}_{p_\tau q_\nu})^2 \left[1 - 2c_{p_\tau q_\nu} \cos(\gamma_{p_\tau q_\nu}) + c^2_{p_\tau q_\nu} \right],$$

$$D = \sum_{p_\tau > q_\nu}\sum Z_{p_\tau q_\nu} (R^{(0)}_{p_\tau q_\nu})^2 \left[c_{p_\tau q_\nu} \cos(\gamma_{p_\tau q_\nu}) - c^2_{p_\tau q_\nu} \right], \tag{12}$$

$$B = \sum_{p_\tau > q_\nu}\sum Z_{p_\tau q_\nu} (R^{(0)}_{p_\tau q_\nu})^2 \, c^2_{p_\tau q_\nu}.$$

Let $Z_{p_\tau q_\nu}$, $R^{(0)}_{p_\tau q_\nu}$, $\gamma_{p_\tau q_\nu}$, $c_{p_\tau q_\nu}$ be random variables, $0 < Z < \infty$; $0 < R < 1$; $0 < \gamma < 2\pi$, $0 < c < \infty$; $\Phi(Z)$ the functions of distribution of chains according to the number Z of the segments, $G(Z, R)$ is the probability density of the chains having Z segments, by R; $P_1(y)$ is the probability density relative to the angle $y = \gamma/2\pi$; $P_2(c)$ is the probability density of the random variable c. Then introducing these functions into the expression (12), we have

$$\langle K \rangle = \int_0^\infty dZ \int_0^1 \Phi(Z)G(Z, R)ZR^2 \, dR \int_0^\infty \int_0^1 [\, 1 - 2\cos(2\pi y)\, c +$$

$$+ c^2 \,] P_1(y)P_2(c) dy dc; \ \langle B \rangle = \int_0^\infty dZ \int_0^1 \Phi(Z)G(Z, R)ZR^2 \, dR \int_0^\infty c^2 P_2(c) dc\,;$$

$$\langle D \rangle = \int_0^\infty dZ \int_0^1 \Phi(Z)G(Z,R)ZR^2\,dR \int_0^\infty \int_0^1 [\cos(2\pi y)c - c^2]\,P_1(y)P_2(c)\,dy\,dc\,. \quad (13)$$

The probability density

$$G\,(Z,R) = 4\left(\frac{3}{2}Z\right)^{3/2} R^2\,\pi^{-0,5} \exp\left(-\frac{3}{2}ZR^2\right) \quad (14)$$

is proposed by James and Guth [2]. By substitution of the function of distribution (14) into the relation (13) we obtain after integration and elementary transformations

$$\langle K \rangle = G_0\,(1-2\langle c \rangle\langle \cos\gamma \rangle + \langle c^2 \rangle);\ \langle D \rangle = G_0\,(\langle c \rangle\langle \cos\gamma \rangle - \langle c^2 \rangle)$$

$$\langle B \rangle = G_0 \langle c^2 \rangle,\ G_0 = \langle G \rangle = \int_0^\infty \Phi(Z)\,dZ \int_0^1 G(Z,R)ZR^2\,dR = 1, \quad (15)$$

where G_0 is the number of chains of the elastic matrix per unit volume. Let us analyse the limitations on the coefficients $\langle K \rangle, \langle D \rangle, \langle B \rangle$, which correspond to the conditions of unstressed microinclusions. The stresses P_i^* transmitted to the microinclusions are equal to

$$p_i^* = -\frac{\partial W^*_{(1)}}{\partial \varphi_i} = -kT\,C_0^3\,(\langle D \rangle \lambda_i + \langle B \rangle \varphi_i);\ W_1{}^* = C_0^{-3}\,W_1,\ 0 < C_0 < (0{,}5)^{1/3}, \quad (16)$$

where C_0 is the coefficient of heterogeneity, which is equal to the ratio of $\langle D \rangle$, the dimension of the microinclusion, to $\langle r_0 \rangle$, the distance between the centres of mass of microinclusions. In the unstressed state $p_i^* = 0$, $\lambda_i = 1$, $\varphi_i = 1$. If we put the expression (16) equal to zero, and using (15), we obtain

$$\langle D \rangle + \langle B \rangle = 0,\ \langle C \rangle\langle \cos\gamma \rangle = 0,\ \langle \cos\gamma \rangle = 0,\ \langle c^2 \rangle = c^2. \quad (17)$$

After substitution of (15, 17) into the eq. (10) which expresses the free energy of the elastic matrix, we find

$$W_1 = f_1(T) + 0{,}5\,G_0\,kT\,[\,(\lambda_i\,\lambda_i - 3) + c^2(\lambda_i - \varphi_i)\,(\lambda_i - \varphi_i) - 2\ln J\,]. \quad (18)$$

For the microinclusions whose nuclei do not depend on the temperature, the functionals W_2 are accepted to be

$$W_2 = f_2(T) + C_0^3 \left[0{,}5 \int_0^t \int_0^t \Pi(t-\tau,\, t-\eta) \mathrm{d}e_i^*(\tau) \mathrm{d}e_i^*(\eta) + 0{,}5\, K_0 (v^* - 3\alpha_0\, \Theta)^2 \right]$$

$$e_i^* = (\varphi_i - 1) - \frac{1}{3} v^*, \quad v^* = \varphi_1 + \varphi_2 + \varphi_3 - 3\ , \tag{19}$$

where $\Pi(t)$ is the function of relaxation, α_0 is the coefficient of heat dilatation, and $K_0 = \lambda_0 + \frac{2}{3}\mu_0$, λ_0, μ_0 are the constants of Lamé. When $\varphi_i - 1 \simeq \varepsilon_i^*$, the functional (19) is transformed into the functional of free energy for the thermoviscoelastic body having finite strain or small strain. Substitution of (19) and the work of exterior forces (16), (17)

$$\delta A_2 = -\, C_0^3 p_i^*\, \mathrm{d}\varphi_i = G_0\, k\, T\, c^2 \left[(s_i - e_i^*)\, \mathrm{d}e_i^* + \frac{1}{3}(v - v^*) \mathrm{d}v^* \right],$$

$$s_i = (\lambda_i - 1) - \frac{1}{3} v, \quad v = (\lambda_1 + \lambda_2 + \lambda_3 - 3)\ ,$$

$$(\lambda_i - \varphi_i)\,(\lambda_i - \varphi_i) = (s_i - e_i^*) + \frac{1}{3}(v - v^*)^2\ , \tag{20}$$

into the eq. (1) written for the microinclusions

$$\mathrm{d}W_2 + S_2\, \mathrm{d}T = \delta A_2 - \Lambda\, \mathrm{d}t$$

and after equating, in the relation so obtained, the coefficients of $\mathrm{d}e_i^*$, $\mathrm{d}v^*$, $\mathrm{d}T$, $\mathrm{d}t$, we obtain the system of equations

$$3k_0 (v^* - 3\alpha_0\, \Theta) = G_0\, k\, T\, c^2\, C_0^{-3}\, (v - v^*)\ ,$$

$$2\mu_0 \left[e_i^*(t) - \int_0^t \Gamma(t - \tau)\, e_i^*(\tau) \mathrm{d}\rho \right] = G_0\, k\, T\, c^2\, C_0^{-3}\, (s_i - e_i^*)\ , \tag{21}$$

and the expression for the entropy S_2 and the dissipative function Λ. The solution of these equations is

$$e_i^*(t) = A_0\, s_i(t) + \int_0^t K(t, \tau) A_0(\tau) s_i(\tau) \mathrm{d}\tau\ , \tag{22}$$

$$v^*(t) = A_1 \left[v(t) + 3\, (a_0 T)^{-1} \alpha_0\, \Theta \right], \tag{23}$$

where $K(t, \tau)$ is the resolvent of the nucleus $\Gamma(t-\tau)/(1+a_0T)$

$$A_0 = \frac{a_0T}{1+a_0T},\ A_1 = \frac{a_1T}{1+a_1T},\ a_0 = \frac{G_0\,k\,c^2}{2\mu_0\,C_0^3},\ a_1 = \frac{G_0 k\,c^2}{3k_0\,C_0^3}.$$

The eqs. (22), (23) determine the relation between strains $\varphi_i(t)$ of the microinclusions and strains $\lambda_i(t)$ and the temperature $T(t)$ of the macroparticle. Integrating the expression (19) by parts, eliminating $e_i^*(t)$, $v^*(t)$ by means of the eqs. (20)-(23), and using (2) and (12) we obtain

$$W = f(T) + p[J - F(T)] + 0{,}5\,G_0\,k\,T\Big\{(\lambda_i\,\lambda_i - 3) - 2\ln J +$$

$$+ c^2\Big[(1+a_0T)^{-1}\int_0^t\int_0^t P(t,\tau)P(t,\eta)\,ds_i(\tau)ds_i(\eta) + \frac{(v - 3\alpha_0\Theta)^2}{3(1+a_1T)}\Big]\Big\}, \qquad (24)$$

$$\partial P(t,\tau)/\partial\tau = (1+a_0T)A_0(\tau)K(t,\tau),\quad P(t,t) = 1,$$

$$f(T) = f_1(T) + f_2(T) = W_0 - S_0\,\Theta + C_\varepsilon\,[\,\Theta - T\ln(T/T_0)\,],$$

where C_ε, is the specific heat. Substituting the free energy (24) into the thermodynamic eq. (1), taking into account the work of exterior forces $\delta A = F(T)\,\sigma_i\,\lambda_i^{-1}\,d\lambda_i$, and identifying in the obtained expression the coefficients of $d\lambda_i$, dT, dt, dp, we find the expression (with no summation)

$$\sigma_i = \frac{G_0 k\,T}{F(T)}\Big\{(\lambda_i^2 - 1) + \lambda_i\,c^2\Big[(1+a_0T)^{-1}\int_0^t P(t,\tau)\alpha_i(\tau) + \frac{(v - 3\alpha_0\Theta)}{3(1+a_1T)}\Big]\Big\} + P$$

and the expression for entropy S, the dissipative function Λ and $J = F(T)$ - the conditions of mechanical non-compressibility. Using the same method we can evaluate the free energy for compressible elastomers as

$$W = f(T) + 0{,}5\,G_0\,k\,T\Big\{(\lambda_i\,\lambda_i - 3) - \frac{2}{1-2\nu^*}\ln J + \frac{4\nu^*}{1-2\nu^*}(J-1) +$$

$$+ c^2\Big[(1+a_0T)^{-1}\int_0^t\int_0^t P(t,\tau)P(t,\eta)ds_i(\tau)ds_i(\eta) + \frac{(v - 3\alpha_0\Theta)^2}{3(1+a_1T)}\Big]\Big\}, \qquad (25)$$

where ν^* is the Poisson coefficient of the matrix. In the particular case when $c^2 = 0$ the expressions (24), (25) give the potentials of Treloar and Blatz, accordingly.

3 Conclusion

The theory of thermoviscoelasticity that has been proposed in this work, can be useful in solving the problems of the synthesis of elastomers having desired properties. The general case of the elastomers with "n" thermoviscoelastic microinclusions, was examined in Ref. [1]. Using the same method, we can study the elastomers having elastoplastic microinclusions, those with microinclusions having piezoelectric properties, etc.

Acknowledgements

This paper was written while the author was a visitor at Laboratoire de Modélisation en Mécanique, Université de Pierre et Marie Curie, Paris, France, within the France-U.S.S.R. Scientific Exchange Program.

References

[1] Dounaev, I.M., A variant of nonlinear theory of thermoviscoelasticity of elastomers, *Mechanics of Solids*, **1**, 110-120 (1985) (in Russian).

[2] James, H.M. and Guth, E., Theory of elastic properties of rubber, *J. Chem. Phys.* **11**, 455-481 (1943)

[3] Treloar, L.R.G., The mechanics of rubber elasticity, *Proc. R. Soc. London*, Ser. A, **351**, 301-330 (1976).

[4] Volkenstein, M.V., Configurational Statistics of Polymeric Chains, S.N. Timasheff and M.J. Timasheff, Trans: Wiley-Interscience, New York (1963).

Dounaev I.M.
Department of Strength of Materials
Polytechnic Institute of Krasnodar
Moskovskaya 2
350042 Krasnodar
U.S.S.R.

R. ELLINGHAUS, C. PAPENFUSS AND W. MUSCHIK

Electrostriction in Liquid Crystals

1 Introduction

Starting with a special trigonal ansatz for the alignment tensor field of a cholesteric liquid crystal of positive dielectric anisotropy we prove, by using Landau theory, that this ansatz has a lower free-energy density than the cholesteric phase and the O^5-phase of Grebel, Hornreich and Shtrikman [3] in a certain region of the reduced temperature-chirality-plane. In this domain the trigonal phase may be the thermodynamically stable one. Using this ansatz we get an implicit equation for the interplanar spacing of certain lattice planes as a function of the external electric field. This equation is solved for the weak-field limit, yielding an expression for the electrostriction.

2 Alignment tensor field

As usual, we choose the free-energy density

$$F = \frac{1}{2}\left(aA_{ij}^2 + c_1 A_{ij,l}^2 + c_2 A_{ij,i} A_{kj,k} - 2\, d\varepsilon_{ijk} A_{in} A_{jn,k}\right)$$

$$- \beta A_{ij} A_{jk} A_{ki} + \gamma (A_{ij}^2)^2 - \frac{\chi}{2} E_i A_{ij} E_j , \qquad (1)$$

with the alignment tensor field (A_{ij}). We assume that the anisotropic electric susceptibility is positive $(\chi > 0)$. When applying an external electric field parallel to one special direction of translational symmetry the length of periodicity in this direction is shortened, whereas the length of periodicity in the two perpendicular directions is elongated, because no change of the volume of the unit cell is observed in electrostrictive effects [1,4]. Consequently we assume that the product of these lengths is invariant under the application of the electric field.

The electric field E and the direction of the z-axis are chosen parallel. The ansatz for (A_{ij}) consists in the superposition of a cholesteric helix, propagating along the z-axis and of two distorted cholesteric helices, propagating along the x- and y-axis of a Cartesian frame. These distorted helices are described by a function φ with $\varphi(n\, g_E) = 2n\pi$ for all integers n. g_E is the length of periodicity perpendicular to E. φ is a solution of the differential equation

$$\varphi' = \pm w \sqrt{1 - \frac{\mathcal{E}^2}{w^2} \sin^2(\varphi)}\,, \quad w > 0, \quad \mathcal{E} := 2E \sqrt{\frac{\chi_r}{\mu}}\,. \tag{2}$$

Later the parameter w will be determined by minimizing the free energy. For appropriate w it is known that the vector field given by $(\cos \varphi(y), 0, \sin \varphi(y))$ represents the director of a cholesteric helix distorted by the electric field E parallel to the z-axis [2]. Our ansatz reads explicity

$$(A_{ij}) := \alpha \begin{pmatrix} \cos^2(\varphi(y)) + \sin^2(q_z z) - 1 & \dots \\ \sin(q_z z)\cos(q_z z) & \dots \\ \sin(\varphi(y))\cos(\varphi(y)) & \end{pmatrix}$$

$$\begin{matrix} \sin(q_z z)\cos(q_z z) \\ \dots \quad \cos^2(q_z z) + \sin^2(\varphi(x + g_E/4) - \pi/2) - 1 \quad \dots \\ \sin(\varphi(x + g_E/4) - \pi/2)\cos(\varphi(x + g_E/4) - \pi/2) \end{matrix} \tag{3}$$

$$\begin{pmatrix} & \sin(\varphi(y))\cos(\varphi(y)) \\ \dots & \sin(\varphi(x + g_E/4) - \pi/2)\cos(x + g_E/4) - \pi/2) \\ & \cos^2(\varphi(x + g_E/4) - \pi/2) + \sin^2(\varphi(y)) - 1 \end{pmatrix}$$

For symmetry reasons the following implication holds:

$$E = 0 \Rightarrow \{\varphi(x) = q_z x \wedge w = \pm q_z \wedge g_E = 2\pi/q_z\}\,, \tag{4}$$

so that (A_{ij}) has trigonal (C_3^1) symmetry in the field free case. In reduced quantities, defined by the relations

$$\mathcal{F} = 36F \frac{\gamma^3}{\beta^4}, \quad \alpha = s\mu\,, \quad s = \frac{\beta}{\sqrt{6\gamma}}, \quad t = 12a \frac{\gamma}{\beta^2}, \quad \xi_R^2 = 12c_1 \frac{\gamma}{\beta^2},$$

$$q_c = \frac{d}{c_1}, \quad \kappa = q_c \xi_R \text{ and } \chi_r = 3\sqrt{6\chi}\, \frac{\gamma^2}{\beta^3} \tag{5}$$

this ansatz results in the mean free-energy density (with $E = 0$)

$$\bar{\mathcal{F}}_{C_3^1} := \frac{\int_V \mathcal{F}_{C_3^1}\,\mathrm{d}z\mathrm{d}y\mathrm{d}x}{V} = \frac{1}{16}\left[\left(24(q_z/q_c)^2 + 24(q_z/q_c)\right)\mu^2\kappa^2 + 6\mu^2 t + 39\mu^4\right], \quad (6)$$

for any volume of periodicity V. This expression is minimized by

$$q_z = -\frac{q_c}{2} \text{ and } \mu = \sqrt{\frac{\kappa^2 - t}{13}}, \quad (7)$$

consequently we get

$$\bar{\mathcal{F}}_{C_3^1} = -\frac{3(t-\kappa^2)^2}{208}, \quad \kappa^2 > t. \quad (8)$$

The free-energy density for the O^5-phase is [3]

$$\bar{\mathcal{F}}_{O^5} = \mathcal{F}_{O^5} = \frac{1}{4}(t-\kappa^2)\,\mu^2 - (23\sqrt{2}/32)\mu^3 + (499/384)\mu^4, \quad (9)$$

which is minimized by

$$\mu = \frac{\sqrt{47904(\kappa^2 - t) + 42849} + 207}{499\sqrt{2}}. \quad (10)$$

Introducing (10) into (9) and using (8) we get

$$\bar{\mathcal{F}}_{C_3^1} - \bar{\mathcal{F}}_{O^5} = -\frac{249}{103792}\Phi^2 + \frac{69}{996004}\Phi\left(2\sqrt{47904\Phi + 42849} + 621\right) +$$
$$+ \frac{985527}{7952095936}\left(\sqrt{47904\Phi + 42849} + 207\right), \quad (11)$$

where $\Phi := \kappa^2 - t$. By numerical computation we get from (11)

$$\Phi > 196.31 \Rightarrow \bar{\mathcal{F}}_{C_3^1} < \bar{\mathcal{F}}_{O^5}. \quad (12)$$

In the framework of Landau theory the phase boundary for the isotropic-cholesteric transition is given by [5]

$$t_{IC} = \frac{3}{4}\kappa^2 + 1 \,. \tag{13}$$

Therefore the alignment tensor field (3) results in a smaller free-energy than the alignment tensors for the O^5 or the cholesteric phase [3], at least in the region

$$t > \frac{3}{4}\kappa^2 + 1 \wedge \kappa^2 > 196.31 + t \,. \tag{14}$$

Detailed considerations yield

$$\kappa > B \Rightarrow \mathcal{F}_{col} > \mathcal{F}_{O^5} \tag{15}$$

for some $B < 2.52$, therefore the free-energy of the C_3^1-phase is lower than the free-energy for the O^5, I- and cholesteric-phases in the domain

$$\kappa^2 > 196.31 + t, \quad t > 0 \,. \tag{16}$$

The smallest chirality κ in this area is given by 14.01. The resulting phase diagram is shown in Figure 1.

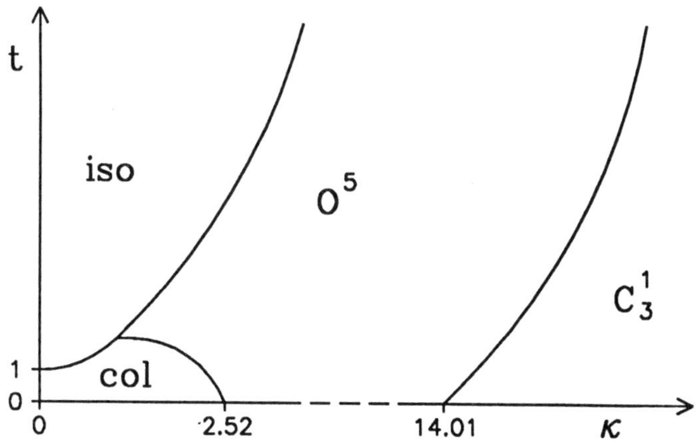

Figure 1

3 The free-energy density for non-vanishing electric field

In the following we need some equations and definitions ($\mathbf{K}(\cdot)$ is the complete elliptic integral of the first kind):

$$g_E = \int_0^{g_E} \mathrm{d}x = \pm\int_0^{2\pi} \frac{1}{w\sqrt{1-\frac{\mathcal{E}^2}{w^2}\sin^2(\varphi)}}\mathrm{d}\varphi = \pm\frac{4}{w}\,\mathbf{K}(\mathcal{E}/w)\,, \tag{17}$$

$$E_1(x) := \int(\varphi'(x))^2\mathrm{d}x\,, \tag{18}$$

$$E_1(g_E) = \pm\int_0^{2\pi} w\sqrt{1-\frac{\mathcal{E}^2}{w^2}\sin^2(\varphi)\mathrm{d}\varphi}\,, \tag{19}$$

$$E_2(x) := \int\cos(2\varphi(x))\mathrm{d}x\,, \tag{20}$$

$$E_2(g_E) = \pm\int_0^{2\pi} \frac{\cos(2\varphi)}{w\sqrt{1-\frac{\mathcal{E}^2}{w^2}\sin^2(\varphi)}}\mathrm{d}\varphi\,, \tag{21}$$

$$E_3(x) := \int\cos(4\varphi(x))\mathrm{d}x\,, \tag{22}$$

and

$$E_3(g_E) = \pm\int_0^{2\pi} \frac{\cos(4\varphi)}{w\sqrt{1-\frac{\mathcal{E}^2}{w^2}\sin^2(\varphi)}}\mathrm{d}\varphi\,. \tag{23}$$

Using the expressions for the volume V of the unit cell of periodicity along the x–, y– an z–axis ($V = (4\pi/q_c)^3$ for $E = 0$ and $V = |2\pi g_E^2/q_z|$, for non-vanishing E), the experimental verified invariance of this volume [1,4] yields the identity

$$q_z = -\frac{g_E^2 q_c^3}{32\pi^2}. \tag{24}$$

(1), (3) and (17) to (24) yield the free-energy per unit cell divided by its volume

$$\frac{\int_0^{g_E}\int_0^{g_E}\int_0^{2\pi/q_z} \mathcal{F}_{C_3^1}\,\mathrm{dzdydx}}{2\pi g_E^2/q_z} =$$

$$= \frac{\kappa^2\mu^2 g_E^4 q_c^4}{2048\pi^4} - \frac{\kappa^2\mu^2 g_E^2 q_c^2}{64\pi^2} + g_E^{-1}\,\kappa^2\,\mu^2\,q_c^{-2}\,(E_1(g_E) + 2\pi q_c) +$$

$$+ \frac{3}{8}t\mu^2 + \frac{1}{8}g_E^{-2}\,t\mu^2\,E_2^2(g_E) + \frac{39}{16}\mu^4 + \frac{1}{4}g_E^{-1}\,\mu^4\,E_3(g_E) +$$

$$+ g_E^{-2}\,\mu^4\,\frac{20E_2^2(g_E) + E_3^2(g_E)}{16} + g_E^{-1}\,\mu E^2\chi_r\,. \tag{25}$$

(25) has to be minimized by substituting the functions E_1, E_2, E_3 and g_E according to (17) to (23) and minimizing with respect to w. Doing so, is of course a matter of (probably hopeless) numerical computation. Here we give an approximation to this problem in the weak field limit. We assume that E is small enough, that cubic and higher powers of $\mathcal{E}/w$ can be neglected. This allows us to expand the roots in (17) to (23). This approximation results in

$$g_E = -\pi\left(\frac{2}{w} + \frac{\mathcal{E}^2}{2w^3}\right) \tag{26}$$

$$E_1(g_E) = -\pi w\left(2 - \frac{\mathcal{E}^2}{2w^2}\right) \tag{27}$$

$$E_2(g_E) = -\pi\frac{\mathcal{E}^2}{4w^3} \text{ and } E_3(g_E) = 0\,. \tag{28}$$

It holds $w = q_c/2$ if $E = 0$, as mentioned earlier. Therefore we set

$$w = q_c\left(\frac{1}{2} + \Delta_w\right), \tag{29}$$

with Δ_w small compared to 1. We neglect powers of cubic and higher order of Δ_w. With these assumptions and approximations and with $\varepsilon := \mathcal{E}/q_c$ we compute from (25)

$$\overline{\mathcal{F}}_{C_3^1} = \Delta_w^2 \left(3(8\varepsilon^2 + 1)\, \mu^2 \kappa^2 - 6E^2 \varepsilon^2 \chi_r \mu\right) + $$

$$+ \Delta_w \left(2E^2 \mu\chi_r \varepsilon^2 - 3\mu^2 \kappa^2 \varepsilon^2\right) -$$

$$- \frac{1}{2} E^2 \mu\chi_r \varepsilon^2 + \frac{39}{16}\mu^4 + \frac{3}{8}(t - \kappa^2)\, \mu^2 . \tag{30}$$

Minimizing this expression with respect ot Δ_w yields

$$\Delta_w = \varepsilon^2 \left(\frac{1}{2} - \frac{E^2 \chi_r}{3\mu\kappa^2} \right), \tag{31}$$

again by neglecting higher-order terms. Introducing this relation into (26) yields (keeping in mind, that ε is of the same order of magnitude as $\mathcal{E}/w$

$$g_E = -\pi \left(\frac{4}{q_c (1 + 2\Delta_w)} + \frac{4\mathcal{E}^2}{q_c^3 (1 + 2\Delta_w)^3} \right). \tag{32}$$

From the last equation we get

$$g_E = -\frac{4\pi}{q_c} \left(1 + \frac{2\varepsilon^2 \mathcal{E}^2 \chi_r}{3\mu\kappa^2} \right), \tag{33}$$

which together with (24) gives the result for the length of periodicity ($g_{\parallel E}$) parallel to the electric field

$$g_{\parallel E} = -\frac{2\pi}{q_z} = \frac{4\pi}{q_c} \left(1 - \frac{4\varepsilon^2 \mathcal{E}^2 \chi_r}{3\mu\kappa^2} \right) = \frac{4\pi}{q_c} \left(1 - \frac{64 E^4 \chi_r^3}{3 q_c^2 \mu^3 \kappa^2} \right). \tag{34}$$

Introducing (31) into (30) and minimizing the resulting expression with respect to μ we get

$$\mu = \sqrt{\frac{\kappa^2 - t}{13}} + O(\varepsilon)\,, \tag{35}$$

which finally yields

$$g_{\|E} = \frac{4\pi}{q_c}\left(1 - \frac{832\sqrt{13}\,E^4\,\chi_r^3}{3q_c^2\sqrt{(\kappa^2 - t)^3}\;\kappa^2}\right),\quad E \ll q_c\,\frac{(\kappa^2 - t)^{1/4}}{\sqrt{\chi_r}}. \tag{36}$$

4 Summary

We predict the possibility, an alignment tensor field of C_3^1 symmetry may be thermodynamically stable in a material characterized by sufficiently high chirality. An equation describing the deformation of the periodic structure induced by the alignment tensor field is derived and approximately solved for weak fields. In this limit electrostriction occurs with a leading term in E^4. Its coefficient is an explicitly calculated expression in the reduced material parameters of Landau theory.

References

[1] V.E. Dmitrienko. Electro-optic effects in blue phases. In *Proceedings of the 12. International Liquid Crystal Conference, Freiburg* (1988).

[2] P.G. de Gennes. Calcul de la distorsion d'une structure cholestérique par un champ magnétique. *Sol. State Comm.* **6** (1968).

[3] H. Grebel, R.M. Hornreich, and S. Shtrikman. Landau theory of blue phases. *Phys. Rev. A* **28** (1983).

[4] G. Heppke, B. Jérôme, H.-S. Kitzerow, and P. Pieranski. Electrostriction of BPI and BPII for blue phase systems with negative dielectric anisotropy. *J. Phys. France* **50** (1989).

[5] R.M. Hornreich, M. Kugler, and S. Shtrikman. Localized instabilities and the order-disorder transition in cholesteric liquid crystals. *Phys. Rev. Lett.* **48** (1982).

Ellinghaus R., Papenfuss C. and Muschik W.
Institut für Theoretische Physik
Technische Universität Berlin
Hardenbergstrasse 36
D-1000 Berlin 12
F.R.G.

D. LHUILLIER

Dilute suspensions with non-homogeneous particle concentrations

1 Introduction

Transforming results obtained for one particle (in a infinite fluid) to average equations for a suspension of particles seems an easy task for a dilute suspension: since the interaction between particles is negligible, the transformation is likely to be a mere multiplication by the number of particles per unit volume. In fact, things are not that simple when the particle concentration, dilute as it may be, is not uniform over the suspension. Upon averaging, there are more particles on one side of the averaging volume than on the other side. This problem was stressed by many authors and in particular by Nigmatulin [10]. In this contribution, we detail the way the averaging problem must be handled for spherical particles with non-uniform concentration.

There is also a second riddle to solve: one usually considers a velocity field $\mathbf{V}_\infty$ in which the particle is introduced and the problem is to express $\mathbf{V}_\infty$ and its spatial derivatives in terms of the average velocity $\mathbf{V}$ of the suspension and its spatial derivatives. Similarly, one must relate the relative velocity $\mathbf{V}_p - \mathbf{V}_\infty$ of the single particle approach to the relative velocity $\mathbf{V}_p - \mathbf{V}$ of the continuum description. We propose a solution to this problem, again in the case of spherical particles.

2 A single particle in a prescribed velocity field

A. The fluid stress on the particle surface

Let us consider a fluid with a stationary but non-uniform velocity field $\mathbf{V}_\infty(\mathbf{x})$, and a stress field

$$\Pi_\infty = - p_\infty \mathbf{I} + 2\eta_0 \mathbf{D}_\infty . \tag{1}$$

Let us suppose at the outset that $\mathbf{V}_\infty$ is a quadratic function of position (like in a Poiseuille flow for instance) so that the strain rate $\mathbf{D}_\infty$ varies linearly with $\mathbf{x}$ while any second-order derivative of $\mathbf{V}_\infty$ is a constant. We also suppose that the fluid flow is slow enough to be governed by the Stokes equation

$$\nabla \cdot \Pi_\infty + \rho_f^0 \mathbf{g} = 0.$$

A rigid sphere of radius a with velocity $\mathbf{V}_p$ and rotation velocity $\boldsymbol{\omega}_p$ is introduced in the above-defined fluid. All around the particle, the flow is perturbed and the fluid stress becomes Π_f with

$$\nabla \cdot \Pi_f + \rho_f^0 \, \mathbf{g} = 0 \,. \tag{2}$$

The detailed behaviour of $\Pi_f(\mathbf{x})$ stems from the knowledge of the perturbations $\mathbf{V}_f(\mathbf{x}) - \mathbf{V}_\infty(\mathbf{x})$ and $p_f(\mathbf{x}) - p_\infty(\mathbf{x})$, which are solutions of the Stokes equation with prescribed boundary conditions at the particle surface. The general solution of the problem was found by Lamb in terms of a development in spherical harmonics [6]. The relevant number of spherical harmonics depends on the complexity of the field $\mathbf{V}_\infty(\mathbf{x})$ and the complexity of the particle motion. We have obtained (is this an original result?) Lamb's solution for a rigid particle in a quadratic velocity field. It involves eight spherical harmonics (as compared to five for a linear velocity field and three for a constant one). We will not write the complete solution here for the velocity and pressure fields, first because we want to save some space, but most importantly, because the only interesting quantity, as far as the particle motion is concerned, is the force $\Pi \cdot \mathbf{n}$ at a point $\mathbf{x}_s = \mathbf{R} + a\mathbf{n}$ of the particle surface. In fact, we found

$$\Pi^f_{ij} \, n_j \big|_s = \Pi^\infty_{ij} \, n_j \big|_s - \frac{3\eta_0}{2a} P_i + 3\eta_0 \, (D^\infty_{ij} - \omega_{ij}) n_j + a\eta_0 \, \varphi_{ijk} \, n_j \, n_k \tag{3}$$

where $\omega_{ij} = \omega^p_{ij} - \omega^\infty_{ij}$ is the antisymmetric tensor associated with the relative rotation, P_i is a generalized relative velocity defined as

$$\mathbf{P}(\mathbf{R}) = \mathbf{V}_p - \mathbf{V}_\infty(\mathbf{R}) - \frac{a^2}{6} \nabla^2 \, \mathbf{V}_\infty \tag{4}$$

while φ_{ijk} is a constant tensor built from second-order derivatives of $\mathbf{V}_\infty$ and verifying

$$\varphi_{ijk} = \varphi_{jik} \text{ and } \varphi_{ikk} = 0 \,.$$

B. The particle stress

The particle stress Π_p is a symmetric tensor which must verify the continuity of forces at the particle surface (surface tension is neglected)

$$\Pi^p_{ij} \, n_j \big|_s = \Pi^f_{ij} \, n_j \big|_s \,. \tag{5}$$

If we want to have an expression compatible both with this requirement and with (3), we must write at any point $\mathbf{x} = \mathbf{R} + \mathbf{r}$ inside the particle

$$\Pi^{p}_{ij} = \Pi^{\infty}_{ij} + 3\eta_0 D^{\infty}_{ij} - \frac{3\eta_0}{2a^2}(P_i r_j + P_j r_i - P_k r_k \delta_{ij}) + \eta_0 \varphi_{ijk} r_k$$

$$- \frac{3\eta_0}{2a^2}(\omega_{ik} r_k r_j + \omega_{jk} r_k r_i) + h_{ij}(\mathbf{r})$$

where h_{ij} is a (yet undetermined) symmetric tensor vanishing at the particle surface.

C. The particle motion

The particle stress is not constant over the particle and, in particular, its divergence is

$$\nabla \cdot \Pi_p = - \rho^0_f \, \mathbf{g} - \frac{9\eta_0}{2a^2}\mathbf{P} - \frac{15\eta_0}{a^2}\boldsymbol{\omega}\cdot\mathbf{r} + \nabla \cdot \mathbf{h} \, .$$

The above expression is nothing but the representation as a force per unit volume of the force exerted by the fluid on the particle surface. But we know that the actual volume force is due to gravity and is a constant over the particle volume if the mass repartition is homogeneous; in this case,

$$\nabla \cdot \Pi_p + \rho^0_p \, \mathbf{g} = 0 \, , \tag{6}$$

and comparing with the former expression we deduce

$$(\rho^0_p - \rho^0_f)\mathbf{g} = \frac{9\eta_0}{2a^2}(\mathbf{V}_p - \mathbf{V}_\infty(\mathbf{R}) - \frac{a^2}{6}\nabla^2 \mathbf{V}_\infty) \tag{7}$$

while

$$\boldsymbol{\omega} = 0 \quad \text{and} \quad \mathbf{h} = 0 \, . \tag{8}$$

One can think of (7) as a balance of forces on the particle while the other two results express the balance of couples. In Ref. [9], we consider the extension of result (8) to the case of particles acted upon by external multipolar forces. To end with these one-particle results, we take (8) into account to write the particle stress as

$$\Pi^{p}_{ij} = \Pi^{\infty}_{ij} + 3\eta_0\, D^{\infty}_{ij} - \frac{3\eta_0}{2a^2}(P_i\, r_j + P_j\, r_i - P_k\, r_k\, \delta_{ij}) + \eta_0\, \varphi_{ijk}\, r_k\,. \tag{9}$$

3 The average stress of a suspension

A. General expression

Let us consider a mixture of fluid and particles with a particle stress obeying (6) and a fluid stress obeying (2). For our future averaging operations, it is convenient to introduce a phase function $\chi(\mathbf{x}, t)$ with value 1 in the particles and 0 in the fluid [3,5]. The spatial derivative of χ is singular on the interface where

$$\nabla\chi = -\,\mathbf{n}\,\delta_I\,, \tag{10}$$

δ_I being a Dirac function localized on the interface. Taking into account the continuity of forces at any point on the interfaces, one gets from (2), (5) and (6)

$$\nabla\cdot[\,\chi\,\Pi_p + (1-\chi)\Pi_f\,] + [\,\chi\rho^0_p + (1-\chi)\rho^0_f\,]\,\mathbf{g} = 0 \tag{11}$$

which upon averaging over a volume containing many particles becomes

$$\nabla\cdot\Pi + \rho\,\mathbf{g} = 0 \tag{12}$$

where

$$\rho = (1-\Phi)\rho^0_f + \Phi\,\rho^0_p$$

is the average mass per unit volume of the suspension (Φ being the particle volume fraction) while the suspension stress is defined as

$$\Pi = \langle\, \chi\Pi_p + (1-\chi)\,\Pi_f \,\rangle\,. \tag{13}$$

B. Dilute limit

For a dilute suspension with an unperturbed stress Π_∞, the average stress (13) can be rewritten as

$$\Pi = \Pi_\infty + \langle\, \chi(\Pi_p - \Pi_\infty)\,\rangle + \langle\, (1-\chi)(\Pi_f - \Pi_\infty)\,\rangle\,.$$

The last term of the right-hand side is always negligible and with (1) and (9) the above expression amounts to

$$\Pi_{ij} = -p_\infty \delta_{ij} + 2\eta_0 (1 + \frac{3\Phi}{2}) D^\infty_{ij} - \frac{3\eta_0}{2a^2} < \chi(P_i r_j + P_j r_i - P_k r_k \delta_{ij}) > +$$

$$+ \eta_0 \varphi_{ijk} < \chi r_k > \qquad (14)$$

The problem of the average stress is thus linked to the evaluation of terms like $<\chi\mathbf{r}>$ and more generally $<\chi\mathbf{r}f(\mathbf{R})>$ where $f(\mathbf{R})$ is constant over a particle but may vary from one particle to another, depending on the particle position in the volume of averaging.

4 The average of $\chi\mathbf{r}$ and related quantities

The averaging volume contains many particles completely inside but some of them are cut by the boundary (see Figure 1). The former are entirely bounded by the interface

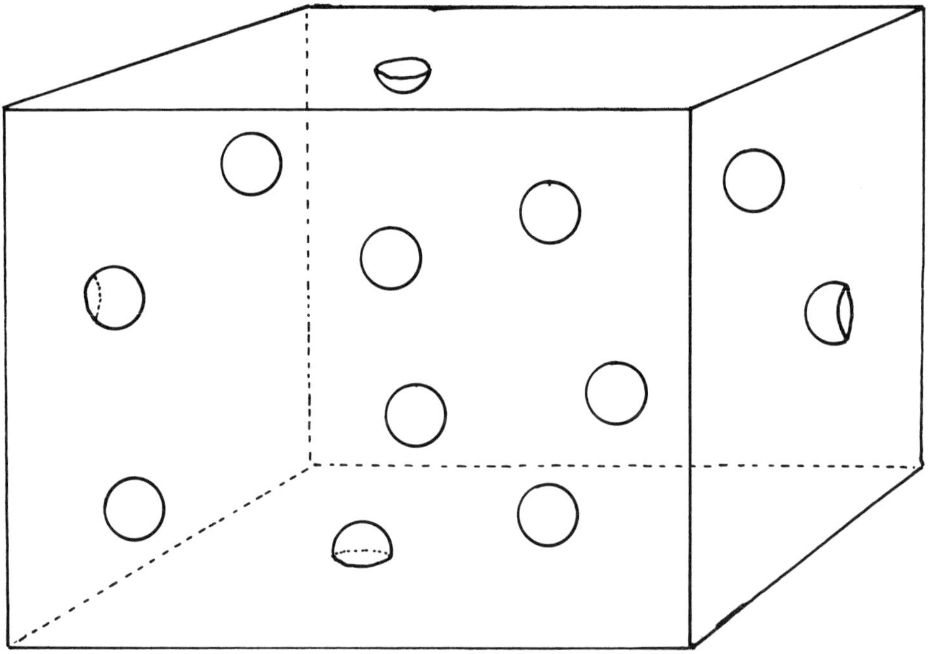

Figure 1. The volume of averaging, with many particles completely inside, and some others cut by the boundary

S_I, while the latter are bounded partly by S_I and partly by S_b (Figure 2). In any case, if V_p stands for the part of the particle volume included in the averaging volume, one can write

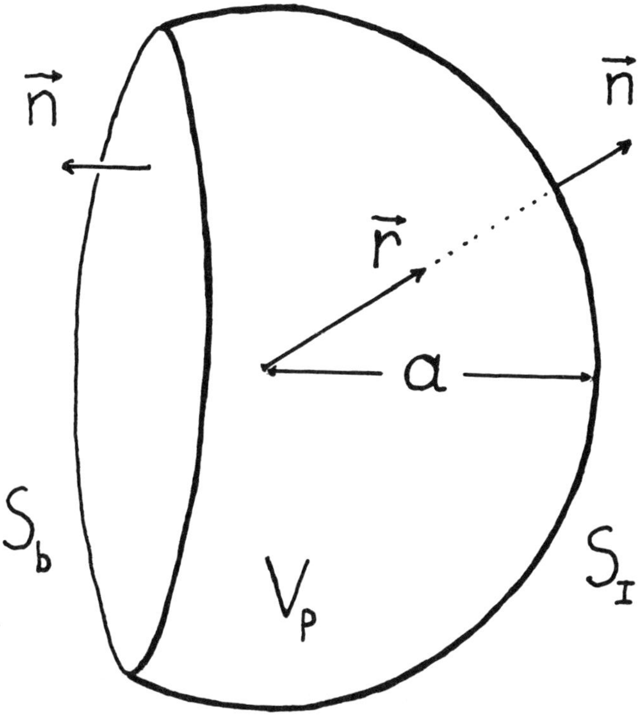

Figure 2. A particle cut by the boundary of the volume of averaging.

$$\int_{V_p} \mathbf{r}\, \mathrm{d}V = \int_{S_I + S_b} \frac{r^2}{2}\, \mathbf{n}\, \mathrm{d}S = \int_{S_b} \frac{r^2}{2} \mathbf{n}\, \mathrm{d}S + \frac{a^2}{2} \int_{S_I} \mathbf{n}\, \mathrm{d}S$$

and since

$$\int_{S_I + S_b} \mathbf{n}\, \mathrm{d}S = 0 \tag{15}$$

one gets

$$\int_{V_p} \mathbf{r}\, \mathrm{d}V = \int_{S_b} \frac{r^2 - a^2}{2} \mathbf{n}\, \mathrm{d}S\,.$$

This result, valuable for any particle (cut or not) can be expressed with the phase function as

$$\chi\, \mathbf{r} = \boldsymbol{\nabla}\left(\chi \frac{r^2 - a^2}{2}\right) \tag{16}$$

and one can easily verify that for any function f which is a constant over a given particle

$$\chi\,\mathbf{r} f(\mathbf{R}) = \nabla\left(\chi \frac{r^2 - a^2}{2} f(\mathbf{R})\right). \tag{17}$$

The same approach permits one to derive (10) from (15) and to get

$$5\chi\, r_i r_j\, g(\mathbf{R}) = a^2 \chi\, g(\mathbf{R})\, \delta_{ij} + \frac{\partial}{\partial x_k}\left(\chi\, r_j (r_k r_i - a^2 \delta_{ik}) g(\mathbf{R})\right). \tag{18}$$

Now, in a continuum approach, any averaged quantity A varies on a scale much larger than the particle radius; consequently

$$a\,|\nabla A| \ll A \quad \text{and} \quad a^2 \nabla^2 A \ll A\,. \tag{19}$$

Taking this fact into account, one gets from (17) and (18)

$$\langle \chi\,\mathbf{r} f(\mathbf{R}) \rangle = -\frac{a^2}{5} \nabla \langle \chi f(\mathbf{R}) \rangle = -\frac{a^2}{5} \nabla(\Phi f(\mathbf{x}))$$

$$\langle \chi\, r_i r_j\, g(\mathbf{R}) \rangle = \frac{a^2}{5} \langle \chi g(\mathbf{R}) \rangle \delta_{ij} = \frac{a^2}{5} \Phi g(\mathbf{x}) \delta_{ij}\,.$$

As a result, the expression (14) for the suspension stress can now be expressed as

$$\Pi_{ij} = -p_\infty \delta_{ij} + 2\eta_0\left(1 + \frac{3\Phi}{2}\right) D^\infty_{ij} + \frac{3\eta_0}{10}\left(\frac{\partial \Phi P_i}{\partial x_j} + \frac{\partial \Phi P_j}{\partial x_i} - \frac{\partial \Phi P_k}{\partial x_k}\delta_{ij}\right). \tag{20}$$

Note that we neglected the contribution coming from φ_{ijk} which is of order $a^2 |\nabla D^\infty_{ij}|\,|\nabla\Phi|$ and is thus negligible as compared to the D^∞_{ij} term.

5 Connection with the suspension variables

In the last step we must relate quantities belonging to the velocity field $\mathbf{V}_\infty$ to their counterparts in the suspension. For instance we must link the strain rate $\mathbf{D}_\infty$ and the strain rate $\mathbf{D}$ of the suspension defined as

$$D_{ij} = \frac{1}{2}\left(\frac{\partial V_i}{\partial x_j} + \frac{\partial V_j}{\partial x_i}\right),$$

where

$$V = \Phi \, V_p + (1 - \Phi) \, V_f$$

is the volume-average velocity. Since the particle motion is strain-free and $u_p = u_f$ at the interfaces one gets [2,10]

$$D_{ij} = (1 - \Phi) \, D_{ij}^{\infty} . \tag{21}$$

The calculation of second-order derivatives is a bit more complicated. For instance

$$\nabla^2 V = \langle (1 - \chi) \nabla^2 \mathbf{u}_f + (\mathbf{n} \cdot \nabla)(\mathbf{u}_f - \mathbf{u}_p)\delta_I \rangle .$$

Note that the velocity gradients at the interface will give a contribution to $\nabla^2 \mathbf{V}$ as soon as there is a concentration gradient, i.e., whenever $\langle \mathbf{n} \, \delta_I \rangle \neq 0$. The result of the averaging happens to be

$$\nabla^2 \mathbf{V} = (1 - \Phi) \nabla^2 \mathbf{V}_{\infty} - 3 \mathbf{D}_{\infty} \cdot \nabla \Phi . \tag{22}$$

The suspension is characterized not only by a convective motion represented by $\mathbf{V}$ but also by the particle sedimentation represented by $\mathbf{W} = \mathbf{V}_p - \mathbf{V}$. Now, what is the equivalent of this relative velocity in the one-particle approach? Although we do not have a crystal-clear proof, we have many indications that the correct definition of $\mathbf{W}$ in terms of the unperturbed velocity field is

$$\mathbf{W} = \frac{\langle \chi(\mathbf{u}_p - \mathbf{V}_{\infty}) \rangle}{\langle \chi \rangle} = \mathbf{V}_p - \langle \mathbf{V}_{\infty} \rangle_p \tag{23}$$

where $\langle \; \rangle_p$ is an average over the volume occupied by the particle. In other words, for our quadratic velocity field

$$\mathbf{W} = \mathbf{V}_p - \mathbf{V}_{\infty}(\mathbf{R}) - \frac{a^2}{10} \nabla^2 \mathbf{V}_{\infty} .$$

It is clear that one could have proposed $\mathbf{W} = \mathbf{V}_p - \mathbf{V}_{\infty}(\mathbf{R})$. This point is a matter of debate but we have a preference for definition (23). If we take it for granted, the vector $\mathbf{P}$ defined in (4) can be expressed in terms of the suspension velocities $\mathbf{V}$ and $\mathbf{W}$ as

$$\mathbf{P} = \mathbf{W} - \frac{a^2}{15} \left(\nabla^2 \mathbf{V} + 3 \, \mathbf{D} \cdot \nabla \Phi \right) .$$

As a consequence, eq. (7) can be rewritten (to lowest order in Φ) as

$$(\rho_p^0 - \rho_f^0)\mathbf{g} = \frac{9\eta_0}{2a^2}\mathbf{W} - \frac{3\eta_0}{10}(\nabla^2 \mathbf{V} + 3\mathbf{D} \cdot \nabla \Phi) \tag{24}$$

and the suspension stress (20) will appear as

$$\Pi_{ij} = -p\,\delta_{ij} + 2\eta_0\left(1 + \frac{5\Phi}{2}\right)D_{ij} + \frac{3\eta_0}{10}\left(\frac{\partial \Phi W_i}{\partial x_j} + \frac{\partial \Phi W_j}{\partial x_i} - \frac{\partial \Phi W_k}{\partial x_k}\,\delta_{ij}\right) \tag{25}$$

Equation (24) describes the relative motion between particles and fluid while eq. (12) completed by (25) describes the convective motion. Clearly, there is a coupling between sedimentation and convection as first noticed by Nozières [12]: New forces appear in convection whenever the velocity $\Phi\mathbf{W}$ is not constant over the flow. And new forces appear in sedimentation whenever Φ, or $\mathbf{D}$, is not constant.

6 Final remarks

According to the above results, many new phenomena should occur in a non-homogeneous suspension but, together with the equations of motion, one must derive boundary conditions for the two velocity fields $\mathbf{V}$ and $\mathbf{W}$. One of these boundary conditions, proposed by Nozières [12], expresses the proportionality of the components of $\mathbf{V}$ and $\mathbf{W}$ parallel to the boundaries ($\mathbf{V} = -3\Phi\,\mathbf{W}$) and this allowed Nozières to give a convincing explanation of why the sedimentation velocity was a function of the shape of the container [1].

It is to be stressed that another kind of $\nabla\Phi$ force can appear in (24), as a result of Brownian motion. The influence of Brownian forces on the relative motion is analysed in [7] and [8]. Here we neglected them, as well as the role of surface tension. The presence of the same quantity $3\eta_0/10$ in both (24) and (25) suggests some kind of Onsager symmetry in the coupling between convection and sedimentation. This symmetry was guessed at by Nozières [12] and confirmed by microscopic calculations in [4] and [11], and with a more macroscopic approach by the present author [9].

We can say that the slow motion of a non-homogeneous suspension (dilute or not) is now rather well understood, but a lot of work remains to be done if one is to take particle inertia or particle velocity fluctuations into account.

References

[1] Beenakker, C.W.J. and Mazur, P., *Phys. Fluids*, **28** (1985), 3203-3206.

[2] Brenner, H., *Ann. Rev. of Fluid Mech.*, **2** (1970), 137-176.

[3] Drew, D.A., *Ann. Rev. of Fluid Mech.*, **15** (1983), 261-291.

[4] Felderhof, B.U., *Physica A*, **153** (1988), 217-233.

[5] Fitremann, J.H., These d'Etat Université P. et M. Curie, 1977.

[6] Happel, J. and Brenner, H., *Low Reynolds Number Hydrodynamics*, Prentice-Hall, Englewood Cliffs, N.J. (1965), Ch. 3.

[7] Kops-Werkhoven, M., Vrij, A. and Lekkerkerker, H., *J. Chem. Phys.*, **78** (1983), 2760-2763.

[8] Lhuillier, D., *J. Physique (Paris)*, **47** (1986), 1687-1696.

[9] Lhuillier, D., Transport phenomena in moderately concentrated suspensions of rigid spheres, to be published in Physica.

[10] Nigmatulin, R.I., *Int. J. Multiphase Flow*, **5** (1979), 353-385.

[11] Noetinger, B., *Physica A* **157** (1989) 1139..

[12] Nozières, P., *Physica A*, **147** (1987), 219.

Lhuillier D.
Laboratoire de Modélisation en Mécanique
Université Pierre-et-Marie Curie
Tour 66, 4 place Jussieu
F-75252 Paris Cédex
FRANCE

A. MORRO

Continuum description of polyelectrolyte solutions

1 Introduction

Polyelectrolytes, as well as proteins and nucleic acids, are made of macromolecules which in aqueous solution (or in other polar solvents) deliver macro-ions. Macro-ion solutions always contain small ions. There must be at least those arising from the dissociation of macromolecules called counter-ions.

The electrostatic interaction between the macro-ion and the small ions has important effects on the configuration of macro-ions and on their chemical combination with other ions, including their acidic and basic properties. That is why great attention has been devoted to the distribution of small ions around a single macro-ion.

Mathematically, determining the distributiion of small ions around the macro-ion is of interest in two respects at least. First, the small ions are taken to obey Poisson's equation along with the Boltzmann distribution law, which makes the problem under consideration non-linear. Second, the high electric field around the macro-ion demands a model of dielectric permittivity which depends on (the square of) the electric field. Besides this, for any species of small ion, the overall number in the solution is fixed. Then the spatial distribution of small ions is to be determined under global constraints.

The first feature is at the basis of the well-known Debye-Hückel theory for spherical, impenetrable ions [4]. The second one arose in [2] and led to interesting developments for the permittivity function [1]. Accounting for (global) constraints usually results in awkward procedures. The use of Lagrange multipliers at least introduces additional unknowns which very rarely have a physical meaning. Furthermore, the occurrence of Lagrange multipliers always hinders the characterization of the sought field as the extremum of an appropriate functional.

In this paper a scheme is elaborated where the constraints are accounted for in a direct way which leads to a genuinely variational formulation for the problem under consideration. Next a finite element method is developed which makes the search for the sought field a straightforward and efficient procedure.

2 Continuum description

In its simplest, but reasonable, scheme a polyelectrolyte solution is regarded as a lattice of cylindrical elementary cells (cf. [3] and references therein). A cylindrical rod, whose axis coincides with the axis of the cell, models the polyelectrolyte macro-ion. The solvent carries the small ions and occupies the ring $R_1 \leq \mathbf{r} \leq R_2$ around the macro-ion. The height of the cell, denoted by l, is assumed to be very large as compared to the other geometric dimensions and hence any corner effects at the bases of the cell are disregarded. The charged groups of the macro-ion are taken to be distributed uniformly and then modelled by a uniform charge density, per unit area, λ at $r = R_1$.

Let φ be the potential of the (time-independent) electric field $\mathbf{E}$, q the proton charge, z_i the valence of the ith species of ion, k the Boltzmann constant, and θ the absolute temperature. Following the Boltzmann distribution law, we assume that the ratio of the concentration (number density) of the ith species of ion at the potential φ to its concentration at $\varphi = 0$ is equal to $\exp(-z_i\beta\varphi)$, where $\beta = q/k\,\theta$. Then we write the concentration c_i of the ith species as

$$c_i = c_{i0} \exp(-z_i\,\beta\varphi)\,;$$

the quantity c_{i0} is the value of c_i at $\varphi = 0$. For the sake of convenience we let φ vanish at the outer surface $r = R_2$. Letting ε be the dielectric permittivity and adopting rationalized MKS units we can write Coulomb's law as

$$\nabla\cdot(\varepsilon\nabla\varphi) + q\sum_i c_{i0}\, z_i \exp(-z_i\,\beta\varphi) = 0\,.$$

Depending on the situation we describe, the concentrations c_i, and then φ, may be subjected to the constraint that the overall charge of any species of ion is fixed, namely

$$|z_i|\, c_{i0} \int_\Omega \exp(-z_i\,\beta\varphi)dx = N_i\,, \qquad i = 1,2, \ldots$$

where Ω is the region occupied by the solution. Hence it follows that c_{i0} is determined by the potential φ through

$$|z_i|\, c_{i0} = \frac{N_i}{\int_\Omega \exp(-z_i\,\beta\varphi)dx}$$

and the charge density ρ is given by the functional

$$\rho(\varphi) = q\sum_i \frac{z_i}{|z_i|}\, \frac{N_i \exp(-z_i\,\beta\varphi)}{|z_i|\int_\Omega \exp(-z_i\,\beta\varphi)\,dx}\,.$$

Then Coulomb's law becomes

$$\nabla \cdot (\varepsilon \nabla \varphi) + q \sum_i \frac{z_i}{|z_i|} \frac{N_i \exp(-z_i \beta \varphi)}{|z_i| \int_\Omega \exp(-z_i \beta \varphi)\, dx} = 0\,. \tag{1}$$

To (1) we have to adjoin appropriate boundary conditions. We already chose

$$\varphi = 0 \text{ at } \mathbf{r} = R_2\,. \tag{2}$$

Moreover, by Gauss law we conclude that

$$\varepsilon \frac{\partial \varphi}{\partial r} = -\lambda \text{ at } r = R_1\,. \tag{3}$$

Concerning the function $\varepsilon(E^2)$ we require that the differential permittivity be positive, namely

$$2\varepsilon' E^2 + \varepsilon > 0 \tag{4}$$

a prime denoting the derivative with respect to the argument E^2. A simple function meeting (4) is

$$\varepsilon = \left(\varepsilon_\infty^2 + \frac{\varepsilon_0^2 - \varepsilon_\infty^2}{1 + \gamma E^2}\right)^{1/2} \tag{5}$$

where ε_0 and ε_∞ are the limit values of ε as $E \to 0$ and $E \to \infty$, respectively. The parameter γ is determined via a comparison with experimental data on $\varepsilon(E^2)$. For water, the function (5) shows a satisfactory agreement with experimental data.

3 Variational formulation

It is rather uncommon to find in the literature variational formulations for differential equations which involve functionals of the unknown function(s). Equation (1) is a non-linear equation, for the unknown function φ, which involves a non-linear functional of φ. Nevertheless it is not difficult to prove that a variational formulation exists for the eq. (1) along with the boundary conditions (2), (3). Based on [1] consider the functional

$$f(\varphi) = \frac{1}{2} \int_\Omega \Big(\int_0^{|\nabla\varphi|^2} \varepsilon(\zeta) d\zeta dx + k\,\theta \sum_i \frac{N_i}{|z_i|} \ln \int_\Omega \exp(-z_i \beta\varphi) dx + \int_{\partial\Omega_m} \lambda \varphi da$$

where $\partial\Omega_m$ is the macro-ion surface $r = R_1$. A straightforward calculation shows that, for any admissible function h,

$$df(\varphi \mid h) = \int_{\partial\Omega} \varepsilon(|\nabla\varphi|^2)\nabla\varphi \cdot \mathbf{n}\, h da + \int_{\partial\Omega_m} \lambda h\, da$$

$$- \int_\Omega \left\{ \nabla \cdot [\varepsilon(|\nabla\varphi|^2)\nabla\varphi] - q \sum_i \frac{z_i \quad N_i \exp(-z_i\, \beta\varphi)}{|z_i| \int_\Omega \exp(-z_i\, \beta\varphi) dx} \right\} h dx$$

where $\mathbf{n}$ is the outward unit normal to $\partial\Omega$. The arbitrariness of h allows us to say that $df(\varphi)h = 0$ implies the validity of (1), (3). Then the extremum of $f(\varphi)$ is the solution to (1)-(3). Look now at the second-order differential. A direct calculation yields

$$\frac{d^2 f}{d\alpha^2}(\varphi + \alpha h) = \int_\Omega \{2\varepsilon' [(\nabla\varphi + \alpha\nabla h) \cdot \nabla h]^2 + \varepsilon \nabla h \cdot \nabla h\} dx + \beta q \sum_i N_i |z_i| J_i$$

where

$$J_i = \frac{\int_\Omega \exp[-z_i\beta(\varphi + \alpha h)]\, h^2 dx \int_\Omega \exp[-z_i\beta(\varphi + \alpha h)]\, dx - \{\int_\Omega \exp[-z_i\beta(\varphi + \alpha h)]\, h dx\}^2}{\left\{\int_\Omega \exp[-z_i\beta(\varphi + \alpha h)] dx\right\}^2}$$

By use of the Cauchy-Schwarz inequality it follows that, for any non-negative function g,

$$\left(\int_\Omega h\, g dx\right)^2 \leq \int_\Omega h^2\, g dx \int_\Omega g dx \, .$$

Then, letting $g = \exp(-z_i\beta\varphi)$ we have

$$\left\{\int_\Omega \exp[-z_i\beta(\varphi + \alpha h)\,]\, h\, dx\right\}^2 \leq \int_\Omega \exp[-z_i\beta(\varphi + \alpha h)\,]\, h^2\, dx \int_\Omega \exp[-z_i\beta(\varphi + \alpha h)\,]\, dx \, ,$$

which proves that the second part of $d^2 f/d\alpha^2$ is positive definite. As to the first part, the validity of

$$2\varepsilon'(E^2)(\mathbf{E} \cdot \mathbf{v})^2 + \varepsilon(E^2)\mathbf{v} \cdot \mathbf{v} > 0 \qquad (6)$$

for any pair of non-vanishing vectors $\mathbf{E}, \mathbf{v}$ ensures the positive definiteness. The validity of (6) is guaranteed by the condition of positive-definite differential permittivity (4). Then the functional f has a strict minimum at the solution of (1)-(3).

On the basis of this remarkable feature we might search approximate solutions through a Ritz-like formulation. Rather we prefer a Galerkin-like approach and proceed as follows. Observe that $df(\varphi \mid w) = 0$ implies

$$(\nabla \cdot (\varepsilon \nabla \varphi) + \rho(\varphi), w) = 0, \qquad w = 0 \quad \text{at} \quad \partial\Omega \tag{7}$$

where $(\cdot,\cdot)$ denotes the inner product in $L_2(\Omega)$. Hence we obtain (1). The function w is required to be of class C^1 in Ω. An obvious integration by parts of (7) allows us to write

$$(\varepsilon(|\nabla\varphi|^2)\nabla\varphi, \nabla w) = (\rho(\varphi), w), \quad w \in C^1(\Omega), \quad w = 0 \text{ at } \partial\Omega\,. \tag{8}$$

We regard (8) as the Galerkin-like formulation of the problem in that a function $\varphi \in C^1(\Omega)$ is sought such that (8) holds with (1)-(3). Of course, by the cylindrical symmetry of the problem the functions φ, w are taken to depend on the radial coordinate r only.

4 Finite-element formulation

Let

$$V^0 = \{\, w : w \in C^1(R_1, R_2),\, w(R_1) = w(R_2) = 0 \,\}\,.$$

Since φ and w are functions of r only, the condition (8) reads

$$\int_{R_1}^{R_2} \varepsilon(\varphi_r^2)\varphi_r\, w_r\, r\, dr = \int_{R_1}^{R_2} \rho(\varphi) w\, r\, dr, \quad w \in V^0\,, \tag{9}$$

where the subscript r denotes the derivative with respect to r.

Let $R_1 = r_0 < r_1 < \ldots < r_M < r_{M+1} = R_2$ be a partition of the interval $[R_1, R_2]$ into subintervals $I_j = (r_{j-1}, r_j)$ of length $\Delta_j = r_j - r_{j-1}, j = 1,\ldots, M+1$, and set $\Delta = \max \Delta_j$. Moreover, let φ_j, $j = 0,1,\ldots,M, M+1$ be a continuous, piecewise linear function that takes the value 1 at the node point r_j and O at the other node points. Let V_Δ be the $M+2$-dimensional subspace of V^0 spanned by the basis $\{\varphi_0, \varphi_1,\ldots,\varphi_M, \varphi_{M+1}\}$ and V_Δ^0 the M-dimensional space spanned by $\{\varphi_1,\ldots,\varphi_M\}$. For ease in calculation, here we let $\Delta_j = \Delta$, $j = 1,\ldots, M+1$, and then $\Delta = (R_2 - R_1)/(M+1)$.

We can now state the following discrete formulation of the problem: Find $\tilde{\varphi} \in V_\Delta$ such that

$$\int_{R_1}^{R_2} \varepsilon(\tilde{\varphi}_r^2)\, \tilde{\varphi}_r (\varphi_j)_r\, r\, dr = \int_{R_1}^{R_2} \rho(\tilde{\varphi})\varphi_j\, r\, dr\,, \quad j = 1,\ldots,M\,. \tag{10}$$

It is convenient to write $\mathbf{r}$ as $R_1 + \eta$, $\eta \in [0, R_2 - R_1]$. Then, because $\tilde{\varphi} \in V_\Delta$ and $\tilde{\varphi}$ vanishes at $r = R_2$, we can write

$$\tilde{\varphi}\,(R_1 + \eta) = \sum_{k=0}^{M} \xi_k\, \varphi_k(\eta)\,.$$

Really, by (3) we can evaluate the (radial) electric field, E_0 say, at $r = R_1$. This means that ξ_0 and ξ_1 are related by

$$\xi_0 = \xi_1 - E_0 \Delta\,.$$

Accordingly, $\tilde{\varphi}$ is determined by the M quantities $\xi_1,\ldots,\xi_M$. Denote by a prime the radial derivative, i.e., $\varphi'_j = (\varphi_j)_r$. Then substitution for $\tilde{\varphi}$ into (10) yields

$$\int_{R_1}^{R_2} \sum_{h=0}^{M} \varepsilon\Big(\big(\sum_{k=0}^{M} \xi_k\, \varphi'_k\big)^2\Big)\xi_h \varphi'_h \varphi'_j\, r\, dr = \int_{R_1}^{R_2} \rho\Big(\sum_{k=0}^{M} \xi_k \varphi_k\Big)\varphi_j\, r\, dr, j = 1,\ldots,M\,. \tag{11}$$

Henceforth, for the sake of simplicity we confine our attention to the significant case when only one species of monovalent positive ion occurs (e.g., H_3O^+), $z = 1$.

Observe that

$$\int_{R_1}^{R_2} \exp(-\beta\tilde{\varphi}) r\, dr = \sum_{k=0}^{M} \int_{k\Delta}^{(k+1)\Delta} \exp\{-\beta\,[\xi_k(1 - \frac{\eta - k\Delta}{\Delta}) +$$

$$+\ \xi_{k+1} \frac{\eta - k\Delta}{\Delta}]\}\,(R_1 + \eta) d\eta$$

whence, by the reasonable approximation $R_1 + \eta \simeq R_1 + k\Delta$ as $\eta \in (k\Delta, (k+1)\Delta)$, we have

$$\int_{R_1}^{R_2} \exp(-\beta\tilde{\varphi}) r\, dr = \sum_{k=0}^{M} (R_1 + k\Delta)\Delta\, \frac{\exp(-\beta\xi_{k+1}) - \exp(-\beta\xi_k)}{\beta(\xi_k - \xi_{k+1})}\,.$$

Similarly we obtain

$$\int_{R_1}^{R_2} \exp(-\beta\tilde{\varphi})\varphi_j \, \mathbf{r} \, dr =$$

$$= [R_1 + (j-1)\Delta]\Delta \frac{\beta(\xi_{j-1} - \xi_j)\exp(-\beta\xi_j) - \exp(-\beta\xi_j) + \exp(-\beta\xi_{j-1})}{\beta^2(\xi_{j-1} - \xi_j)^2}$$

$$+ [R_1 + j\Delta]\Delta \frac{-\beta(\xi_j - \xi_{j+1})\exp(-\beta\xi_j) + \exp(-\beta\xi_{j+1}) - \exp(-\beta\xi_j)}{\beta^2(\xi_j - \xi_{j+1})^2}.$$

The support of $\varphi'_h \varphi'_j$ for a given j is the interval $\eta \in ((j-1)\Delta, (j+1)\Delta)$. Then a straightforward calculation yields

$$\int_{R_1}^{R_2} \sum_{h=0}^{M} \varepsilon\Big(\big(\sum_{k=0}^{M} \xi_k \varphi'_k\big)^2\Big)\xi_h \varphi'_h \varphi'_j \, \mathbf{r} \, dr = [R_1 + (j-1)\Delta]\, \varepsilon\Big(\Big(\frac{\xi_j - \xi_{j-1}}{\Delta}\Big)^2\Big) \frac{\xi_j - \xi_{j-1}}{\Delta}$$

$$+ [R_1 + j\Delta]\, \varepsilon\Big(\Big(\frac{\xi_{j+1} - \xi_j}{\Delta}\Big)^2\Big) \frac{\xi_j - \xi_{j+1}}{\Delta}.$$

In conclusion, the discrete formulation (11) leads to the non-linear system of algebraic equations

$$[R_1 + (j-1)\Delta]\, \varepsilon\Big(\Big(\frac{\xi_j - \xi_{j-1}}{\Delta}\Big)^2\Big) \frac{\xi_j - \xi_{j-1}}{\Delta} + [R_1 + j\Delta]\, \varepsilon\Big(\Big(\frac{\xi_{j+1} - \xi_j}{\Delta}\Big)^2\Big) \frac{\xi_j - \xi_{j+1}}{\Delta}$$

$$= \frac{Nk\,\theta}{2\pi\, l}\Big[\sum_{k=0}^{M} (R_1 + k\Delta) \frac{\exp(\beta\xi_{k+1}) - \exp(-\beta\xi_k)}{\xi_k - \xi_{k+1}}\Big]^{-1} \times$$

$$\times \Big\{ [R_1 + (j-1)\Delta] \frac{\beta(\xi_{j-1} - \xi_j)\exp(-\beta\xi_j) - \exp(-\beta\xi_j) + \exp(-\beta\xi_{j-1})}{(\xi_{j-1} - \xi_j)^2}$$

$$+ [R_1 + j\Delta] \frac{-\beta(\xi_j - \xi_{j+1})\exp(-\beta\xi_j) + \exp(-\beta\xi_{j+1}) - \exp(-\beta\xi_j)}{(\xi_j - \xi_{j+1})^2} \Big\}$$

The next step is to evaluate explicitly the solution $\xi_1, \ldots, \xi_M$ to such a system. Moreover it is of interest to check the validity of the model by contrating this approximate solution with suitable experimental data. Both investigations are now under way.

Acknowledgements

The research leading to this paper has been supported through the Research Project MADESS of CNR.

References

[1] Angelo Morro, Variational methods for nonlinear dielectrics, *Atti Sem. Mat. Fis. Univ. Modena*, **36**, 339-353 (1988).

[2] Angelo Morro and Mauro Parodi, A variational approach to non-linear dielectrics: application to polyelectrolytes, *J. Electrostatics*, **20**, 219-232 (1987).

[3] Fumio Oosawa, *Polyelectrolytes*, M. Dekker, New York, 1971.

[4] Charles Tanford, *Physical Chemistry of Macromolecules*, J. Wiley, New York, 1961; ch. 7.

Morro A
Dept. Ingegneria Biofisica e Elettr.
FacoltàdiEngegneria
Università di Genova
Viale Causa 13
I-16145 Genova
ITALY

A.I. MURDOCH

Remarks on continuum modelling of fluid interfacial regions

1 Introduction

Fluid–fluid interfacial regions are in general extremely thin and characterized by great bulk density changes. For example, liquid–vapour interfacial thicknesses are typically of order 10 Å (= 10^{-9} m) away from near-critical temperatures and liquid : vapour density ratios are of order $10^3 : 1$. In the case of ("immiscible") liquid–liquid interfaces the situation is similar: each density vanishes on one side of the interface.

Mechanical phenomena associated with such interfacial regions in macroscopic equilibrium are well-described in terms of a *surface of tension* in which bulk phases are considered to be separated by a surface which acts as if it were a massless membrane in a state of uniform tension. Further, a satisfactory model of thermostatic behaviour of multicomponent interfaces was established by Gibbs [1]. However, non-equilibrium situations are somewhat complex: even for single-component liquid-vapour interfaces any comprehensive theory must accommodate the possibilities of mass transport both within *and* across the interface (e.g., evaporation), and of bulk motions induced by interfacial behaviour (the *Marangoni* effect). While a number of (bidimensional) continuum theories of fluid–fluid interfaces have been *postulated* [8, 9, 2, 10], it is instructive to derive, or at least motivate, such theories in some way. Here will be outlined an approach which takes account of microscopic considerations and interprets all continuum concepts and relations in terms of space–time averages taken over sums of molecular quantities. This is a natural continuation of earlier work which addressed fundamental aspects of continuum modelling [4], mixtures [3,5], and generalized continua [6].

2 Delineation of interfacial boundaries

The erratic motion of small ("Brownian") particles when immersed in a liquid is vivid evidence of the inhomogeneity of *gross* molecular behaviour in non-solid phases at length scales of 10^{-6} m and below. Accordingly, for *fluids* the macroscopic mass density ρ and velocity $\mathbf{v}$, regarded as purely spatial averages, should be associated with regions of characteristic dimension in excess of 10^{-6} m.

Let fundamental discrete entities (molecules, ions or atoms) be modelled as point masses P_i $(i = 1,2, ...)$ whose masses and velocities are denoted by m_i and $\mathbf{v}_i$. Spatial averaging may be effected using the notion of an "ε*-cell centred at a* (geometrical) *point* $\mathbf{x}$". This is essentially a simply-connected region with centroid $\mathbf{x}$, piecewise-smooth boundary with subsurfaces having principal curvatures of order ε^{-1} or less, and all boundary points $\mathbf{y}$ satisfying $\varepsilon/2 < |\mathbf{y}-\mathbf{x}| < 3\varepsilon/2$. Consider $\sum_i' m_i/V_\varepsilon$, where the sum is taken over all P_i within some ε-cell of volume V_ε centred at $\mathbf{x}$. (All cell sums are indicated by a superposed prime.) If, at instant t, this ratio is essentially insensitive to cell shape and size over a range of ε values macroscopically small yet microscopically large, then its value is denoted by the "pseudo-limit" $\lim_\varepsilon \{\sum_i' m_i/V_\varepsilon\}$ and identified with $\rho(\mathbf{x},t)$. Similarly, $\mathbf{v}(\mathbf{x},t)$ is identified with $\lim_\varepsilon \{\sum_i' m_i\mathbf{v}_i/\sum_i' m_i\}$. It follows that the fields ρ and $\mathbf{v}$ will vary imperceptibly over distances commensurate with associated cell size ε and that continuum theory cannot be invoked at length scales less than ε. Brownian motion indicates that for *fluids* a lower bound on ε is 10^{-5} m.

If an ε-cell intersects or straddles a liquid-vapour interfacial region the ratio $\sum_i' m_i/V_\varepsilon$ is highly shape- and size-dependent and ρ will fail to exist. The interfacial zone I can thus be associated with the region in which ρ, interpreted as a purely spatial average, fails to exist.[1] Hence I is identified with a region of thickness of order 2×10^{-5} m, corresponding to the smallest relevant length scale compatible with continuum theory. This is unsatisfactory since there are compelling reasons for delineating an interface with an accuracy commensurate with its thickness ($\sim 10^{-9}$ m). Greater precision can be gained by reinterpreting ρ as a space-*time* average.

The Δ-time average of a fluctuating quantity φ at instant t is defined by

$$\varphi_\Delta(t) := \frac{1}{\Delta}\int_{t-\Delta}^{t} \varphi(\tau)d\tau .$$

It may happen that over a range of Δ values, microscopically large yet macroscopically small, φ_Δ is essentially constant, with value $\bar{\varphi}(t)$ say. In such case we write

$$\bar{\varphi}(t) = \lim_\Delta \{ \varphi_\Delta(t) \} .$$

Now consider ratios $\sum_i' m_i/V_\varepsilon$, where the sums are over ε-cells with $\varepsilon \sim 10^{-6}$ m or less. Although such sums will fluctuate it may be possible to smooth these by time averaging. If so the smoothed value at instant t is identified with $\rho(\mathbf{x}, t)$ and we write

---

[1] A *multicomponent* interfacial zone is associated with that region in which *at least one* constituent density (defined as ρ but with sum taken only over the relevant molecular species) fails to exist.

$$\rho(\mathbf{x}, t) = \lim_{\Delta} \lim_{\varepsilon} \{ \sum_i{}' m_i / V_\varepsilon \} . \tag{1}$$

The boundary of the ε-cell is here assumed to deform with the motion prescribed by $\mathbf{v}$ and to coincide with some ε-cell centred at $\mathbf{x}$ at instant t. Thus ρ, reinterpreted as a space-time average, may be meaningful at length-time scale pairs $(\varepsilon_0, \Delta_0)$ where $\varepsilon_0 < 10^{-5}$ m, and, if so, will vary insensibly at these scales. Intuitively we expect such averages to exist for a range of $(\varepsilon_0, \Delta_0)$ pairs lying on a monotonic-decreasing curve Γ in $\mathbb{R}^2$ with increasing spatial accuracy (associated with ε_0) achieved at the expense of temporal precision (measured by Δ_0). As before, the interfacial zone can be identified with that region in which [2] ρ, as defined by (1) fails to exist. It follows that the more temporal precision is sacrificed the more precisely may be delineated the interface.

It is also possible to reinterpret $\mathbf{v}$ as a space-time average via

$$\mathbf{v}(\mathbf{x}, t) = \lim_{\Delta} \lim_{\varepsilon} \{ \sum_i{}' m_i \mathbf{v}_i / \sum_i{}' m_i \} , \tag{2}$$

where the sums are taken over an ε-cell whose boundary deforms *with the motion prescribed by* $\mathbf{v}$. Such an implicit definition can be motivated [7] as the limit of an iterative procedure.

For the present purpose, improved interpretations of ρ and $\mathbf{v}$ as space-time averages over ε-wafers[3] are possible. This is treated carefully in [7]: here we only sketch the idea. Suppose ρ fails to exist in the sense of (1) in the region I bounded by *distinct* surfaces S_1 and S_2. It may prove possible to extend fields ρ and $\mathbf{v}$ into I by reinterpreting these as space-time averages taken over ε-wafers based on families of surfaces parallel to S_1 and S_2. By this we mean ρ and $\mathbf{v}$ as given by (1) and (2) make sense when the instantaneous sums are over wafer-like regions whose walls deform with the motion of $\mathbf{v}$ restricted to the surface on which the wafer is based. The consequences of this second reinterpretation of (1), (2) are:

(i) the interfacial region (redefined in terms of the non-existence of the reinterpretation of ρ) is now delineated with an accuracy commensurate with wafer thickness, and

---

[2] This is appropriate to single-component (i.e., liquid-vapour) interfaces. Cf. preceding footnote for *other* interfaces.

[3] Let Σ be a surface, $\mathbf{x} \in \Sigma$, and let Σ_ε denote the intersection of Σ with an ε-cell centred at $\mathbf{x}$. Consider that region W_ε bounded by Σ, a parallel surface to Σ distant 10^{-9} m therefrom, and normals to Σ through the boundary $\partial\Sigma_\varepsilon$ (a curve) of Σ_ε (assuming such normals form, locally, a ruled surface). W_ε is termed an ε-wafer based on Σ and centred at $\mathbf{x}$. The ruled surface is termed the *wall* of the wafer.

(ii) ρ and $\mathbf{v}$ can exhibit pronounced spatial variation normal to interfacial boundaries (on a length scale in excess of 10^{-9} m) but parallel thereto perceptible spatial change will only be manifest at length scales above ε (notionally 10^{-5} m).

We observe that S_1, S_2 are essentially *singular* surfaces.

3 Selection of a model surface and interfacial velocity fields

The intersection with I of an ε-cell ($\varepsilon \sim 10^{-5}$ m) centred at $\mathbf{x} \in I$ is termed an *interfacial* ε-*cell*, described as *"walled"* if its boundary component between S_1 and S_2 is a ruled surface generated by normals to some smooth surface lying strictly between S_1 and S_2. We note that the mass centre of the set of molecules instantaneously within such a cell must be expected to fluctuate, even in conditions of macroscopic equilibrium. It proves possible to motivate, again via an iterative procedure [7], the following modelling assumptions:

M.A. To each instant t corresponds a smooth surface $S(t)$ which is the locus of Δ-time-averaged walled interfacial ε-cell mass-centre locations, where the time-averaging involves wall deformations mandated by the surface motion of S prescribed by the *interfacial surface velocity*

$$\mathbf{u}_S := U\mathbf{n} + \mathbf{P}\mathbf{u} \,. \tag{3}$$

Here $U\mathbf{n}$ denotes the normal velocity field on S ($\mathbf{n}$ is a unit normal field) and [4] $\mathbf{Pu}$ the tangential component of the *interfacial velocity* $\mathbf{u}$, where

$$\mathbf{u} = \lim_{\Delta} \lim_{\varepsilon} \{ \textstyle\sum_i' m_i \mathbf{v}_i / \sum_i' m_i \} \,. \tag{4}$$

In (4) the sums are taken over walled interfacial ε-cells whose walls deform with the motion prescribed by $\mathbf{u}_S$.

Remarks (i) The somewhat complex inter-relationships between S, $\mathbf{u}$ and $\mathbf{u}_S$ arise naturally in the iterative considerations of [7].

(ii) Regarded as a family of surfaces parametrized by time, the trajectory of S has a well-defined *normal* (singular surface) *velocity* field $U\mathbf{n}$. The interfacial velocity $\mathbf{u}$ is defined entirely in terms of molecules within the interface and is that velocity in terms of which individual interfacial molecular *thermal* velocities $\tilde{\mathbf{v}}_i$ should be computed ($\tilde{\mathbf{v}}_i := \mathbf{v}_i - \mathbf{u}(\mathbf{z}_i, t)$, where $\mathbf{z}_i$ is the closest point of $S(t)$ to $\mathbf{x}_i(t)$). However,

---

[4] $\mathbf{P} := \mathbf{1} - \mathbf{n} \otimes \mathbf{n}$ is the perpendicular projection of spatial vectors onto the relevant tangent space to S.

considerations of evaporation indicate that in general $\mathbf{u} \cdot \mathbf{n} \neq U$. In order to follow the motion of the surface yet simultaneously move as closely as possible with the local gross interfacial molecular average velocity (represented by $\mathbf{u}$) then the appropriate velocity field is $\mathbf{u}_s$.

4 Balances of mass and linear momentum

These relations are obtained in similar fashion to their bulk counterparts [4]: starting from the observation that $dm_i/dt = 0$ and writing down Newton's second law for P_i in I, summing over all P_i in a portion of $\hat{I}$ of I subdivisible into very many ε-interfacial cells and then time averaging yields these (integral) balances for $\hat{I}(t)$, omitted here for brevity. All terms are space-time averages taken over molecular quantities. The decomposition of the surface stress tensor into interaction and diffusive contributions makes it clear that the existence of surface *tension* derives from the cohesive nature of molecular interactions at separations in excess of the average associated with liquid phases.

Acknowlegement

This work was initiated with Professor H. Cohen, University of Manitoba.

References

[1] Gibbs, J.W. On the equilibrium of heterogeneous substances. In *The Scientific Papers of J.W. Gibbs*, Vol. I., Dover (1961).

[2] Moeckel, G.P. Thermodynamics of an interface. *Arch. Rational Mech. Anal.*, **57**, 255–280 (1975).

[3] Morro, A. and Murdoch, A.I. Stress, body force, and momentum balance in mixture theory. *Meccanica*, **21**, 184–190 (1986).

[4] Murdoch, A.I. The balance equations of continuum mechanics from a corpuscular viewpoint. Proc. CMDS 5. Balkema (1987).

[5] Murdoch, A.I. and Morro, A. On the continuum theory of mixtures: motivation from discrete considerations. *Int. J. Eng. Sc.*, **25**, 9–25 (1987).

[6] Murdoch, A.I. On the relationship between balance relations for generalised continua and molecular behaviour. *Int. J. Eng. Sc.*, **25**, 883–914 (1987).

[7] Murdoch, A.I. and Cohen, H. On the continuum modelling of fluid interfacial regions I: Kinematical considerations. Submitted for publication.

[8] Scriven, L.E. Dynamics of a fluid interface. *Chem. Eng. Sc.*, **12**, 98–108 (1960).

[9] Slattery, J.C. Surfaces I: Momentum and moment of momentum balances for moving surfaces. *Chem. Eng. Sc.*, **19**, 379–385 (1964).

[10] Napolitano, L.G. Thermodynamics and dynamics of surface phases. *Acta Astronautica*, **6**, 1093–1112 (1979).

Murdoch A.I.
Department of Mathematics
University of Strathclyde
Livingstone Tower
26 Richmond Street
Glasgow G1 1XH
U.K.

B. STRAUGHAN

Oscillatory, non-linear and penetrative convection in an isotropic thermomicropolar fluid

1 Introduction

In this paper we describe results on the linear instability and non-linear stability of a layer of micropolar fluid heated from below (or sometimes from above). The theory of micropolar fluids is due to Eringen [4-7] whose theory allows for the presence of particles in the fluid by additionally accounting for particle motion. In [4] he introduced the theory for a simple microfluid and extended this in [5] to allow deformation of local fluid elements (particles). This was further generalized in [6] where he gave precise meaning to a thermomicropolar fluid and in [7] the theory appropriate to anisotropic fluids was elucidated.

We review here the Bénard problem for the theory of Eringen [6]. This problem was first examined by Datta and Sastry [3] who studied stationary convection in the linear problem under an assumption on the sign of a thermal coupling term in the energy equation. Ahmadi [1] used non-linear energy stability theory although he chose to neglect the coupling term which gave rise to an interesting effect in [3]. Further use of the energy method was made by Lebon and Perez-Garcia [8] who presented non-linear stability results for the problem of Datta and Sastry [3]. Payne and Straughan [11] also looked at the effect of the thermal coupling term but concentrated primarily on the opposite sign to that selected by Datta and Sastry [3] and Lebon and Perez-Garcia [8], while Lindsay and Straughan [9] analyse a non-Boussinesq effect likely to be important in a water-based suspension with part of the layer at a temprature below 4°C.

In [3] it was shown that heating both from above and below could lead to stationary convection instabilities. The analysis of [11] provides an alternative route where stationary convection only occurs in the heated below case. Furthermore, as the magnitude of the thermal coupling term is increased [11] predicts a substantial decrease in the critical Rayleigh number (unlike [3] and [8] where the opposite sign is chosen for the thermal coupling term). Chandra [2] observed in his experiments that adding smoke particles to a layer of gas could lead to such a convective motion commences. Since the particle spin associated with Eringen's [6] theory could possibly be appropriate to the added dust situation described by Chandra [2] there may be a justification for use of the analysis presented here for convection in a fluid fairly evenly interspersed with "particles", which may be dust, dirt, ice or raindrops, or other additives. Thus we believe that heuristically our results give reason to believe that the Eringen [6] micropolar convection model may be applicable to geophysical or industrial convection contexts.

The equations for an incompressible, isotropic thermomicropolar fluid are now recalled, from Eringen [6]. We take the microinertia moment tensor to be $j\mathbf{I}$, with j constant. The continuity, moment, moment of momentum and energy equations are then ([6], pp. 489, 490)

$$v_{i,i} = 0, \tag{1}$$

$$\rho \dot{v}_i = \rho f_i + t_{k\,i,k}\,, \tag{2}$$

$$\rho j \dot{\nu}_i = \rho \ell_i + \varepsilon_{i\,k\,h}\, t_{k\,h} + m_{k\,i,\,k}\,, \tag{3}$$

$$-\rho\dot{\varepsilon} + t_{k\,h}\, b_{k\,h} + m_{k\,h}\, \nu_{h,k} + q_{k,k} + \rho h = 0, \tag{4}$$

where standard indicial notation is used, $b_{k\,h} = v_{h,k} - \varepsilon_{k\,h\,r}\,\nu_r$, a superposed dot denotes the material derivative, $\mathbf{v}$, ν, $\varepsilon(T)$ are fluid velocity, particle spin vector and internal energy, T being temperature; ρ, $t_{k\,i}$, f_i, ℓ_i, $m_{k\,i}$, q_k and h are density (presumed constant except in the body force term), stress tensor, body force, body couple, couple stress tensor, heat flux vector and heat supply.

The constitutive equations are

$$t_{k\,h} = -\,\pi\delta_{k\,h} + \mu(v_{k,h} + v_{h,k}) + \bar{\kappa}(v_{h,k} - \varepsilon_{hk\,h\,a}\,\nu_a), \tag{5}$$

$$m_{k\,h} = \bar{\alpha}\nu_{r,r}\,\delta_{k\,h} + \bar{\beta}\nu_{k,h} + \gamma\nu_{h,k} + \alpha\varepsilon_{k\,h\,m}\,T_{,m}\,, \tag{6}$$

$$q_k = \kappa T_{,k} + \beta\,\varepsilon_{k\,h\,m}\,\nu_{h,m}\,, \tag{7}$$

where π is the pressure and, in general, $\mu, \bar{\kappa}, \bar{\alpha}, \bar{\beta}, \gamma, \alpha, \kappa, \beta$ are functions of T. We here treat $\mu, \bar{\kappa}, \bar{\alpha}, \bar{\beta}, \gamma, \alpha, \kappa$ as constant and in (1)–(4) ρ is assumed constant ($= \rho_0$) except in the body force term in (2) where the equation of state investigated in [1, 3, 8, 11] is

$$\rho = \rho_0\,[1 - A(T - T_0)], \tag{8}$$

A being the coefficient of thermal expansion of the fluid, while Lindsay and Straughan [9] take the non-Boussinesq equation of state

$$\rho = \rho_0\,[1 - A(T - T_0)^2]\,. \tag{9}$$

It is now further assumed that $\ell_i \equiv h \equiv 0$.

The relevant equations are now obtained by inserting (5)–(8) into (1)–(4).

However, some reduction of (4) is necessary to make the convection problem tractable. Equation (4) is

$$- \rho \frac{\partial \varepsilon}{\partial T} \dot{T} + \{(v_{h,k} - \varepsilon_{khr} \nu_r)[\mu(v_{k,h} + v_{h,k}) + \bar{\kappa}(v_{h,k} - \varepsilon_{kha} \nu_a)]$$
$$+ \nu_{h,k}(\bar{\alpha}\nu_{r,r}\, \varphi_{kh} + \bar{\beta}\nu_{k,h} + \gamma\nu_{h,k})\} + \kappa\Delta T$$
$$+ \alpha\varepsilon_{khm} T_{,m}\, \nu_{h,k} + \varepsilon_{khm} \nu_{h,m}\, \beta_{,k} = 0. \tag{10}$$

Three approaches to reducing (10) have been advocated in the literture. All assume the quadratic terms (in { ... } parentheses) are negligible.

Ahmadi [1] further assumes β constant and also neglects the α term. The resulting Bénard problem is then essentially symmetric and so no subcritical instabilities are possible.

The fluid is supposed contained in the layer $z \in (0,d)$ with gravity in the negative z direction. The planes $z = 0, d$ are kept at constant temperatures T_1, T_2 with $T_1 > T_2$ for most of the time in [11]. The stability/instability of the steady solution

$$\bar{T} = - Bz + T_1, \qquad \bar{\mathbf{v}} \equiv \mathbf{0}, \quad \bar{\boldsymbol{\nu}} \equiv \mathbf{0}\,, \tag{11}$$

where B is the temperature gradient, $B = (T_1 - T_2)/d$, is studied.

The reduction of equation (10) by Datta and Sastry [3] takes β constant but retains the term of $\alpha\varepsilon_{khm} T_{,m}\, \nu_{h,k}$. This term gives rise to a non-zero contribution from the basic temperature gradient which makes the linear operator non-symmetric. The final approach by Lebon and Perez-Garcia [8] neglects the α term but they assume $\beta = \hat{\beta} T$ for $\hat{\beta}$ constant. This again leads to a contribution from the basic temperature gradient and hence some skew-symmetry. A much richer structure is obtained by retaining a thermal coupling as do Datta and Sastry [3] or Lebon and Perez-Garcia [8].

The coefficients in (5)-(7) are restricted by (see [6])

$$\begin{gathered} 0 \le 2\mu + \bar{\kappa}, \quad 0 \le 2\mu + \kappa, \quad 0 \le \kappa, \quad 0 \le \bar{\kappa}, \quad \mu \ge 0, \\ 0 \le 3\bar{\alpha} + \bar{\beta} + \gamma, \quad 0 \le \gamma + \bar{\beta}, \quad (\alpha - \beta T^{-1})^2 \le 2\kappa T^{-1}(\gamma - \bar{\beta}). \end{gathered} \tag{12}$$

It would appear there is no reason why $\beta T^{-1} < 0$ and for the most part [9] and [11] investigate the case where $\hat{\beta} > 0$.

The non-dimensional perturbation equations are

$$u_{i,t} + u_k u_{i,k} = R\theta\delta_{i3} - \pi_{,i} + \Delta u_i + K(\Delta u_i - \varepsilon_{kir}\, \nu_{r,k})\,, \tag{13}$$

$$\Pr(\theta_{,t} + u_i\, \theta_{,i}) = Rw + \Delta\theta - bR\varepsilon_{3hm}\, \nu_{h,m} + b\Pr\varepsilon_{khm}\, \nu_{h,m}\, \theta_{,k}\,, \tag{14}$$

$$j(\nu_{i,t} + u_a \, \nu_{i,a}) = K(\varepsilon_{ikh}\, u_{h,k} - 2\nu_i) + G\nu_{a,a\,i} + \Gamma\Delta\nu_i \,, \tag{15}$$

together with the condition that $\mathbf{u}$ is solenoidal, where $\mathbf{u} = (u, v, w)$,. and where R^2 is the Rayleigh number.

Equations (13)–(15) hold on $z \in (0,1)$ with $\theta = \nu_i = 0$ on $z = 0,1$ and either $u_i = 0$ on $z = 0,1$ or stress free boundary conditions there. Periodicity is assumed in the x and y directions.

If instead of (8) the quadratic equation (9) is chosen, then (15) remains unchanged, (14) only changes in that R is replaced by $-R$ while in (13) the term $R\theta\delta_{i3}$ is replaced by $\delta_{i3}[-2R\theta(\xi - z) + \mathrm{Pr}\,\theta^2\,]$.

2 Non-linear energy stability and linear instability

The stationary convection instability boundary is given by

$$R^2 = \left(\frac{\Gamma(1+K)\Lambda + 2K + K^2}{(\Gamma + bK)\Lambda + 2K} \right) \frac{\Lambda^3}{a^2}\,, \tag{16}$$

where a is a wave-number and, for most cases, $\Lambda = \pi^2 + a^2$. For $b > 0$, $R^2 > 0$ and so stationary convection is allowed only by heating from below.

For oscillatory convection Payne and Straughan [11] find

$$R^2 \frac{\Lambda^3}{a^2}(1+K) + [j\,a^2\,]\,\{\Gamma\Lambda^3(1 + \mathrm{Pr}[1+K\,]) + 2k\Lambda^2(1+\mathrm{Pr}) + K^2\,\Lambda^2\,\mathrm{Pr}\,\}$$

$$+ \frac{\Lambda\,\mathrm{Pr}}{a^2[\,j\,(1+\mathrm{Pr}) + K\,\mathrm{Pr}(j - b)\,]}\,\{\Lambda K^2 + K\Lambda^2 b(1+K\,)$$

$$+ j^{-1}(\,[\,K\,b + \Gamma\,]\Lambda + 2k)\,[\,\Gamma\Lambda(1 + \mathrm{pr}\,[1+K\,]) + K^2\,\mathrm{Pr} + 2k\,(1+\mathrm{Pr})\,]\,\}\,. \tag{17}$$

Payne and Straughan [11] then deduce that overstable convection with $b < 0$ will not be possible unless b is suitably large: certainly it is necessary that $b <= \Gamma/K$, a condition which they show is also a possible source of trouble for stationary convection. For positive b, they show that a necessary condition for overstability is that

$$b > j\,[1 + (1+\mathrm{Pr})/K\,] \tag{18}$$

and from this they deduce that for probable situations of practical interest oscillatory convection is only possible in the case where the layer is *heated from above*, and even then the resulting critical Rayleigh numbers are likely to be enormous, possibly of the order of 4 million, and this leads them to argue that such a situation would not be likely to be encountered in a mundane setting.

Non-linear energy stability is investigated by Ahmadi [1], Lebon and Perez-Garcia [8], Payne and Straughan [11] and by Lindsay and Straughan [9]. The penetrative convection analysis of [9] is interesting in that if $b = 0$ they must evidently use a *weighted energy* to obtain unconditional (i.e., for all initial amplitudes) non-linear stability: the actual weighted energy is similar to that suggested by Payne et al. [10]. Apparently the method of [10] fails to work if $b \neq 0$ and in that case we only obtain conditional non-linear stability.

References

[1] Ahmadi, G. Stability of a micropolar fluid layer heated from below. *Int. J. Engng. Sci.*, **14** (1976), 81-89.

[2] Chandra, K. Instability of fluids heated from below. *Proc. Roy. Soc. London*, **A 164** (1938), 231-242.

[3] Datta, A.B. and Sastry, V.U.K. Thermal instability of a horizontal layer of micropolar fluid heated from below. *Int. J. Engng. Sci.*, **14** (1976), 631-637.

[4] Eringen, A.C. Simple microfluids. *Int. J. Engng. Sci.*, **2** (1964), 205-217.

[5] Eringen, A.C. Micropolar fluids with stretch. *Int. J. Engng. Sci.*, **7** (1969), 115-127.

[6] Eringen, A.C. Theory of thermomicrofluids. *J. Math. Anal. Appl.*, **38** (1972), 480-496.

[7] Eringen, A.C. Theory of anisotropic micropolar fluids. *Int. J. Engng. Sci.*, **18** (1980), 5-17.

[8] Lebon, G. and Perez-Garcia, C. Convective instability of a micropolar fluid layer by the method of energy. *Int. J. Engng. Sci.*, **19** (1981), 1321-1329.

[9] Lindsay, K.A. and Straughan, B. Penetrative convection in an isotropic thermomicropolar fluid. To be published.

[10] Payne, L.E., Song, J.C. and Straughan, B. Double diffusive porous penetrative convection; thawing subsea permafrost. *Int. J. Engng. Sci.*, **26** (1988), 797-809.

[11] Payne, L.E. and Straughan, B. Critical Rayleigh numbers for oscillatory and nonlinear convection in an isotropic thermomicropolar fluid. *Int. J. Engng. Sci.* (in press).

Straughan B.
Department of Mathematics
University of Glasgow
Glasgow G12 8QW
U.K.

PART 2 : DEFORMABLE SOLIDS WITH MICROSTRUCTURE

M. BERVEILLER AND H. SABAR

Grain size effects in polycrystalline plasticity

1 Introduction

The granular structure of metallic polycrystals induces an additional hardening (with respect to the hardening of single crystals) due to the presence of grain boundaries.

At least three kinds of effect may be mentioned :

(1) Plastic incompatibilities inducing long range internal stresses produce multiple plastic slips which lead to a supplementary isotropic and kinematic hardening with respect to single crystals. These phenomena are taken into account by the advanced models (like self-consistent or statistical ones) which deal with the internal stresses associated with interfacial or geometrically necessary dislocations. This is the case of models based on the solution of the plastic inclusion problem. Nevertheless, the grain size effect does not appear in such approaches.

(2) The triple junctions between grains induce a non-uniform stress field inside the grains and contribute to the development of inhomogeneous plastic strain inside the grains. This rather complicated effect has to be dependent on the grain size but is not considered here.

(3) At a more local scale, one may expect to observe a grain size effect associated with the fine structure of interfacial dislocations, reflecting the inhomogeneous deformation at grain boundaries (pile up). The self-consistent modelling for which the plastic strain is homogeneous inside the grains, distributes the dislocations exclusively inside the grain boundaries (surface dislocations). This repartition may be acceptable if one wishes to evaluate the long range stress fields but does not take into account phenomena like pile up.

In order to take into consideration the fine structure of dislocations at grain boundaries, we suppose that the average mean free path of geometrically necessary dislocation decreases with increasing plastic strain. This decrease is evaluated using a very simple theory of dislocation pile up.

The stress field for a plastic inclusion with mobile interface is given and used to build a model which incorporates both long range stresses (Kröner's model) and the (long range) stress due to pile up.

In this manner, the effects of grain size associated with the fine structure of the geometrically necessary interfacial dislocations are taken into account and give a polycrystalline hardening depending on grain size.

2 The plastic inclusion with a moving interface

An ellipsoidal inclusion V_I is considered in an infinite body with homogeneous elastic constant C.

Plastic or eigenstrains in the medium are assumed to be piece-wise uniform, equal to ε^{PI} inside the inclusion and $\mathbf{E}^P$ in the matrix.

Using the indicator function $\theta^I(r)$

$$\theta^I(r) = \begin{cases} 1 & \text{if } r \in V_I, \\ 0 & \text{if } r \notin V_I, \end{cases} \tag{1}$$

the plastic strain field can be written as

$$\varepsilon^P_{ij}(r) = E^P_{ij} + \left(\varepsilon^{P_I}_{ij} - E^P_{ij}\right)\theta^I(r)\,. \tag{2}$$

Using the Green tensor G for the infinite medium with elastic constant C_{ijkl}, the

total strain field $\varepsilon^T_{ij}(r)$ is given by

$$\varepsilon^T_{ij}(r) = E^P_{ij} + \int_{V_I} \Gamma_{ijkl}(r - r')\, C_{klmn}\left(\varepsilon^{P_I}_{mn} - E^P_{mn}\right) dV_I'\,, \tag{3}$$

where

$$\Gamma_{ijkl} = \frac{1}{2}\left(G_{ik,jl} + G_{jk,il}\right) \tag{4}$$

is the first strain Green tensor defined by Kröner [7]. As is well known from Eshelby's [4] results, the field $\varepsilon^T_{ij}(r)$ is uniform inside the inclusion and equal to ε^{T_I}.

Setting

$$S_{ijmn} = \int_{V_I} \Gamma_{ijkl}(r - r')\, C_{klmn}\, dV_I' \quad (\text{for } r \in V_I) \tag{5}$$

then

$$\varepsilon^{T_I}_{ij} = E^P_{ij} + S_{ijkl}\left(\varepsilon^{P_i}_{kl} - E^P_{kl}\right). \tag{6}$$

The elastic strain ε^{e_I} inside the inclusion is by definition

$$\varepsilon^{e_I}_{ij} = \varepsilon^{T_I}_{ij} - \varepsilon^{P_I}_{ij} \tag{7}$$

or from (6)

$$\varepsilon^{e_I}_{ij} = \left(I_{ijkl} - S_{ijkl}\right)\left(E^{P}_{kl} - \varepsilon^{P_i}_{kl}\right) \tag{8}$$

which is the well-known Kröner formula [8].

From this state, we suppose that both $\mathbf{E}^{P}$ and ε^{P_I} suffer a change dE^{P} and $d\varepsilon^{P_I}$ and that simultaneously the boundary moves by a displacement field $du_p(r)$ leading to a new inclusion $(V_I + dV_I)$. From a theorem given for instance by Germain [6], the increase $d^{T}(r)$ is

$$d\varepsilon^{T}_{ij}(r) = dE^{P}_{ij} + \int_{V_I} \Gamma_{ijkl}(r - r')\, C_{klmn}\left(d\varepsilon^{P_I}_{mn} - dE^{P}_{mn}\right) dV'_I$$

$$+ \int_{V_I} \left[\Gamma_{ijkl}(r - r') C_{klmn}\left(\varepsilon^{P_I}_{mn} - E^{P}_{mn}\right) du_p(r')\right]_{,p'} dV'_I . \tag{9}$$

If the new inclusion is also an ellipsoid, it is easy to verify from (3) that the field $\varepsilon^{T_I} + d\varepsilon^{T_I}$ is also uniform inside the new inclusion $(V_I + dV_I)$. In this case, for $r \in (V_I + dV_I)$, the field $d\varepsilon^{T_I}_{ij}$ is given by

$$d\varepsilon^{T_I}_{ij} = dE^{P}_{ij}\, S_{ijkl}\left(d\varepsilon^{P_I}_{kl} - dE^{P}_{kl}\right) + dS_{ijkl}\left(\varepsilon^{P_I}_{kl} - E^{P}_{kl}\right), \tag{10}$$

where we introduce the tensor dS defined by

$$dS_{ijkl} = \int_{V_I} \left[\Gamma_{ijmn}(r - r')\, C_{mnkl}\, du_p(r')\right]_{,p'} dV_I . \tag{11}$$

Let a_1, a_2 and a_3 be the initial semi-axes of the ellipsoid and their variations are da_1, da_2, da_3.

From geometrical arguments, the displacement field of the boundary is given by

$$du_p(r) = \frac{x_p}{a_p}\, da_p \quad \text{(no summation on } p\text{)}. \tag{12}$$

From (11), one obtains for dS

$$dS_{ijkl} = \sum_{P} \frac{da_p}{a_p} \int_{V_I} \left(\Gamma_{ijmn}(r - r') C_{mnkl}\, x'_p\right)_{,p'} dV'_I \tag{13}$$

or by partial integration

$$dS_{ijkl} = \sum_P \frac{da_p}{a_p}\Big\{ \int_{V_I} (\Gamma_{ijmn}(r - r')C_{mnkl}\, dV_I' + \int_{V_I} \Gamma_{ijmn,p'}(r - r')C_{mnkl}\, x_p' dV_I' \Big\}. \tag{14}$$

The first integral is equal to S and finally

$$dS_{ijkl} = \sum_P \frac{da_p}{a_p} \Big\{ S_{ijkl} + \int_{V_I} \Gamma_{ijmn,p'}(r - r')C_{mnkl}\, x_p' dV_I' \Big\}. \tag{15}$$

From (15) and (7) the stress increment $d\sigma_{ij}$ inside the inclusion is given by

$$d\sigma_{ij} = C_{ijkl}\left(I_{klmn} - S_{klmn}\right)\left(dE^{P}_{mn} - d\varepsilon^{P_I}_{mn}\right) + C_{ijkl}\, dS_{klmn}\left(E^{P}_{mn} - \varepsilon^{P_I}_{mn}\right). \tag{16}$$

The components of dS have been calculated for different forms of inclusions and by different methods [2].

For an isotropic solid with shear moduli μ and Poisson's ratio ν, the component dS_{1212} is equal to

$$dS_{1212} = \frac{-5 + 7\nu}{105(1 - \nu)} \frac{da}{a}. \tag{17}$$

Here, we have assumed an initial spherical form for the inclusion and a change of its axis defined by

$$a_1 = a \rightarrow a + da\,,$$
$$a_2 = a \rightarrow a\,, \tag{18}$$
$$a_3 = a \rightarrow a\,.$$

The above results are now used to evaluate the grain size effects in the polycrystalline plasticity.

3 Evaluation of the mean free path of dislocations by single slip

In the following, we restrict ourselves to the microplasticity state where the plastic deformation inside the matrix is zero and only a few grains start to deform plastically by single slip γ on a slip system with unit normal $n = (0, 1, 0)$ and a unit vector $m = (1, 0, 0)$ parallel to slip direction.

The plastic distortion β^{P}_{ij} is given by

$$\beta^{P}_{ij}(r) = \gamma\, m_j\, n_i\, \theta^{I}(r)\,. \tag{19}$$

From Kröner's [9] formula, the dislocation density α is equal to

$$\alpha_{ij} = -\,\varepsilon_{ikl}\,\beta^{P}_{lj,k}$$

$$= \varepsilon_{ikl}\,\gamma\, m_j\, n_l\, N_k\, \delta(s) \tag{21}$$

where ε_{ikl} is the permutation tensor, and N_k is the unit vector normal to the grain boundary.

Equation (21) means that all the dislocations are located in the grain boundary in the form of surface dislocations. These results do not take into account the local interactions between dislocations like pile up or tangles eventually associated with a local multislip.

As a matter of fact, the density α is partially distributed in front of the grain boundary, so that a new "interface" is built which moves toward the centre of the grain as the slip progresses. Here the term interface is taken as the surface in front of the pile up.

In the plane $x_3 = 0$, the dislocation density is given by

$$\alpha_{31} = \gamma\,\frac{x_1}{a}\,\delta(S)\,. \tag{22}$$

We can now assume that the displacement u_1 of the interface is proportional to the dislocation density so that

$$u_1 = -\,K(\gamma, a)\,\frac{x_1}{a}\,\gamma\,, \tag{23}$$

where K is an unknown function depending on the local interaction between dislocations in a pile up.

The increment du_1 is obtained from (23) by differentiation

$$du_1 = -\left\{\frac{\partial K}{\partial \gamma}\,d\gamma\,\gamma + K\,d\gamma\right\}\frac{x_1}{a} \tag{24}$$

or

$$du_1 = -\,F(\gamma, a)\,\frac{x_1}{a}\,d\gamma\,. \tag{25}$$

By comparison with (12) the corresponding changes in $\mathrm{d}a_p$ are

$$\mathrm{d}a_2 = \mathrm{d}a_3 = 0\,,$$

$$\mathrm{d}a_{1=} = -\,F\,(\gamma, a)\,\mathrm{d}\gamma\,. \tag{26}$$

The evaluation of F is done using the results from the pile up theory [3].

The length of a pile up with N dislocations under stress σ is

$$u = \frac{\mu\, b}{\pi(1-\nu)}\,\frac{N}{\sigma}. \tag{27}$$

The number of dislocations N is related to the slip amplitude γ and the distance P between slip lines by [5]

$$\gamma = \frac{N\,b}{P}. \tag{28}$$

From (27) and (28), one obtains

$$u = \frac{\mu\, P}{\pi(1-\nu)\,\sigma}. \tag{29}$$

For F.C.C. metals, μ/σ is close to 10^3 and P is about $0.1\ \mu\mathrm{m}$. Then, F is given by

$$F = \frac{10^3}{\pi(1-\nu)}\,P\,. \tag{30}$$

For dislocations with low stacking fault energy, their dissociation is higher and F can be one order greater than in (30).

In the next section the hardening on the single slip system due to statistical dislocation density, the long range stress appearing in (9) and the new term depending on grain size are discussed.

4 Grain size influence on intragranular hardening

From eq. (16) and the hypothesis made at the beginning of Section 3, the shear stress $\mathrm{d}\tau$ on the slip system due to the applied stress $\mathrm{d}\Sigma$ and internal stresses is given by

$$\mathrm{d}\tau = \mathrm{d}\Sigma - 2\,\mu(1-\beta)\,\frac{\mathrm{d}\,\gamma}{2} + 2\,\mu\,\gamma\,\frac{\partial\, S}{\partial\, a}\mathrm{d}a\,, \tag{31}$$

where $\beta = (2/15)\,(4-5\nu/1-\nu) \sim 1/2$ is Eshelby's constant (for $\nu \sim 1/3$) and

$$\frac{\partial S}{\partial a} = \frac{\partial S_{1212}}{\partial a} = \frac{\partial S_{1221}}{\partial a} = \frac{\partial S_{2112}}{\partial a} = \frac{\partial S_{2121}}{\partial a}.$$

Assuming a hardening due to the statistical dislocation distributions inside the grain equal to H ($d\tau = H\, d\gamma$), one obtains from (31)

$$d\Sigma = \left\{ H + \frac{\mu}{2} - 0.08\,\mu\,\gamma\,\frac{F(\gamma, a)}{a} \right\} d\gamma \tag{32}$$

or

$$d\Sigma = \mu \left\{ \frac{H}{\mu} + \frac{1}{2} - 0.08\,\gamma\,\frac{F(\gamma, a)}{a} \right\} d\gamma\,. \tag{33}$$

The first term is about 1/250 for F.C.C. metals and the second one is overevaluated; both terms are grain size-independent. The third term depends on the grain size a and his sign depends on the difference $(\mathbf{E}^p - \varepsilon^p)$ (see (10)). Figure 1 shows the evolution of the three terms of (33) with the grain size. The intergranular term $(\mu/2)$ has been corrected with an accommodation factor $\alpha \sim 1/100$ proposed by Berveiller and Zaoui [1] in order to take into account the evolution of the elastoplastic stiffness of the polycrystal.

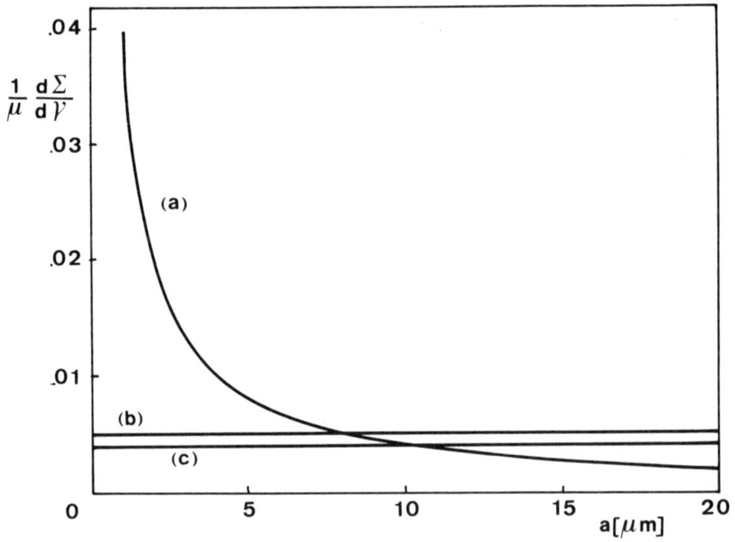

Figure 1. Influence of the grain size on the three hardening terms of equation (33): (a) grain size effect, (b) single crystal hardening, (c) intergranular hardening due to Kröner's long range stress.

A very important influence of the grain size on the global hardening is observed for grain diameter less than some value (in this case ~ 10 μm) depending on the dissociation of dislocations. For grain diameter greater than this value, the single crystal hardening and the interactions between grains are preponderant. In this case it may be expected that the polycrystalline hardening is grain size-independent. Such effects are experimentally well confirmed by A.W. Thomson et al. [11]. The above theory is now introduced into a self-consistent model [10] in order to describe the effects of grain size on the overall behaviour of polycrystals.

References

[1] Berveiller, M. and Zaoui, A., An extension of the self-consistent scheme to plastically flowing polycrystals, *J. Mech. Phys. Solids*, **26** pp. 325-344, 1979.

[2] Berveiller, M. and Sabar, H., To be published.

[3] Eshelby, J.D., Frank, F.C. and Nabarro, F.R.N., The equilibrium of linear arrays of dislocations, *Phil. Mag.*, **42**, pp. 351-364, 1951.

[4] Eshelby, J.D., Elastic Inclusions and Inhomogeneties, *Progress in Sol. Mech.*, **2**, pp. 89-140, 1961.

[5] Friedel, J., *Les dislocations*, Ed. Gauthier-Villars, Paris, 1964.

[6] Germain, P., *Cours de Mécanique des milieux continus*, Ed. Masson, Paris, 1973.

[7] Kröner, E., *Statistical Continuum Mechanics*, CISM Lecture notes no. 92, Springer-Verlag, Wien, 1972.

[8] Kröner, E., Zur plastischen verformung des Vielkristalls, *Acta Met.*, **9**, pp. 155-161, 1961.

[9] Kröner, E., *Kontinuumstheorie der Versetzungen und Eigenspannungen*, Springer-Verlag, Wein, 1958.

[10] Lipinski, P. and Berveiller, M., Elastoplasticity of micro inhomogeneous metals at large strains, to appear in *Int. J. Plasticity.*

[11] Thompson, A.W., Baskes, H.I. and Flanagan, W.F., The dependence of polycrystal work-hardening on grain size, *Acta Met.*, **21**, pp. 1017-1028, 1973.

Berveiller M.
L.P.M.M.,
Faculté des Sciences
Université de Metz
Ile du Saulcy
F-57045 Metz Cédex
FRANCE

Sabar H.
L.P.M.M.,
Faculté des Sciences
Université de Metz
Ile du Saulcy
F-57045 Metz Cédex
FRANCE

T. BRETHEAU, E. HERVE AND A. ZAOUI

On the influence of the phase connecteness on the yield point and the plastic flow of two-phase materials

1 Introduction

It is well known, from a theoretical point of view, and also widely accepted, that the overall mechanical behaviour of inhomogeneous random materials depends not only on the volume fractions of their constituent phases but also on their space distribution. Nevertheless, such a dependence has rarely been shown experimentally and, except for very special situations, it may be commonly considered that it leads only to negligible quantitative deviations from one distribution or the other. We intend to show in the following that several prominent features of the plastic behaviour of two-phase materials can only be accounted for if one of the most important of their morphological properties, namely the connectedness of the phase, is taken into account. This will be done from the results of an experimental investigation of the tensile response of an iron-silver blend which, as well as for what concerns the yield-point characteristics as for the analysis of the plastic flow regime, strongly depends on the connectedness of one phase or of the other phase. This will lead to a tentative characterization of such a connectedness by using a stereological model and image analysis techniques and to some suggestions concerning the way of modelling the mechanical influence of the morphological properties.

This study is motivated by the fact that the space distribution of the phases of composite materials is a basic parameter for the optimization of their microstructural constitution with respect to expected overall mechanical properties, which is likely to be a major aspect of tomorrow's materials engineering, aiming at the production of "à la carte" materials. In this view, on the one hand, usual models are only valid for very specific, extremely ordered or perfectly disordered models, e.g., those using correlation functions of material moduli [9], look hardly tractable and inadequate to express simply some major primary morphological characteristics (such as the occurrence of matrix/inclusion configurations or of imbricated arborescences). This paper is a contribution to the development of practical tools in view of improved morphological analysis and mechanical modelling.

2 Experimental background

The iron-silver composite has been chosen as a model material for this study (monotonic tensile tests at room temperature) for two main reasons: iron and silver are not miscible in any proportion, so that the mechanical behaviour of each phase of such composites is unchanged whatever their volume fraction; in addition, the mechanical plastic properties of these metals are quite different from one another, so that the influence of the constitution of the two-phase material on its overall response is expected to be stronger and then easier to be studied. The first argument has led to the use of a powder metallurgy technique (hot isostatic pressing) in order to elaborate the composite in the whole range of volume fractions, from pure iron to pure silver. The second argument is illustrated by two facts: first, the flow stress level of iron is roughly four times larger than the one of silver; second, iron exhibits the Lüders bands phenomenon which is responsible for an inhomogeneous plastic deformation of the sample after the yield point, corresponding to the initial plateau on the stress-strain curve of Figure 1, whereas silver does not do so. It was expected that the possible propagation of Lüders bands in the two-phase material could be strongly influenced by the space distribution of the iron phase in the composite while the overall response of the composite in the plastic regime could especially depend on the morphological characteristics of the silver phase.

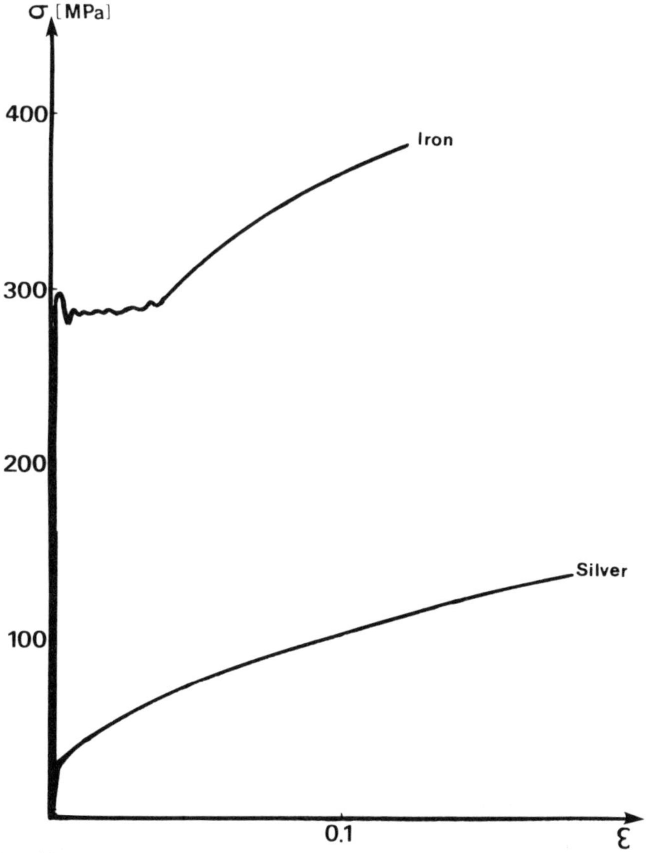

Figure 1. Tensile stress-strain curves of the pure constituent phases.

As a matter of fact, the main experimental results showed a clear sensitivity to the geometrical constitution of the composites:

- beyond about 70% iron, Lüders bands can nucleate and propagate, even if in a more difficult way than in pure iron (multiple initiation, bent propagation fronts, increase of the flow stress during the first plastic stage, early necking formation in the sample, etc.). More precisely, between 70% and 85% iron, silver enters the plastic regime first and then iron does so by Lüders bands propagation, whereas for more than 85% iron, Lüders bands propagate first and then silver begins to flow plastically. Below 70% iron, no Lüders bands can be observed, though iron is flowing according to a homogeneous mode which starts later and later.

- below 60% iron, the two-phase stress-strain curve is significantly and systematically "softer" than expected from a self-consistent analysis: such an analysis [2], which considers successively the iron and the silver phases as a spherical inclusion embedded in an infinite matrix constituted with the effective (homogeneous equivalent) medium and derives this effective behaviour from an averaging process on the solution of both of these situations, needs a theoretical stress-strain curve of pure iron exhibiting no Lüders plateau; this has been performed by an "inversion" of the self-consistent scheme deriving from the experimental curves of pure iron and a 65% iron composite, considered as a perfectly disordered one, such a theoretical iron behaviour (Figure 2).

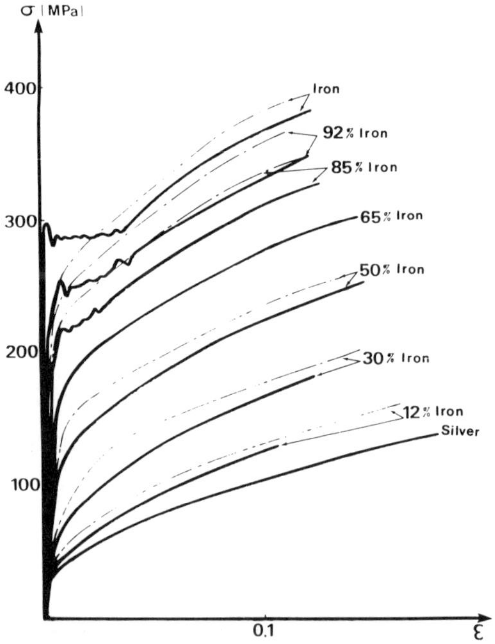

Figure 2. Experimental (full line) and predicted (dotted line) stress-strain curves (according to the self-consistent scheme) (from [5]).

At the same time, microscopic observations of the materials with a lower and lower iron content indicate that they grdually evolve from a silver inclusion/iron matrix configuration to the reverse one, going through intermediate situations where both phases are likely to be continuous. Taking account of the fact that the self-consistent scheme makes both phases play symmetrical roles, we were led to ask a couple of questions, namely:

- do Lüders bands disappear in the composite as soon as the iron matrix is no more continuous, so that their development in a whole cross-section of the sample and over a significant part of its length becomes impossible?

- is the iron inclusion/silver matrix morphology responsible for the softer than expected response of the composite for a sufficiently large silver content, if so, can any adequate model take such a situation into account in a better and more realistic way than the self-consistent scheme?

Unfortunately, the main part of such questions cannot be answered directly from the simple two-dimensional observation of a few micrographs since the determination of the connectedness of the phases of our materials is obviously a three-dimensional problem. That is the reason why we first developed a morphological analysis on a stereological basis; we then performed a mechanical study in order to model the plastic behaviour of inclusion/matrix composites.

3 Morphological analysis

Matheron's Boolean model was used in order to get a three-dimensional modelling of our materials from two-dimensional measurements on photographs obtained by optical microscopy and to calculate the adequate connectivity parameters which could not be measured directly. According to this model [10], iron germs are first randomly scattered in a silver matrix by a Poisson point process with density θ. Random primary iron grains, which can overlap, are then implanted on each germ: we used spherical grains with a given sphere diameters distribution; as a matter of fact, we checked only a uniform and a linear distribution, which both revealed a reasonably adequate range for our materials. The model parameters were then identified from the measurement of the covariance $Q(h)$ of the silver phase, i.e., the probability that a segment of length h lie in the silver phase completely. The linear variation of log $Q(h)$ with h (Figure 3) yields the proof that our materials obey such a model fairly well.

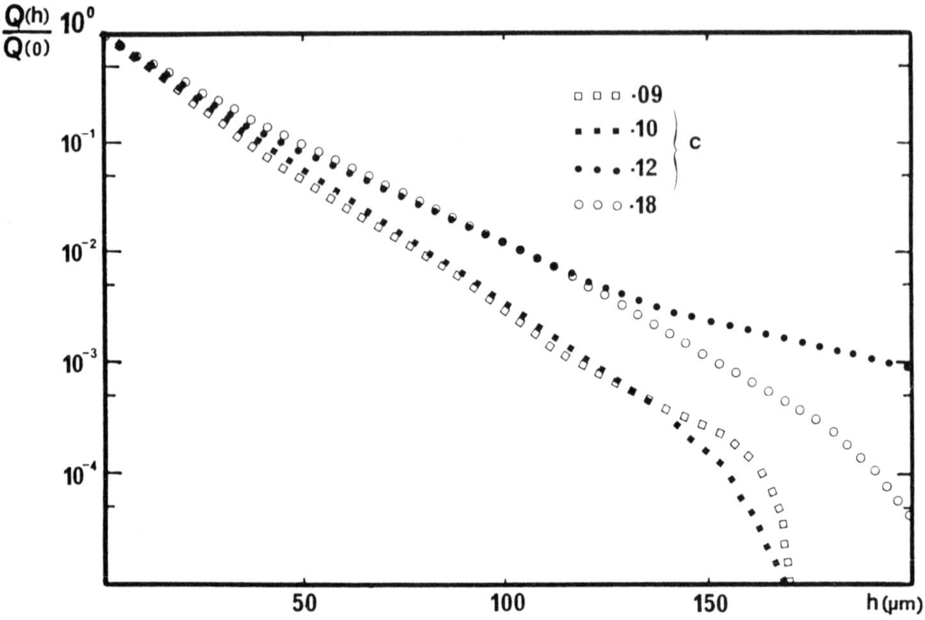

Figure 3. The almost linear variation of log $Q(h)$ indicates that the Boolean scheme used is adequate for our materials.

We can then calculate the specific number of connexity of our materials for various iron volume fractions; this number is known to give pertinent information on the phases connectivity. Let us briefly recall that the connexity number C of a domain X is defined as the integral of the total (or Gauss) curvature, namely:

$$C = \frac{1}{4\pi} \int_{\partial X} \frac{ds}{R_1 R_2}. \tag{1}$$

It is known that [8]:

$$C = N - G, \tag{2}$$

where N is the number of independent closed surfaces of X and G (the "genus") is the maximum number of cuts which may be performed without separating the body into independent parts ($C = 1-0 = 1$ for a full sphere; $C = 1-1 = 0$ for a torus; $C\, 2 - 0 = 2$ for a hollow sphere; C may be negative for a multiply-connected body, for instance when several spheres have been aggregated in an imperfect manner, etc.). There specific number of connectivity C_V of a body with the volume V is simply C/V.

Figure 4 shows the variation of C_V with the iron volume fraction, as derived form the quantitative morphological analysis of seven silver-iron samples with 82-91% iron (twenty contiguous micrographs have been analysed for each sample). The specific number of connectivity C_v is varying in a complex manner when the iron content is increasing [3]:

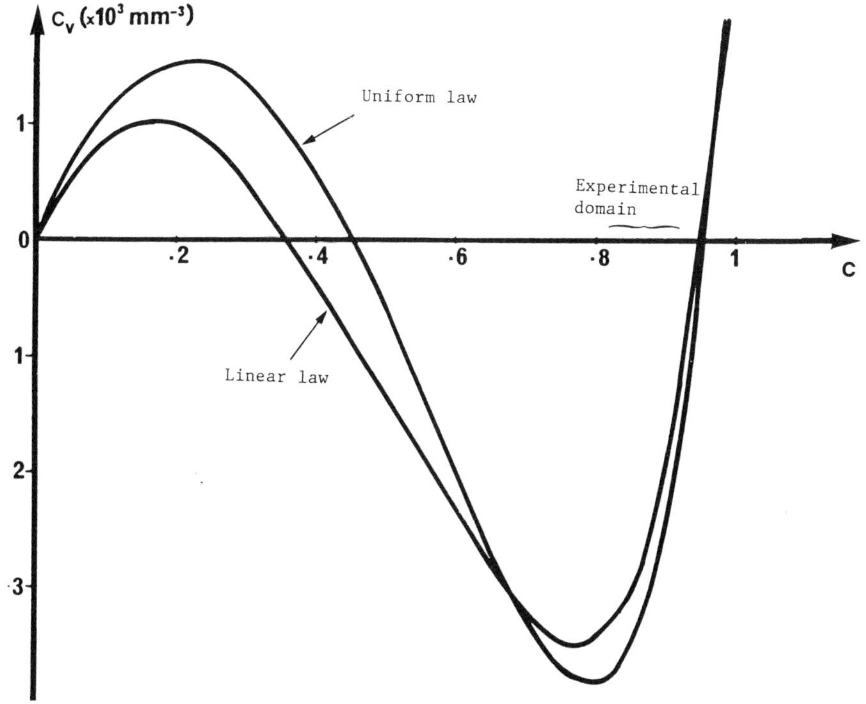

Figure 4. The three-dimensional connectivity specific number C_v dependence versus the iron volume fraction c.

- first, it is positive and increasing while more and more isolated iron spheres are implanted in the silver matrix;

- beyond about 20% iron, overlapping of iron grains is more frequent so that N is decresing and quasi-torical configurations are developed, which makes the genus G increase: C_v is then decreasing more and more up to large negative values;

- from about 70% iron, this decrease is slowed down and beyond 80% iron C_v is increasing: it can be concluded that a continuous iron skeleton has been constituted, first with an only partial interconnection and then with a well-connected configuration. C_v reaches positive values and then abruptly falls towards zero due to the resorption of silver residual islets.

Qualitative correlations with the observed Lüders bands behaviour may be settled: beyond 70% iron the constitution of an iron skeleton makes the propagation of the

Lüders bands easier and easier, despite the presence of the inhibiting silver phase. First the iron connections are seldom and weak and the preliminary plastification and strain-hardening of the silver phase is necessary to allow the bands propagation; for a larger iron content this is not the case and the strongly connected matrix commands the overall composite behaviour according to its own way of yielding.

Obviously, such an interpretation is more intuitive than rigorous. It is reported here mainly in order to illustrate two basic remarks:

- first, morphological characteristics may be responsible for sharp mechanical effects of a "critical" nature which, similarly to percolation phenomena, suddenly vanish when some geometrical threshold has been crossed;

- second, stereological modelling and image anlaysis techniques are likely to become an essential ingredient of any practical development of new materials design and engineering.

4 Mechanical modelling

Whereas any quantitative mechanical modelling of the Lüders bands regime of our materials appears unapproachable at the moment, the more regular and homogeneous response of our samples with less than 60% iron seem more tractable. At variance with the self-consistent approach, which is definitely inadequate to inclusion/matrix two-phase configurations, the classical Hashin composite spheres model [6] offers an interesting way of dealing with such situations since the phase which is disposed in the spherical shells forms a continuous matrix, while the other phase inside the inner spheres consists of well-isolated inclusions. Nevertheless, this model is restricted to isotropic linear elasticity and bounds can only be derived for the effective shear modulus.

The three-phase model proposed by Christensen and Lo]4], which is based upon a similar idea and then allows us to simulate inclusion/matrix configurations, yields well-defined estimates of the overall moduli thanks to an equivalence condition for elastic strain energy. The main drawbacks of this model lie in the fact that it cannot be extended to plasticity in a straightforward manner. This difficulty was resolved thanks to the demonstration, even for much more general situations, that, in the elastic case, the energy condition is strictly equivalent to an average strain (or stress) condition according to which the average strain (or stress) of the composite inclusion is equal to the macroscopic applied one [7]. This basic result permits a quite simple and direct extension of the three-phase model to the non-linear case, the principle of which is analogous to what is now classical for the self-consistent scheme [1].

From that it was not possible to go back to the basic stress-strain curves of pure iron and silver and to derive from them predictions for the two-phase metal at any volume fraction of iron (less than 60%). This resulted in a significant "softening" of the curves with respect to the self-consistent prediction, which led to a much better agreement with the experimental data (see Figure 5 for a typical example of such an improvement).

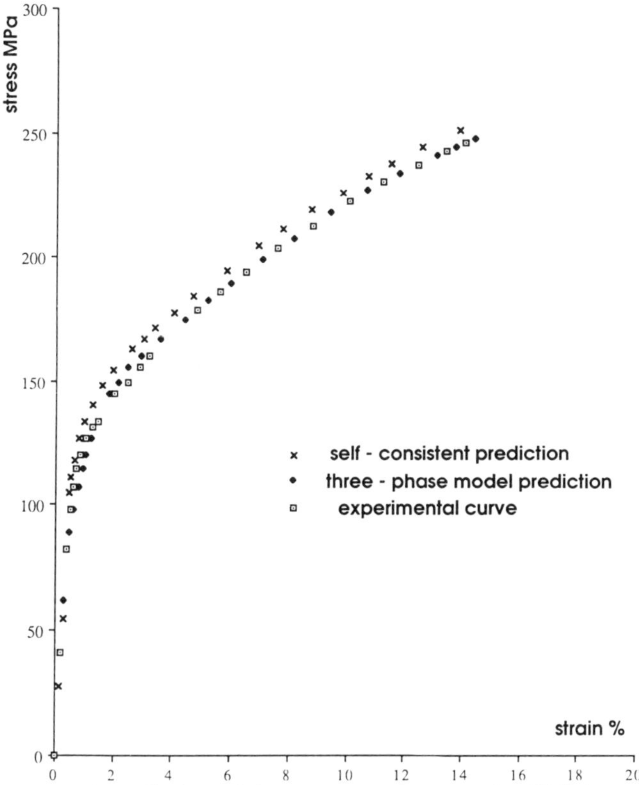

Figure 5. Improved prediction of the stress-strain curve for 50% iron (from [5]).

Consequently it may be concluded that, in addition with the critical sharp mechanical effects discussed hereabove, the phase connectivity is also responsible for smoother consequences on the effective overall behaviour of composite materials, which may be modelled by simple but efficient new tools using composite inclusion patterns embedded in an homogeneous equivalent matrix. The development of such models is now in progress in order to deal with more general morphological situations and more elaborate and varied mechanical properties.

Acknowledgements

Dr Dominique Jeulin, from the "Centre de géostatistique de l'Ecole des Mines de Paris" (Fontainebleau - France) is gratefully acknowledged for his prominent part in the morphological analysis reported hereabove. The mechancial aspects of this study (experimental tests and part of the three-phase modelling) have been reported in the doctoral thesis of A. Feylessoufi.

References

[1] Berveiller, M. and Zaoui, A. (1979). An extension of the self-consistent scheme to plastically flowing polycrystals, *J. Mech. Phys. Solids*, **26**, 325.

[2] Berveiller, M. and Zaoui, A. (1981). A simplified self-consistent scheme for the plasticity of two-phase metals, *Res. Mech. Let.*, **1**, 119.

[3] Bretheau, T. and Jeulin, D. (1989). Caractéristiques morpjologiques des constituants et comportement à la limite élastique d'un matériau biphasé Fe/Ag, *Rev. Phys. App.* (in press).

[4] Christensen, R.M. and Lo, K.H. (1979). Solutions for effective shear properties in three phase sphere and cylinder models, *J. Mech. Phys. Solids*, **27**, 315.

[5] Feylessoufi, A. (1988). Doctoral thesis, Université Paris-Nord, Villetaneuse, France.

[6] Hashin, Z. (1962). The elastic moduli of heterogeneous materials, *J. Appl. Mech.*, **29**, 143.

[7] Herve, E. and Zaoui, A. (1989). Modelling the effective behavior of nonliner matrix-inclusion composites, submitted to *Eur. J. Mech. (A)*.

[8] Jeulin, D. (1979). Morphologie mathématique et propriétés physiques des agglomérés de minerais de fer et de coke métallurgique. Doctoral thesis, Ecole des Mines de Paris, France.

[9] Kroner, E. (1977). Bounds for effective elastic moduli of disordered materials, *J. Mech. Phys. Solids*, **25**, 137.

[10] Matheron, G. (1967). "Eléments pour une théorie des milieux poreux", Masson, Paris.

Bretheau Th., Hervé E., and Zaoui A.
L.P.M.T.M.,
Université de Paris-Nord
Avenue J.B. Clément
F-93430 Villetaneuse
FRANCE

G. CANOVA AND L.P. KUBIN

Dislocation microstructures and plastic flow: a three-dimensional simulation

1 Introduction

Plasticity is a dissipative process, far from thermodynamic equilibrium. As a consequence, dislocations, the defects responsible for plasticity at the atomic scale, very often exhibit self-organization in plastically strained materials. Basically, one can distinguish between two prominent types of dislocation populations: (i) Mobile dislocations carry the plastic strain rate and shear the crystal along crystallographic slip planes. They leave strain markings (slip lines, slip bands) at the specimen surfaces and these traces build up static patterns or may, in some cases, propagate with a well-defined velocity. (ii) Immobile dislocations, which are mutually trapped or blocked by other defects, are also frequently organized in the form of walls, defining open channels of three-dimensional structures. Such dislocation patterns are either in dynamic equilibrium or evolve during straining, being then responsible for strain hardening.

The physical definition of constitutive equations for the bulk material necessarily involves, then, a study of the collective arrangements of dislocation populations. This question becomes even more important when the macroscopic response of the material is affected by strain heterogeneity processes whose charcteristic length scales are in general linked to the ones describing the organization of dislocations. To date, however, this problem has received no convincing theoretical solution because of its analytical complexity (cf. [5] for a review). The number of coupled populations to take into account is quite large since dislocations may have different charcters (in practice two only are considered, screw and edge) and several Burgers vectors, each of them being active along a preferential glide plane. The interactions between dislocations take place via their long range $(1/r)$ elastic stress fields, leading to the build-up of a non-local internal stress field. There are also short range interactions, which are responsible for the mechanisms of intersection, reaction, multiplication, mutual annihilation and trapping. Finally each possible configuration of mutually trapped dislocations (dipole, junctions etc.) must be considered as a distinct population since it has its own creation and destruction rate.

It appears, therefore, that the question of the spontaneous organization of dislocation structures and of its relation to mechanical properties can, at present, only be approached through numerical simulations. To date, only two-dimensional simulations have been developed, which necessarily involve some arbitrariness in the treatment of some processes (such as dislocation multiplication) and which deal with arrays of infinite, straight, dislocations of identical character (edge or screw) and of the same Burgers vector. Starting with random configurations, evidence for self-organization was unambiguously identified after motion under the effect of applied and/or internal stresses. Two simulation techniques were used, molecular dynamics [3] and a technique deriving from cellular automata [5,6], that is involving a discretisation of both space and time, but still with long range interactions. Although the two methods yield rather comparable results, the second one appears easier to implement and is better suited for an extension to three dimensions.

The aim of the present work is to describe a simulation technique which is able to realistically deal with the full complexity of three-dimensional dislocation structures. It accounts for the coupling between screw and edge characters, between different slip systems, for the influence of crystallographic structure and of crystal orientation with respect to it and it includes all the effects bound to line tension and dislocation bending under stress. Measured values of both local and average quantities can be extracted in order to clarify the relations between microstructure and the macroscopic response. The simulated material is a single crystal of pure copper (FCC structure) at room temperature, but the method described in Section 2 is able to deal with dislocations in virtually any type of crystal structure. Some preliminary results are reported in Section 3.

2 Simulation technique

In the FCC structure, slip planes are usually {111} and there are three 1/2 <110> Burgers vectors per slip plane, i.e., 12 possible slip systems. Two dislocation characters are defined, screw and edge, and the line and slip directions of all the possible Burgers vectors can conveniently be represented on a FCC lattice. In Figure 1, one can see a portion of a (square) dislocation loop with a screw segment of line direction and Burgers vector 1/2 $[0\bar{1}1]$ and an edge segment of the same Burgers vector and of line direction $1/2[\bar{2}1]$. The direction of motion of the screw segment is the line direction of the edge one and vice-versa.

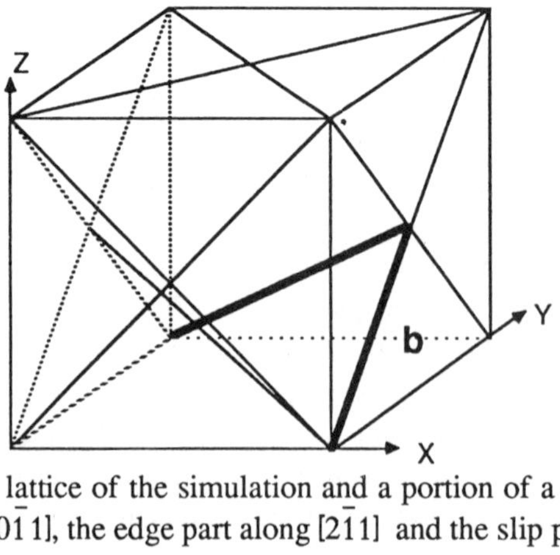

Figure 1. The FCC lattice of the simulation and a portion of a dislocation loop. The screw part is along $[0\bar{1}1]$, the edge part along $[2\bar{1}1]$ and the slip plane is $(11\bar{1})$.

Because parallel dislocations repel each other or attract each other and mutually annihilate, it is not possible to find two parallel dislocations at a separation smaller than a critical value which depends on the Burgers vectors and signs of the dislocations considered. The smallest of these critical distances is the natural basis for the discretization in space. As in the two-dimensional simulations, it is equal to $2y_e$, where y_e is the critical annihilation distance for an edge dipole under the influence of its self-stresses. In copper and at room temperature, $2y_e \sim 3$ nm [1]. Therefore, inserting elementary dislocation segments along the crystallographic directions defined in Figure 1, we must have $| a/2 \langle 110 \rangle | = a\backslash\sqrt{2} = 2y_e$ with $a \sim 4.23$ nm being the parameter of the lattice on which the simulation is based. It follows that $2y_e$, the elementary translation distance of edge segments is also the length of elementary screw segments, while the elementary translation distance of screw segments, i.e., the length of elementary edge segments, is equal to $| a/2 \langle 211 \rangle | = a\sqrt{6}/2 \sim 1.73$ nm. Then, the shape of any dislocation segment or loop can be represented by a succession of straight elementary dislocation segments (EDSs) of edge and screw character. These EDSs move by elementary steps in their slip direction, which are multiples of the elementary translations of the underlying lattice.

The simulated volume has a size of $15 \times 15 \times 15\ \mu m^3$ and contains about 4.56×10^7 elementary FCC cells with periodic boundary conditions. The three-dimensional simulation method consists of following the EDSs themselves rather than the nodes of the lattice (this last procedure being closer to a classical automaton technique), and of coding the state of each segment (line direction, Burgers vector, position of its end points) in the form of one number. It is then possible to deal with dislocation densities up to $10^{15}\ m^{-2}$, which are representative of large plastic strains.

The local stress tensor on an EDS is computed as the sum of three terms. (i) The applied stress tensor. (ii) The internal stress tensor which results from the mutual elastic interaction of the considered EDS with all the other EDSs within a cut-off radius of 5μm. The stress fields associated with EDSs are necessarily those of segments of finite length (cf., e.g. [4]) and in such conditions edge and screw EDSs mutually interact, which is not the case with segments of infinite length. As a result, the configurations which simulate a dislocation loop or a curved segment exhibit a line tension, a feature which was absent from two-dimensional simulations. Figure 2 shows a plot of the resolved stress τ, necessary to maintain a loop of diameter $D = 2R$ in equilibrium, versus D. The theoretical value is $\tau = \Gamma/R$, where Γ is the line energy and it follows form dislocation theory that $\Gamma \sim (\mu b/4\pi \mathrm{Ln}(R/b)$, μ being the shear modulus (4.2×10^{10} Pa for copper). Further, when an EDS moves, it remains connected with the neighbouring EDSs and new EDSs are created, if necessary, which simulates bending under local stresses and complies with the condition that a dislocation line cannot end within a crystal. (iii) A friction stress which accounts for the inherent resistance of the crystal to the motion of dislocation cores. In FCC metals this friction stress is low and the value selected is within the commonly accepted range: $3 \times 10^4\ \mu \sim 1.2$ MPa. The resulting resolved shear stresses on all potential slip planes are then computed.

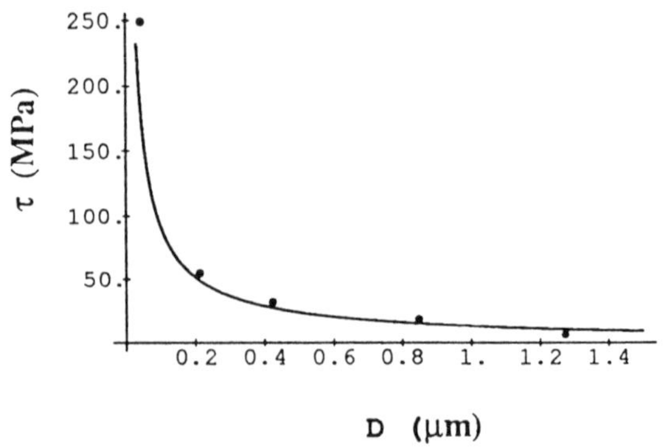

Figure 2. The shear stress τ needed to maintain a loop of diameter D in equilibrium versus D. • : Measured on the simulation, --- : Theoretical value.

The slip plane of an EDS is, among the possible ones, the one where the resolved local stress is maximum. This allows for cross-slip (i.e., a change of active slip plane during dislocation motion) if suitable conditions are locally met by the resolved shear stresses. Cross-slip can be impeded, to simulate the influence of a low stacking fault energy leading to dislocation dissociation. For copper, the local criterion to obtain cross-slip is, then, that the stress resolved on the cross-slip plane should be 10% larger than that on the normal slip plane (in further implementations of the present simulation, the thermally activated nature of cross-slip will be accounted for by introducing a waiting time). Two screw dislocations of opposite sign gliding past each other may then annihilate by cross-slip, leaving cross-slipped segments of edge character in the cross-slip plane.

Most of the elementary mechanisms can, then, be naturally accounted for: glide, cross-slip, mutual trapping and dipole or miltipole formation, the mutual annihilation of screw dislocations by cross-slip, the mutual annihilation of closely spaced edge dipoles, the mutual annihilation of two segments of opposite sign gliding in the same elementary "slice" of slip plane. In three-dimensions, one has also to take into account the interactions between dislocations of different Burgers vectors which are of two kinds according to the sign of their mutual interaction: cutting of repulsive segments or reaction between attractive segments, leading to the formation of attractive junctions. In FCC metals at room temperature intersection processes and the formation of atomic jogs do not significantly contribute to the flow stress [2] and have been neglected. Intersecting dislocations, however, still interact through their (repulsive) long range stress fields at distances larger than $2y_e$. Attractive junctions are fundamental in the "forest mechanism" defining the yield and flow stresses of

FCC metals at room temeprature [2,9]. They provide anchoring points for the dislocation lines which are responsible for strain hardening since their density increases with the total dislocation density, i.e., with strain. Together with the cross-slipped segments mentioned above, these anchoring points can also serve as pinning points for the operation of dislocation sources. Finally, junction dislocations can be destroyed according to a local stress criterion [2,9]. The essence of these properties was simply accounted for by imposing that the meeting point of two attractive EDSs should have a zero velocity as long as the local stress remains below a threshold value. With these conditions, multiplication mechanisms as well as the athermal hardening typical of pure FCC metals at room temeprature can be modelled.

In its present stage, the three-dimensional simulation includes all the main dislocation-dislocation interactions. Most of these processes are athermal at 300 K, and one difficulty resides in the fact that the simulation is based on local stress criteria and cannot distinguish easily between changes in local stresses originating from, e.g., one neighbouring dislocation or from a group of dislocations at some distance. This difficulty will not be met with further developments involving additional fixed obstacles, possibly thermally activated ones, such as clusters of solute atoms, or impurity atoms, or precipitates.

The most difficult problem consisted, however, in correctly accounting for dislocation velocities. These velocities are associated with the free-flight of EDSs between anchoring points or local obstacles, under the influence of resolved local stresses, τ^*. The velocity v was therefore taken in the form of an expression for viscous drag: $\tau^* b = Bv$, with $B = 5 \times 10^{-5}$ Pa/s [8]. Because the spectrum of instantaneous velocities may extend over several orders of magnitude, the histogram of local stresses was chopped into four different classes, defined through the value of the ratio of local to applied stress. Given the elementary time step of the simulation, an average value of the instantaneous free-flight distance was defined for each class, which ranges between zero and 16 elementary translation steps. EDSs are moved one by one towards their final position, the local stresses being periodically checked and updated so that the actual final position of each EDS takes into account the variations in local stresses during its motion.

Any kind of initial dislocation configuration can be prescribed, the most usual one being a Frank net, typical of well-annealed crystals, i.e., an ensemble of dislocation segments of all possible characters and Burgers vectors, pinned at their ends, in total density $\sim 10^{12}$ m^{-2} and with an average length about $\rho^{-1/2} \sim 1$ µm. The orientation of the simulated crystal with respect to the applied stress tensor is an input parameter and various types of external loading conditions can be simulated. The most convenient ones and the most useful for an investigation of the three-dimensional dislocation structures are creep conditions (constant applied uniaxial stress) and alternate tensile-compressive stresses for the simulation of fatigue experiments. An homogeneous lattice rotation can also be introduced to account for the lattice rotations associated with glide processes. The output of the simulations contains both microscopic and macroscopic features. The stress or velocity histograms are periodically recorded, together with the values of total and mobile dislocation densities on each active slip system, from which all relevant quantities (internal stresses, density of immobile dislocations, average values of the dislocation densities and velocities and of the local and internal stresses) can be deduced. In addition, simulated micrographs can be obtained in the form of slices of thickness 500 nm, cut along prescribed orientations, on which the projected features of the spatial arrangement of the microstructure can be examined. Finally the average value of strain rate and/or stress is also periodically computed, from which the deformation curve of the simulated specimen can be deduced.

Preliminary results are presented below. Small scale simulations were performed on one Apollo 3500 workstation at Metz Univeristy, with a simulated crystal size of $3.5 \times 3.5 \times 3.5\ \mu m^3$. Full scale simulations will be performed on the Cray 2 computer of the CCVR (Ecole Polytechnique, Palaiseau).

3 Results

Figure 3 illustrates the mechanism of dislocation multiplication. The initial configuration consists of one dislocation segment pinned at its ends and a constant stress is applied to the simulated crystal. Some steps of operation of the Frank-Read source are reproduced in Figure 3. Figure 4 shows that the corresponding total dislocation density increases linearly with time, resulting in the exponential dependence of strain on time of Figure 5. Such dependences are currently used in the modelling of the mobile density but it is not certain that they still hold for larger values of time and strain, i.e., when internal stresses achieve values comparable to that of the applied stress.

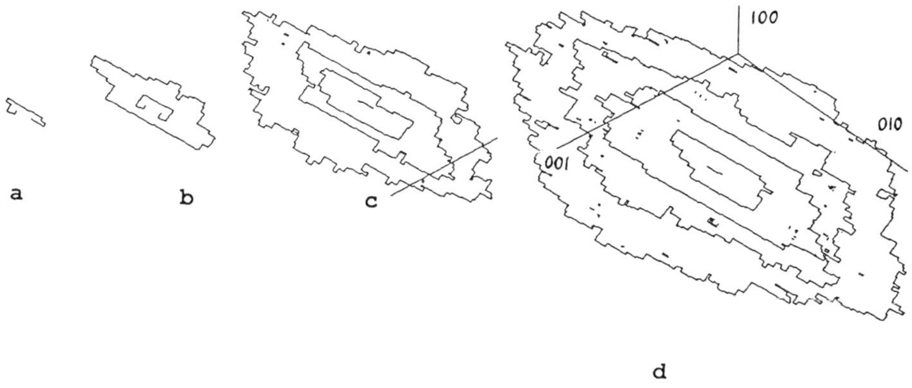

Figure 3. Successive stages of operation of a Frank-Read source from an initial segment pinned at its ends. Number of time steps: (a): 1, (b): 5, (c): 8, (d): 12.

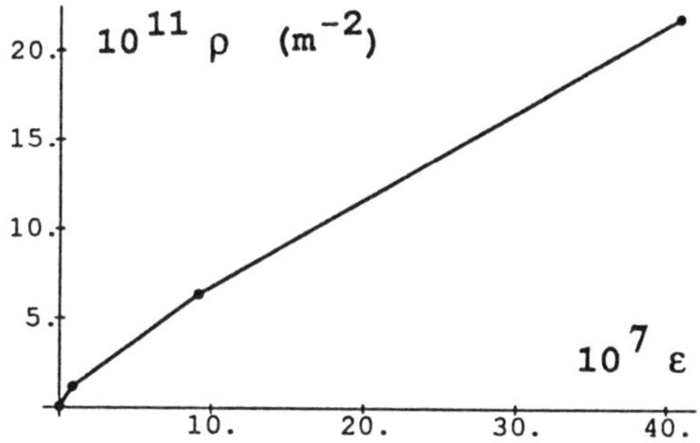

Figure 4. Dislocation density, ρ, versus strain, ε, for the Frank-Read source of Figure 3.

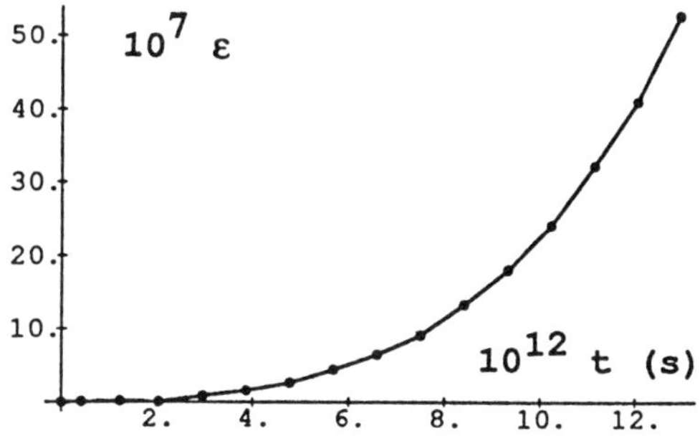

Figure 5. Strain, ε, versus time, t, for the Frank-Read source of Figure 3.

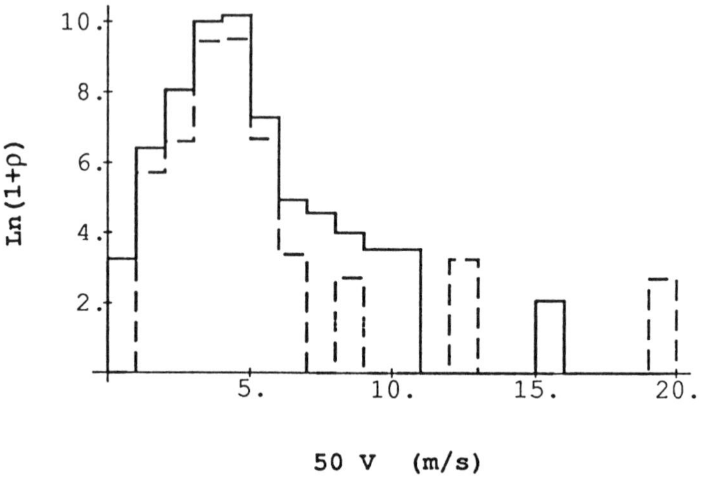

Figure 6. Two histograms of dislocation velocities after one simulated step in creep deformation (σ = 100 MPa).

Figure 6 shows two histograms of dislocation velocities obtained after one simulated step in creep conditions, under a constant uniaxial stress of 100 MPa. The initial configurations consisted of two Frank nets with dislocation densities 5×10^{11} and 10^{12} m^{-2} (50 and 100 segments respectively, with average length 1 μm). The two distributions of Figure 6 are centred around the value yielded by the velocity law when the stress is taken equal to the resolved creep stress. This means that the average internal stress on dislocations is zero or, in other words that dislocations do not mutually interact. Indeed, the strain value is fairly low, typically 10^{-6} and no organization of the microstructure is expected to occur before strains of the order of 10^{-2}.

Another interesting feature of Figure 6 is the very high value of the average velocities, which are in the 100 m/s range. Such values are common at the beginning of plastic flow and they are associated with the initial multiplication stage, leading to the quasi-instantaneous formation of the first slip lines [7].

4 Conclusion

The preliminary results presented here indicate that it is possible to take into account the full complexity of plasticity processes at dislocation scale. By taking into account both long and short range interactions, we expect to be able to reproduce the organizational properties of dislocation ensembles in typical situations like creep and fatigue deformation. Further, the introduction of solute atoms, clusters and precipitates will allow for a study of alloying effects. A major goal will be to use the information extracted from these simulations to build up constitutive forms for the mechanical behaviour of the bulk material, which include the basically heterogeneous character of the microstructures developed during plastic flow.

References

[1] Essmann, U. and Mughrabi, H., *Phi. Mag. A*, **40**, 731 (1979).

[2] Friedel, J., *Dislocations,* Pergamon Press, Oxford (1969).

[3] Ghoniem, N.M. and Amodeo, R., in *Non Linear Phenomena in Materials Science*, Ed. by L. Kubin and G. Martin, Trans. Tech. Publications Ltd. Switzerland, p. 377 (1988).

[4] Hirth, J.P. and Lothe, J., *Theory of Dislocations*, McGraw-Hill, New-York (1968).

[5] Kubin, L.P. and Lépinoux, J., in *Strength of Metals and Alloys* (ICSMA 8), Ed. by P.O. Kettunen et al., Pergamon Press, Oxford, Vol. 1, p. 35 (1988).

[6] Lépinoux, J. and Kubin, L.P., *Scripta Met.*, **21**, 833 (1987).

[7] Neuhäuser, H., in *Mechanical Properties and Behaviour of Solids: Plastic Instabilities*, Ed. by V. Balakrishnan and C.E. Bottani, World Scientific, Singapore, p. 209 (1986).

[8] Philibert, J., in *Dislocations et Déformation Plastique*, Ed. by P. Groh et al., Les Editions de Physique, Orsay, p. 101 (1979).

[9] Saada, G., *Acta Metall.*, **8**, 841 (1960).

Canova G.
L.P.M.M.
Faculté des Sciences
Université de Metz
Ile du Saulcy
F-57045 Metz Cédex
FRANCE

Kubin L.P.
Laboratoire d'Etude des
Microstructures CNRS/ONERA
B.P. 72
F-92322 Chatillon Cédex
FRANCE

C. DAVINI AND G.P. PARRY

Invariants for defective crystals

1 Introduction

One of us (Davini [1], [2]) has proposed a continuum model of crystals where fields of lattice vectors and mass density are regarded as the sole primitive quantities. To interpret the notion of defectiveness in terms of these quantities, we make a couple of assumptions:

(1) based on the atomistic idea of counting defects like vacancies, certain types of dislocations, assume that any object which measures defectiveness is additive over subregions of the crystal.

(2) based on what one might call Taylor's conjecture, that elastic deformation does not create or destroy defects, assume that these objects are invariant under elastic deformation.

Bürgers integral is one object which fulfils requirements (1) and (2), so are various integrals associated with the dislocation density tensor. However we show in [3], [4] that: (a) there is an infinite number of objects with properties (1) and (2), (b) Bürgers integral, etc., are just (some of) the "lowest order" objects, in a precise sense, (c) knowledge of Bürgers integral, etc., is *insufficient* to describe the defectiveness of the crystal, (d) there is a finite list of objects which does describe the defectiveness, (e) the geometric connection between states of the same defectiveness corresponds to *slip* in certain surfaces, (f) there are compatibility conditions which allow, or disallow, slip – this can be seen as an entirely geometric "locking" mechanism controlling the passage of vacancies or dislocations through the crystal.

So it seems that assumptions (1) and (2) provide a simple basis for the assessment of the defectiveness of any particular state of the crystal. One might quibble about assumption (1), for there are indeed exotic non-additive defects (Dzyaloshinskii [5]), but we judge that assumption (2) finds general acceptance in the community. We find it striking that the mechanism of slip emerges naturally in this framework, and hope that this work will provide a rigorous basis for the assessment of various phenomenological theories of plasticity. Here we expand briefly upon the results outlined above, confining attention to special cases, and refer the interested reader to [3, 4] for details of the calculations and discussion of the general case.

2 The model, defect measures

Let the *state* Σ of the crystal correspond to the prescription of three linearly independent lattice vector fields $\mathbf{d}_1(\cdot), \mathbf{d}_2(\cdot), \mathbf{d}_3(\cdot)$ defined over a region of space

$B \subset \mathbf{R}^3$, and write

$$\Sigma = \{ \mathbf{d}_a(\cdot), B \} \tag{1}$$

with the understanding that the subscript a takes the values 1, 2 and 3. Also

$$\mathbf{d}_1 \cdot \mathbf{d}_2 \wedge \mathbf{d}_3 > 0 \ \ in \ \ B. \tag{2}$$

This definition of state is not so general as that introduced in [1], and employed in [3], [4], but it is sufficient to illustrate the nature of our results. Suppose that Σ^* is a different state,

$$\Sigma^* = \{ \mathbf{d}_a^*(\cdot), B^* \}. \tag{3}$$

One can imagine that points of B are mapped to points of B^* by an invertible function

$$\mathbf{x}^*: B \to B^* \tag{4}$$

called the *macroscopic deformation*, but generally there is no reason to presuppose any connection between $\mathbf{d}_a(\mathbf{x})$ and $\mathbf{d}_a{}^*(\mathbf{x}^*)$, or any connection between the fields $\mathbf{d}_a(\cdot)$ and $\mathbf{d}_a{}^*(\cdot)$. However in *elastic deformation*, one supposes that lattice vectors are embedded in the macroscopic deformation, behaving as material vectors, so that

$$\mathbf{d}_a{}^* (\mathbf{x}^*) = F(\mathbf{x})\, \mathbf{d}_a (\mathbf{x})\,, \tag{5}$$

where we have introduced the matrix F of deformation gradients $\nabla\mathbf{x}^* = \left(\dfrac{\partial \mathbf{x}^*}{\partial \mathbf{x}}\right)$, evaluated at the point $\mathbf{x} \in B$. Let $\mathbf{d}^a(\mathbf{x})$, $a = 1,2,3$, denote the dual lattice vectors at $\mathbf{x} \in B$ which are such that

$$\mathbf{d}^a(\mathbf{x}) \cdot \mathbf{d}_b(\mathbf{x}) = \delta^a_b \tag{6}$$

Then one might write (5) loosely as

$$\mathbf{d}_a \to F\, \mathbf{d}_a\,, \tag{7}$$

and calculate that

$$\mathbf{d}^a \to F^{-T}\, \mathbf{d}^a\,, \tag{8}$$

where F^{-T} is the inverse transpose of F. The quantities $\mathcal{S}^{ab}$, n defined by

$$\mathcal{S}^{ab} = \nabla \wedge \mathbf{d}^a \cdot \mathbf{d}^b$$

$$n = \mathbf{d}^1 \cdot \mathbf{d}^2 \wedge \mathbf{d}^3 , \tag{9}$$

will play a crucial role in the analysis. In elastic deformation

$$\mathcal{S}^{ab} \to (\det F)^{-1} \mathcal{S}^{ab}$$

$$\mathcal{S}^{ab} \to (\det F)^{-1} n , \tag{10}$$

where det denotes the determinant. If we bear in mind that line and volume elements $\mathbf{d}x$, dV transform like

$$\mathbf{dx} \to F \, \mathbf{dx}$$

$$dV \to (\det F) \, dV \tag{11}$$

in the change of variables $\mathbf{x} \to \mathbf{x}^*$, it is easily seen from (10) and (11) that

$$\oint_C \mathbf{d}^a \cdot \mathbf{dx} , \int_V \mathcal{S}^{ab} \, dV , \int_V n \, dV \tag{12}$$

are *additive elastic invariants*, in the sense that they are such that 1 holds (with the obvious rules for composition of circuits, volumes) and 2 holds. Define

$$\mathbf{b}^a = \nabla \wedge \mathbf{d}^a , \tag{13}$$

so that by Stokes theorem

$$\int_{\partial S} \mathbf{d}^a \cdot \mathbf{dx} = \int_S \mathbf{b}^a \cdot \mathbf{d}S . \tag{14}$$

Bürgers vector and the (lattice components of the) dislocation density tensor are $\mathbf{b}^a$, $\mathcal{S}^{ab}$ respectively. We refer to $\mathbf{b}^a$, $\mathcal{S}^{ab}$, n as *densities associated with the invariants* (12).

To see that, generally, there is an infinite number of integral invariants with properties 1 and 2, note that from (10)

$$\nu \equiv \frac{\mathcal{S}^{ab}}{n} \to \frac{\mathcal{S}^{ab}}{n} , \tag{15}$$

calculate that

$$\nabla \nu \to F^{-T} \nabla \nu \tag{16}$$

and use (7) to deduce that

$$\mathbf{d}_a \cdot \nabla v \to \mathbf{d}_a \cdot \nabla v \,. \tag{17}$$

So the scalar v generates a different scalar $\mathbf{d}_a \cdot \nabla v$ (provided that v is not constant), and in fact *any* scalar has this property. But clearly

$$\oint_C v\mathbf{d}^a \cdot \mathbf{d}x \tag{18}$$

is an additive elastic invariant for any choice of v, and there is generally an infinite number of such choices. Densities corresponding to (12) and (18) are

$$\mathbf{b}^a,\, S^{ab}\,,\, n,\, \nabla v \wedge \mathbf{d}^a \,. \tag{19}$$

In fact the first three densities of (19) provide a functional basis for those densities which depend just on the lattice vectors and their first gradients, [2], and this is the sense of assertion (b).

3 Neutral deformations

The functionals (12), (19), amongst others, are invariant under elastic deformation, possibly they are also invariant in more general changes of state. In fact the non-elastic changes of state which preserve these elastic invariants turn out to be *slips* of rather general form, more or less, cf. (e). This section is devoted to an illustration of basic ideas in the derivation of this result, in a particular case.

Firstly let us agree to call the non-elastic changes of state which preserve the elastic invariants (12), (18) *neutral deformations* (n.d.)

$$\Sigma = \{\, \mathbf{d}_a(.), B \,\} - \text{n.d. } g^* \to \Sigma^* = \{\, \mathbf{d}_a^*(.), B^* \,\}$$

n.d.g elastic deformation f

$$\Sigma' = \{\, \mathbf{d}_a'(.), B \,\}$$

Clearly one can factorize a general n.d. $g^*\colon \Sigma \to \Sigma^*$ into a n.d. $g\colon \Sigma \to \Sigma'$ where the corresponding macroscropic deformation is just the identity, together with an elastic deformation $f\colon \Sigma^* \to \Sigma'$, since g, g^*, f all preserve the invariants. Thus

$$g^* = f \cdot g \,. \tag{20}$$

For the same reason, knowledge of the n.d. corresponding to a state Σ determines the n.d. of all states Σ^+ obtained by elastic deformation of Σ.

$$\begin{array}{ccc} \Sigma & \text{— elastic} \longrightarrow & \Sigma' \\ \text{n.d.} & & \text{n.d.} \\ \Sigma' & \text{— elastic} \longrightarrow & (\Sigma^+)' \end{array}$$

To calculate n.d. corresponding to a state Σ, then, it is sufficient to calculate the n.d. which leave the region B fixed, and one can choose to simplify expressions for $\mathbf{d}_a(\cdot)$ by appropriate choice of elastic deformation. Thus invariants (12), (18) coincide over all circuits in, and parts of, B, so that *the densities* (19) *coincide in* Σ, Σ'. It follows that, for example,

$$\mathbf{b}^a = \nabla \wedge \mathbf{d}^a = \nabla \wedge \mathbf{d}^{a'} = \mathbf{b}^{a'} , \tag{21}$$

so that there exist potentials τ^a such that

$$\mathbf{d}^{a'} = \mathbf{d}^a + \nabla\tau^a . \tag{22}$$

We consider, here, the case where

$$\text{rank}\,\{\mathbf{b}^a\} = 2, \;\; \nabla\left(\frac{\overset{ab}{S}}{n}\right) \neq 0 . \tag{23}$$

Since

$$\mathcal{S}^{ab'} = \nabla \wedge \mathbf{d}^{a'} \cdot \mathbf{d}^{b'} = \nabla \wedge \mathbf{d}^a \cdot \mathbf{d}^{b'} = \nabla \wedge \mathbf{d}^a \cdot \mathbf{d}^b = \mathcal{S}^{ab} , \tag{24}$$

it follows that

$$\mathbf{b}^a \cdot \nabla\tau^b = 0 . \tag{25}$$

By simplifying choice of elastic deformation, one can assume that τ^a depends on just one variable x^3 – the choice ensures that the three vectors $\mathbf{b}^a$ define a surface $x^3 =$ constant, via $(23)_1$, and then (25) shows that $\tau^a = \tau^a(x^3)$. A somewhat lengthy calculation, given in [4], shows that if n.d. of Σ are non-trivial, then the lattice vector fields $\mathbf{d}_a$ can be taken in the form

$$\mathbf{d}_a = \mathbf{d}_a(x^3) . \tag{26}$$

So from (22)

$$\mathbf{d}^{a'} = \mathbf{d}^a(x^3) + f^a(x^3)\,\mathbf{e}^3\,, \tag{27}$$

where $f^a = \dfrac{d\tau^a}{dx^3}$. The relation $n = n'$ gives, from (27),

$$f^a\,\mathbf{d}_a \cdot \mathbf{e}^3 = 0\,. \tag{28}$$

Hence the "compatibility condition" $(23)_1$ gives strong constraints on the geometry of the lattice fields which allow n.d., (26); moreover the corresponding n.d. are characterized explicitly, (27), (28).

Introduce the macroscopic deformation

$$\mathbf{x} \to \mathbf{x}^* = \mathbf{x} - \int^{x^3} f^a(\theta)\,\mathbf{d}_a(\theta)d\theta\,, \tag{29}$$

with corresponding deformation gradient

$$F = (\nabla\mathbf{x}^*) = \mathbf{I} - f^a\,\mathbf{d}_a \otimes \mathbf{e}^3\,. \tag{30}$$

Note that, via (28)

$$F^{-T} = \mathbf{I} + \mathbf{e}^3 \otimes f^a\,\mathbf{d}_a\,, \tag{31}$$

so that in the induced elastic deformation $\Sigma \to \Sigma^*$

$$\mathbf{d}^a \to \mathbf{d}^a(\mathbf{x}^*) = F^{-T}\,\mathbf{d}^a(\mathbf{x}) = (\mathbf{I} + \mathbf{e}^3 \otimes f^b\,\mathbf{d}_b)\,\mathbf{d}^a(\mathbf{x})$$

$$= \mathbf{d}^a(x) + \mathbf{e}^3 f^b\,\delta^a_b \tag{32}$$

$$= \mathbf{d}^{a'}(\mathbf{x})\,.$$

So there is an elastic deformation $\Sigma \to \Sigma^*$ such that the lattice vectors $\mathbf{d}^{a*}$, $\mathbf{d}^{a'}$ correspond (i.e., are equal) at the *different* points $\mathbf{x}^*, \mathbf{x}$. Said differently, there is a *rearrangement* of the fields $\mathbf{d}_a'(.)$ which produces fields elastically related to $\mathbf{d}_a(.)$. In this case, it is natural to call this rearrangement a slip since, from (29), points in the plane x^3 = constant are displaced by the same amount. So the fundamental mechanism of plasticity theory emerges independently of any presupposition regarding the detailed kinematical nature of defectiveness.

4 Equidefectiveness

In the analysis of Section 3 we have employed (implicitly) just a few of the infinite number of invariants which enjoy the two properties 1 and 2. It is natural to enquire if any more information derives from the equality of the other invariants. Also one might ask if any subset of this infinite list of invariants suffices to describe the defectiveness of the crystal, in any sense. To address these questions in the particular case which we have treated above, note that from (28) and (30),

$$x^{3*} = x^3 , \tag{33}$$

so from (27)

$$\mathbf{d}^{a'} = \mathbf{d}^{a'}(x^3) = \mathbf{d}^{a'}(x^{3*}) \tag{34}$$

and finally from (32)

$$\mathbf{d}_a{}^*(\mathbf{x}^*) = \mathbf{d}'_a(x^{3*}) = \mathbf{d}'_a(\mathbf{x}^*) . \tag{35}$$

The fields $\mathbf{d}^{a'}, \mathbf{d}^{a*}$ are *identical fields*, then. There is an elastic deformation $\Sigma \to \Sigma^*$ such that the fields in Σ^*, Σ' are identical, the rearrangement $\Sigma' \to \Sigma^*$ are *slips which preserve the lattice structure,* locally, they are slips in the level sets of the lattice vector fields; as such one must surely reckon that the defectiveness of states Σ, Σ' is the same, basing this judgement on the local geometric identity of Σ', Σ^* rather than on the densities (19). We give more rigorous analysis and justification of these last assertions in [4], and use the term "equidefective" to signify this local identification of fields, loosely. The main result of [4] is that all neutrally related states are equidefective.

Lastly, note that we have used just the densities (19) to derive these results. In the case that we have treated, the local identity of fields in Σ', Σ^* implies that *all* densities match in the two states. Since densities depend just on x^3, it follows that all densities match in Σ, Σ' too. So the identity of the densities (19) is sufficient that all densities match, and one might regard the list (19) as a functional basis appropriate for the assessment of the defectiveness in a crystal, in the context of this model. We show in [4] that the common descriptors related to $\mathbf{b}^a, \mathcal{S}^{ab}$ are insufficient for this purpose, in fact that $\mathbf{b}^a, \mathcal{S}^{ab}, n$ are likewise insufficient, so that the list (19) is a *minimal* functional basis.

References

[1] Davini, C. A proposal for a continuum theory of defective crystals, *Arch. Rational Mech. Anal.*, **96** (1986) 295-317.

[2] Davini, C. Elastic invariants in crystal theory, in *Proc. Symp. Year on Material Instability in Continuum Mechanics*, (1987), Heriot-Watt University, Edinburgh, J.M. Ball ed., Oxford Press.

[3] Davini, C. and Parry, G.P. On defect-preserving deformations in crystals, to appear in *Int. J. Plasticity*.

[4] Davini, C. and Parry, G.P. A complete list of invariants for defective crystals, in preparation.

[5] I. Dzyaloshinskii. Topological singularities, in *Proc. Summer School on the Physics of Defects*, (1980) Les Houches (France), North-Holland, Amsterdam.

Davini C.
Istituto di meccanica teorica
ed applicata,
Università Udine
Viale Ungheria
Udine
ITALY

Parry G.P.
School of Mathematics
University of Bath
Claverton Down
Bath BA2 7AY
U.K.

D. FAVIER, P. GUELIN, F. LOUCHET, P. PEGON, A. TOURABI AND B. WACK

Microstructural origin of hysteresis and application to continuum modelling of solid behaviour

1 Introduction

The aim of this work may be introduced as follows. One considers first the classical dislocation theory: "The questions raised here, in one form or another, are questions of the kinetics of materials with defects. It is clear from the outset that the new internal variables that are introduced by the kinematic equations of defects require kinetic restrictions. Further, the laws of balance of linear momentum, balance of moment of momentum, and balance of energy of a body without defects require possibly drastic revisions as soon as defects are present. New physical principles are thus required because we depart from the classical arena as soon as defects are allowed" [2]. Whether or not defects are considered, an analysis which is, at the theoretical level, based on classical concepts, should be focused at both micro and macroscopical levels on phenomena dominated by "un caractère hystérétique violent" [7].

Secondly, the study of usual "continuum defect-free idealizations" of rate independent solid behaviour, exhibiting elastic-plastic hysteresis, may be envisioned as a rate form pattern in which constitutive kinetic equations of differential-difference form are found to work without any basic theoretical drawback [3, 4, 5 and 13]. However the proposed pattern is then, from the outset, strongly connected with well founded but non-classical thermodynamics [9]: the discrete memory notion is indeed its basic concept.

Thirdly, the only favourable points one may notice in such an ominous situation are the following: on one hand it is well known, although not overpublished, that continuum modelling is founded, whether or not defects are allowed, on a radical departure from the classical arena: the strain tensor definition indeed implies the discrete memory notion. In short, the Cauchy tensor associated with a past state is dragged along and used in the current state to define the current Almansi strain tensor [14]. On the other hand, and putting aside once and for all the irreversible phenomena of viscous type, one notices that the (isothermal) macroscopic experimental results demonstrate the non-existence of a strictly reversible behaviour and the permanence of hysteretic phenomena [6, 10, 11 and 12]. The *pure hysteresis* behaviour, introduced by the proposed pattern as a stationary behaviour under periodical loading, is then strongly connected with the results of macroscopic experiments.

Finally, in spite of the fact that the price of the classic dislocation theory "proves" to be surprisingly high" [2] and that the proposed pattern is proved to be rather effective, its price may be considered as quite excessive if the material discrete memory assumption is not justified further as physically relevant at the microscopic level. The aim of this work is therefore to provide some remarks regarding this serious question (Section 5). The applications of the material discrete memory pattern are at first briefly illustrated with respect to three questions which are up to now rather unsolved: heat rate supply along an elastic-plastic path (Section 2); isotropic-deviatoric coupling effect exhibited by granular media (Section 3); effect of the simultaneous existence of reversible and hysteretic processes expressed by the shape memory properties (Section 4).

2 Heat rate supply along an elastic-plastic path

One considers the pure hysteresis behaviour under the following stress controlled loading: after a radial path, a spiral-like path is performed in the deviatoric plane, first along 4 turns, secondly along $3+1$ turns with 3 as the previous one but the last one tending closer to the Von-Mises plastic yield than the fourth of the first loading process (Figure 1(a)) [13]. As anticipated by the proposed theory, the strain path is very sensitive to the small perturbation of the stress loading path (Figure 1(b), (d)) but the heat rate supply is not (Figure 1(c), (e)). The heat rate evolution remains of pure hysteresis type and is in agreement with the "Taylor-Farren-Quinney effect".

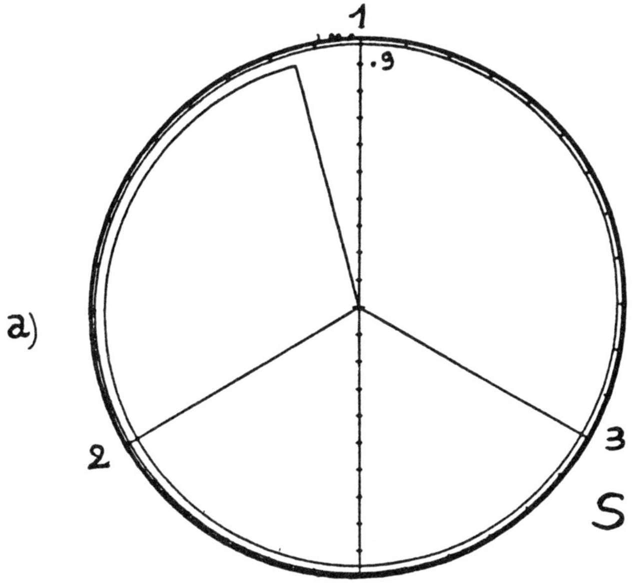

Figure 1(a)

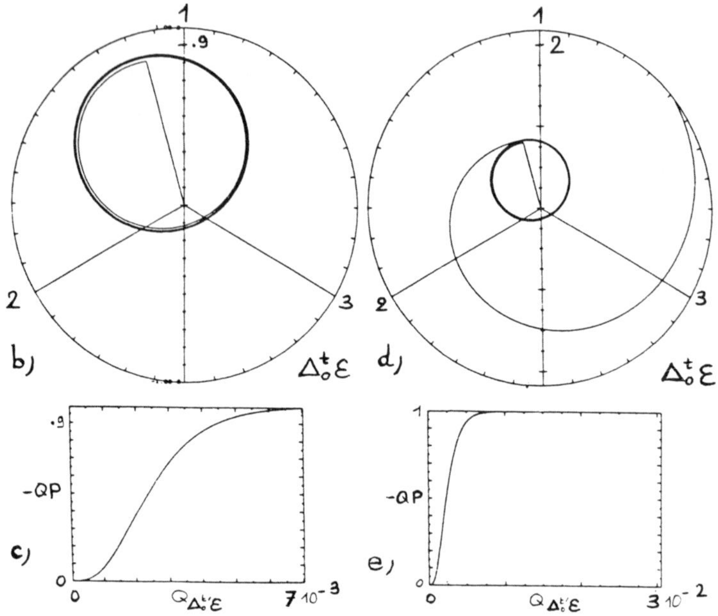

Figure 1(b), (c), (d) and (e)

3 Isotropic-deviatoric coupling effects

The generalized pattern [15] is applied to the case of cyclic strain controlled loading in a deviatoric plane (Figure 2). Both the deviatoric evolution (at left in Figure 2) and the volume evolution are in agreement with the results of recent experiments.

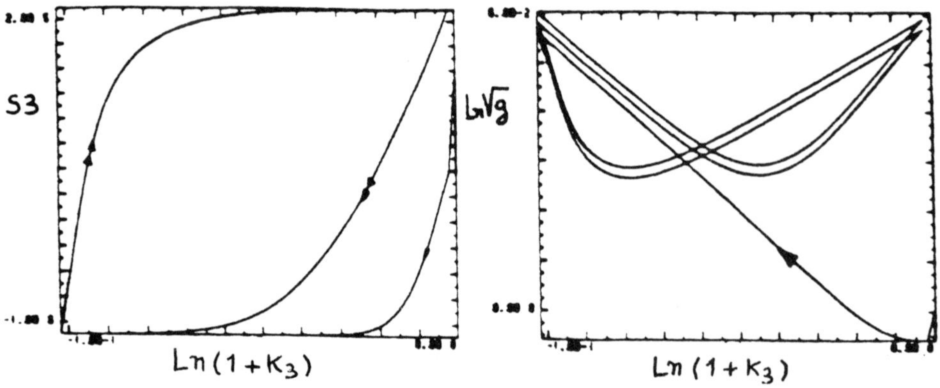

Figure 2

This agreement is found to hold also in the usual stress controlled biaxial test whatever be the amplitudes of the cycles (Figure 3 for small, medium and large cycles from top to bottom).

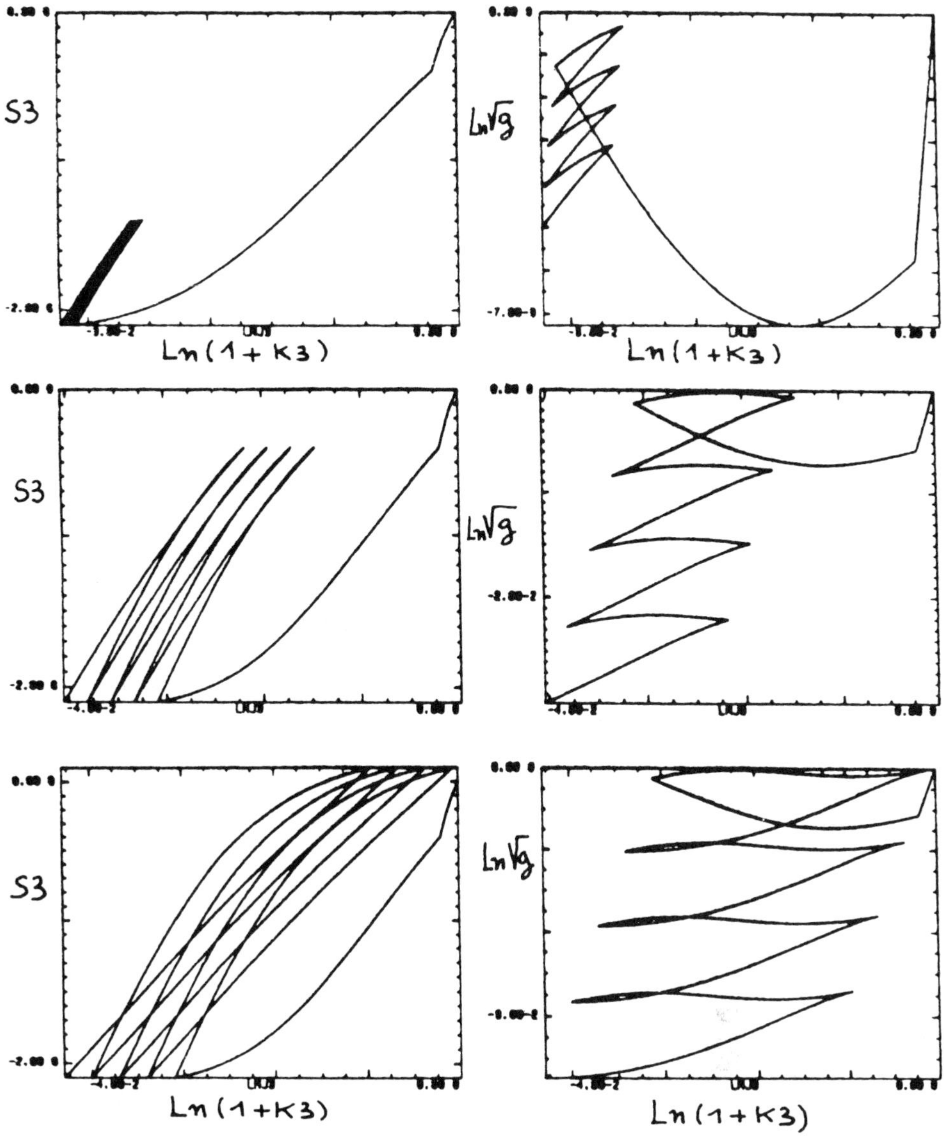

Figure 3

4 Shape memory properties

The usual thermomechanical "training" is defined by the following 6 stages of simple traction type (Figure 4(a), (b)): OAB at "cold" temperature T_1 and cyclic loading between O and F_{max}; BC from T_1 to $T_2 > T_1$ without load; CD at T_2 and cyclic loading between 0 and F_{max}; DE under F_{max} and with temperature varying from T_2 to T_1; EF unloading at cold temperature T_1; FG like BC. Then under a specified constant load the two-way shape memory effect is obtained (Figure 4(e)) for any cyclic evolution of the temperature between T_1 and T_2 (Figure 4(f)). The pattern is also able to describe the gradual establishment of the shape memory properties (Figure 5) with 5 sequences of training followed by shape memory effects under 3 constant loads. These results are in agreement with experiments [5, 6].

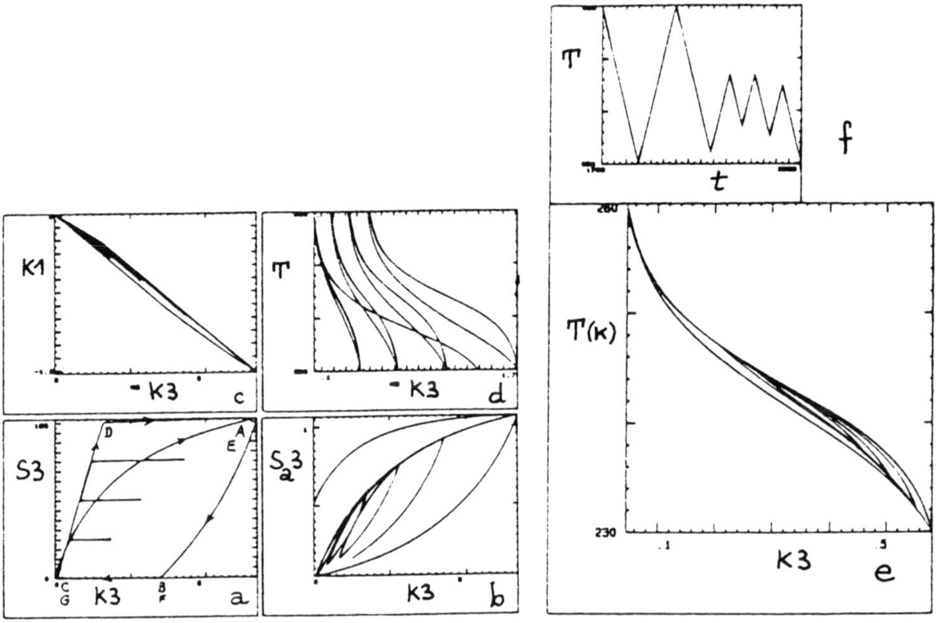

Figure 4. (a) evolutive hysteresis; (b) associated pure hysteresis; (c) axial versus transverse strain; (d) axial strain versus temperature

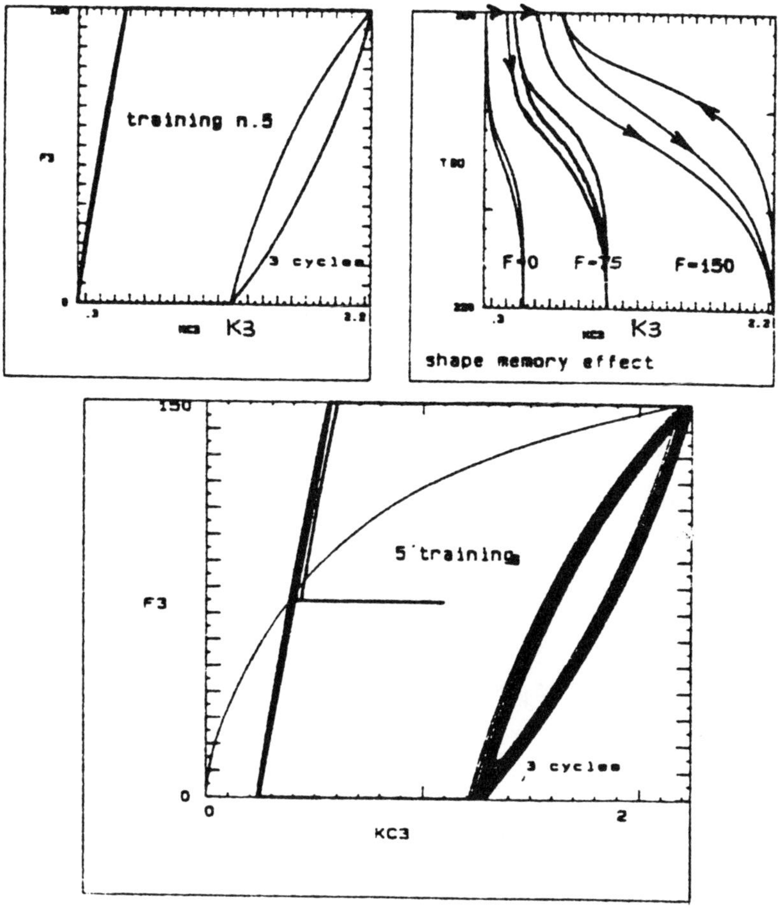

Figure 5

5 Remarks on the microstructural origin of hysteresis

The fact that the pattern of pure hysteresis behaviour cannot be considered as entirely uncomplete in the sense of Bunge [1] can now be illustrated recalling briefly how the associated symbolic models (Figure 6) are heuristic when some particular insight is needed for the study of fundamental microscopic processes based on the dislocation concept.

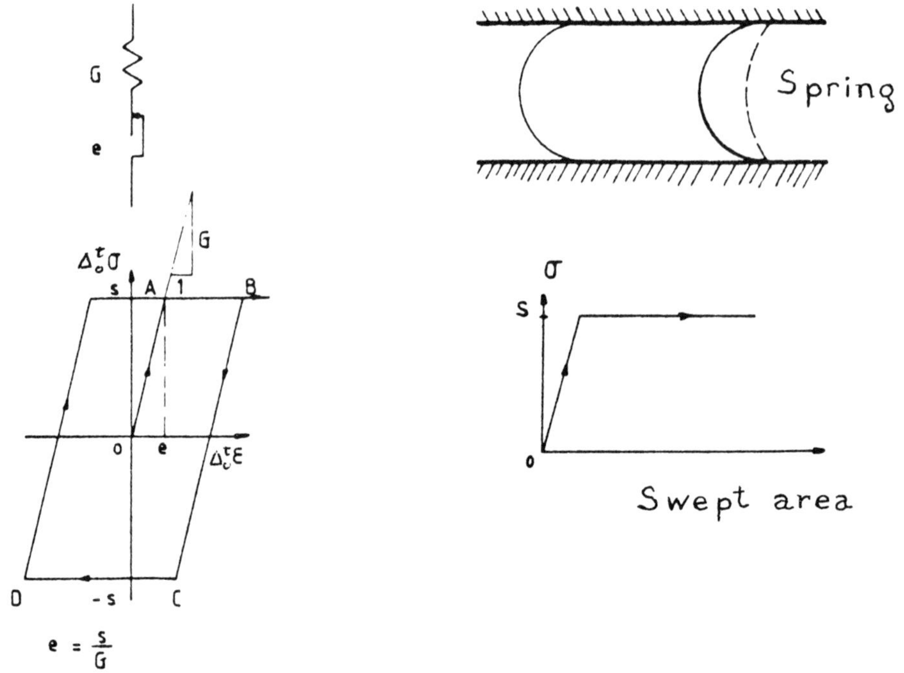

Figure 6

For nearly 20 years it has been possible, with the aid of transmission electron microscopy, to observe dislocation movements occurring during the deformation of small samples. For the study of pure hysteresis, the interesting mechanisms are those related to the case of stationary cycles after stabilization of the microstructural phenomena. Under these conditions a noticeable analogy between gliding microstructural processes and the pure hysteresis symbolic model has recently been established [16, 17].

The simplest case is represented by a *single dislocation* moving between two parallel walls: the process is analogous to the simplest symbolic model (one spring and one friction slider associated in series). Under a small stress the dislocation will bend reversibly, and thus behaves like the spring. For a characteristic value S_0 of the stress, the *dislocation pinning points will move and drag two segments along the walls*: the behaviour is analogous to that of the friction slider (Figure 6). The diagram of the stress versus the area swept by the dislocation is of elastic-perfectly-plastic type. When the stress decreases it is necessary for the stress to reach the opposite value $-S_0$, before the pinning points move in the reverse direction. This behaviour is thus really of pure hysteresis type.

A more complex case is that of a *Frank-Read source* operating between two parallel walls and characterized by two thresholds, S_1 for which the contact points on the walls move and $S_2 > S_1$ for which the source emits a loop (Figure 7). This case is equivalent to a model with two spring-slider couples.

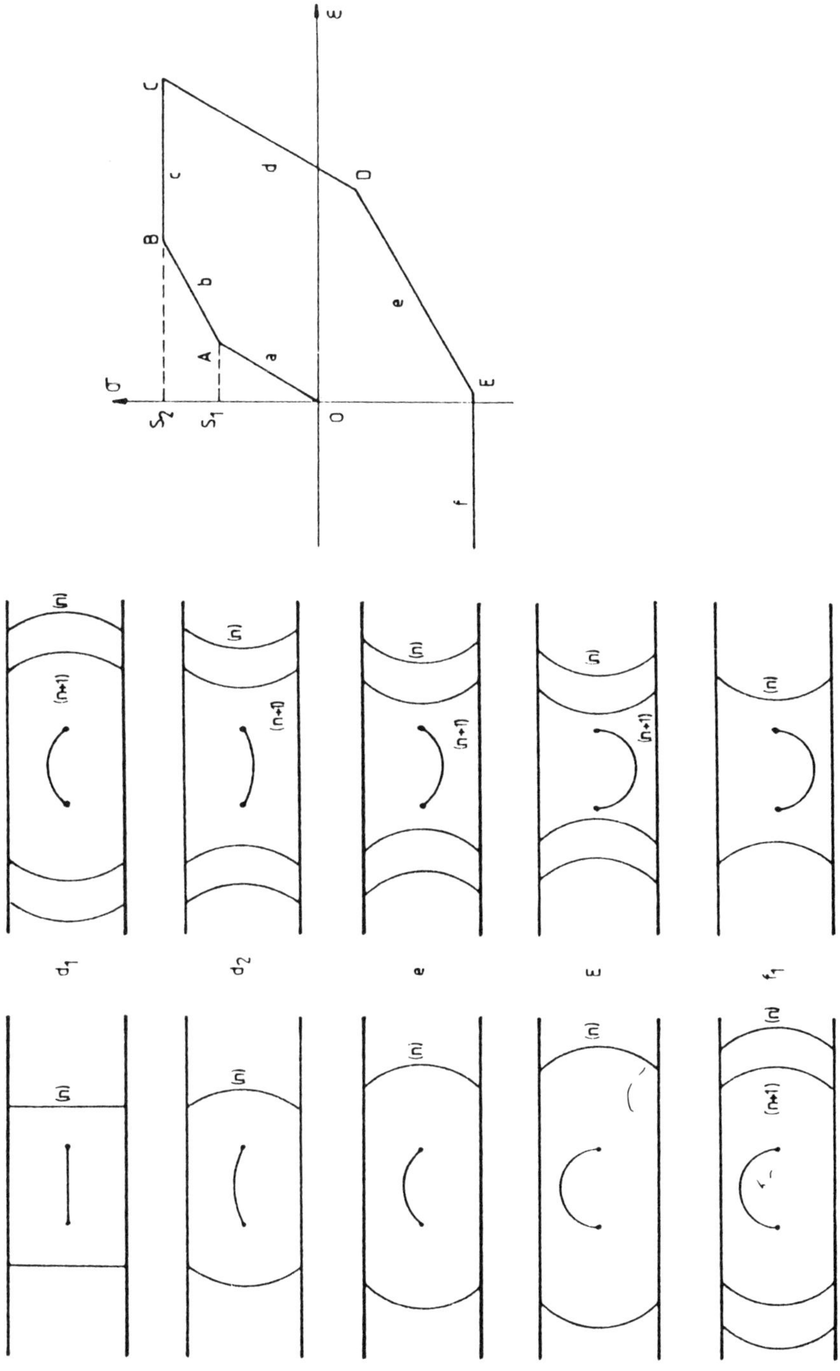

Figure 7

In a real crystalline material the deformation is the result of the activation of a great number of elementary mechanisms like the preceding one. The activation thresholds of these mechanisms is widely and continuously scattered, due to different parameters such as type, characteristic values, orientation with regard to the "external stress", intergranual compatibility, etc. On the basis of this analogy it is possible to give an evident physical interpretation of the *pure hysteresis* properties; this is particularly the case of the stress-strain discontinuity obtained by describing a small cycle inside a large cycle. This is also the case for the properties at the right of an inversion point, for the fundamental cyclic loading properties and for the absence of reversible domain of the real crystalline materials [16, 17].

6 Concluding remark

In spite of the fact that the proposed pattern of *pure hysteresis* must be studied further, it proves from now on to be heuristic regarding the analysis of basic micro-structural processes. If, at more or less long term, the discrete material memory assumption is shown as really physical in character, then continuum idealizations will require possible drastic revisions at the physical level, whether or not defects are considered and even if the application of the catastrophe theory [8] to physical systems is expanded.

References

[1] Bunge, M., *Foundations of Physics*, Vol. 10 of Springer Tracts in Natural Philosophy, Chapter 3, Section 2, p. 143, Section 2.4, p. 156 (1967).

[2] Edelen, D. and Lagoudas, D., *Gauge Theory and Defects in Solids*, Vol. 1 of Mech. and Physics of Discrete Systems, G. Sih, ed., North-Holland, Amsterdam (1988).

[3] Favier, D., Contribution à l'étude théorique de l'élastohystérésis à température variable; application aux propriétés de mémoire de forme. Thèse Doct. d'Etat, Grenoble (1988).

[4] Favier, D., Guelin, P., Nowacki, W.K. and Pegon, P., "Theoretical schemes of thermomechanical coupling", Thermomechanical Coupling in Solids; Ed. by H.D. Bui and Q.S. Nguyen, Elsevier Science Publishers, North Holland, **383** (1987).

[5] Favier, D., Guelin, P., Pegon, P., Tourabi, A. and Wack, B., Ecrouissage, Schémas thermomécaniques et à variables internes: méthodes de définition utilisant le concept d'hystérésis pure, *Arch. Mech. Stos*, **40** 5-6, 611 (1988).

[6] Favier, D., Guelin, P., Pegon, P., Tourabi, A. and Wack, B., Schémas constitutifs à mémoire discrète de la thermomécanique des matériaux métalliques écrouissables, Proc. of Mecamat, Int. seminar on inelastic behaviour of solids: models and utilization, 30 août - 1 sept. 1988, Besançon, France (1988).

[7] Friedel, J., Preface to Yravals Summer School, Groh, Kubin, Martin editors, Editions de Physique (1979).

[8] Gilmore, R., Catastrophe theory: what it is; why it exists; How it works, Math. Analysis of Physical Systems, Mickens editor, Van Nostrand Reinhold, New York (1985).

[9] Guelin, P., Remarques sur l'hystérésis mécanique, *J. de Méca*, **19**, 2, 217 (1980).

[10] Han, S. and Wack, B., Properties of the pure hysteresis behavior of solids: case of stainless steel and superalloy, *Arch. Mech. Stos.*, **38**, 4, 439 (1986).

[11] Han, S. and Wack, B., Discrete memory description of strain hardening and softening with application to stainless steel and superalloy, *Res. Mechanica*, **20**, 1 (1987).

[12] Han, S. and Wack, B., Torsion ratchet associated with mechanical hysteresis, Trans. 9th SMIRT Conf., L, **421** (1987).

[13] Pegon, P. Contribution à l'étude de l'hystérésis élastoplatique, Thèse Doct. d'Etat, Grenoble (1988).

[14] Pegon, P. and Guelin, P., On thermomechanical Zaremba schemes of hysteresis, *Res. Mechanica*, **21**, 21 (1987).

[15] Pegon, P., Nowacki, W.K., Guelin, P., Favier, D. and Wack, B., Hysteresis in cyclic behaviour of granular materials: constitutive scheme, modes of bifurcation and numerical methods, Gdansk Workshop (1989).

[16] Tourabi, A., Contribution à l'étude de l'hystérésis élastoplastique et de l'écrouissage de métaux et alliages réels, Thèse (1988).

[17] Wack, B., Louchet, F. and Tourabi, A., Comportement d'hystérésis pure et passage micro-macro qualitatif, Colloque Plasticité **88**, Grenoble (1988).

Favier D.
G.P.M2.,
Ecole Nationale Supérieure de
Physique de Grenoble, I.N.P.G.,
B.P. 46
F-38402 Saint-Martin d'Hères Cédex
FRANCE

Guelin P.
Institut de Mécanique de Grenoble,
B.P. 53X
F-38041 Grenoble Cédex
FRANCE

Louchet F.
Laboratoire de Thermodynamique
et de Physico-Chimie Métallurgique, I.N.P.G.,
B.P. 75
F-38402 Saint-Martin d'Hères Cédex
FRANCE

Pegon P.
Joint Research Center C.C.E.,
I-21020 Ispra (Varese)
ITALY

Tourabi A., and Wack B.
Institut de Mécanique de Grenoble,
B.P. 53X
F-38041 Grenoble Cédex
FRANCE

M. FÄHNLE, J. FURTHMÜLLER AND R. PAWELLEK

Continuum models of amorphous and polycrystalline ferromagnets: magnetostriction and internal stresses

1 Introduction

It is well known that the excellent soft magnetic properties [11, 17] of amorphous ferromagnets based on 3d-transition metals are determined by structural inhomogeneities occurring on three different length scales [17, 18]: (i) long-range density fluctuations (some μm) responsible for the domain structure, (ii) medium-range density fluctuations (≈ 100 Å) responsible for domain wall pinning; (iii) intrinsic atomic scale inhomogeneities responsible for the intrinsic magnetic properties. For instance, as a result of these atomic scale structural fluctuations the local environment of any atomic site exhibits some departure from high (e.g., cubic) symmetry giving rise to strong local uniaxial magnetic anisotropies [2, 6, 8] with easy axes varying randomly from site to site. It will be shown in the following sections that these local anisotropies are responsible for magnetostriction of amorphous ferromagnets, which is very often comparable to the one of crystalline ferromagnets [11] and which mediates the magnetoelastic coupling of domain walls to internal stresses of type (ii). Furthermore, the intrinsic structural fluctuations cause atomic level self-stresses [15], which have been used by Egami and coworkers [3, 25] to characterize the local topology and symmetry of the amorphous structure and to analyse the microscopic processes of structural relaxation and glass transition in metallic glasses.

Various microscopic approaches have been considered to deal with the effect of atomic scale inhomogeneities. Concerning the magnetostriction, however, a realistic first-principle calculation is far beyond the present computational capabilities. Former attempts in this direction [19, 23, 24] relied on assumptions not valid for amorphous 3d-transition metals [2, 6, 8]. For instance, in all these papers the inhomogeneous character of the magnetostrictive strain field has been neglected although it will be shown in Section 3 that this represents the most important difference between magnetostriction in amorphous and crystalline systems. Concerning the atomic level stresses, a lot of numerical information has been obtained from atomistic computer simulations [3, 25], but it has been pointed out [25] that for an intuitive understanding of the physical meaning of all these numbers, continuum theoretical considerations may be very useful.

In line with the scope of the present symposium we therefore developed continuum models of the discrete amorphous structure which allow a calculation of magnetostriction and atomic level stresses of amorphous ferromagnets on an equal phenomenological footing. We thereby use the so-called incompatibility method, which was introduced by Reissner [22], developed further by Kröner [13] in his theory of dislocations and discussed in Section 2. It is well known that continuum theories can be extended to atomic scale, but in this case in principle a non-local theory should be applied because of the discreteness of the structure and the finite range of interactions between the atoms [14]. However, it was also conjectured [15, 25] that a local approximation suffices for general, qualitative considerations for which no very precise numbers are demanded, and we therefore confine ourselves to the local linearized theory of elasticity in this paper.

2 Description of the formalism

To illustrate the application of the incompatibility method to random media we discuss the states a-f of the gedanken experiment sketched in Figure 1.

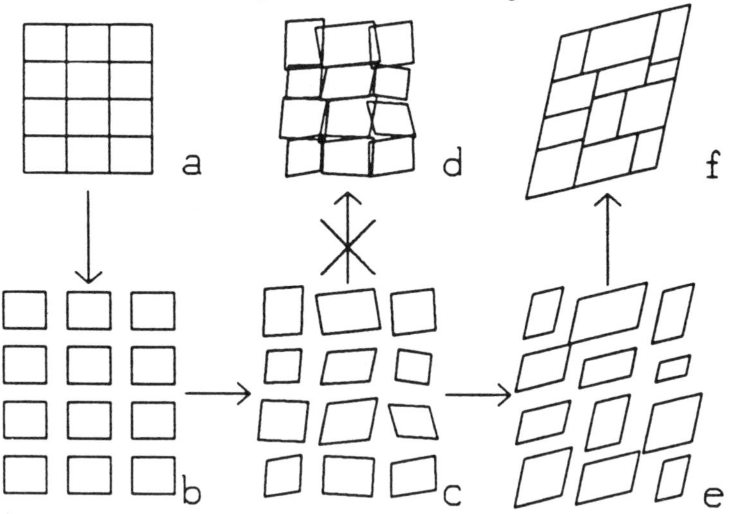

Figure 1. Schematic illustration of the incompatibility method (see text).

State a: We first must define the reference state of our system starting from which we introduce the deformations under consideration. For the calculation of the atomic level stresses it is appropriate to consider the corresponding crystalline state as the reference state. For the calculation of magnetostriction we deal with a hypothetical amorphous reference state with the exchange interactions between the magnetic moments switched off, i.e., without magnetization and magnetostrictive deformations.

State b: We now cut the material into very small structural units ("grains") consisting basically of an atom and its nearest neighbour atoms. In the case of magnetostriction this is motivated by the fact that the local magnetic anisotropy varies randomly from site to site [2, 6, 8]. For simplicity, we can assume [7-10, 20] that all the grains are identical within their own local coordinate systems and exhibit hexagonal symmetry with random grain orientation ("polycrystalline model"). In the case of atomic level stresses we note that the amorphous state is obtained by quenching from the melt and is often regarded as similar to a frozen liquid state (but not identical because of the different temperatures!). In the liquid state small (e.g., icosahedral [5]) clusters of an atom and its nearest neighbour atoms may be considered as more or less independent.

State c: We now allow for so-called quasi-plastic deformations of the units. In the magnetostriction case the quasi-plastic deformations are introduced by switching on the exchange interactions and applying a very strong external field, so that all units have a magnetization in the same direction. If we do not allow for rigid rotations of the units by exposing additional constraints, the units will exhibit spontaneous magnetostrictive strains like small monocrystalline ferromagnets,

$$\varepsilon_{ij}^{Q}(\mathbf{r}) = \lambda_{ijkl}(\mathbf{r})\,\gamma_k\,\gamma_l\,. \qquad (1)$$

Here $\boldsymbol{\gamma}$ is the vector of direction cosines characterizing the orientation of the magnetization. The quantity $\lambda_{ijkl}(\mathbf{r})$ denotes the values of the magnetostriction tensor of the grains at sites $\mathbf{r}$, which in our polycrystalline model has hexagonal symmetry in the internal coordinate systems of the grains, but exhibits spatial fluctuations in a fixed external coordinate system due to the random orientation of the grains. As a result, all units are deformed in a different way, and the statistics of the quasi-plastic strains is determined by the statistics of the grain orientation.

In the case of atomic level stresses we associate [21] the quasi-plastic deformations with the thermal expansions that the isolated units of the liquid state will experience at a characteristic temperature T_c, which may be identified with the temperature just before freezing in the quenching process takes place, probably a temperature between the melting temperature T_s and the glass temperature T_g. Because of thermal fluctuations there is a random distribution of quasi-plastic strains according to

$$p(\boldsymbol{\varepsilon}^{Q}) \sim \exp[-W(\boldsymbol{\varepsilon}^{Q})/(k_B T_c 0]\,. \qquad (2)$$

Here $W(\boldsymbol{\varepsilon}^{Q})$ is the increase of energy of the unit associated with the thermal expansion, described by

$$W(\boldsymbol{\varepsilon}^{Q}) = 1/2\, V_0\, C^{(2)}\, (\mathrm{Tr}\,\boldsymbol{\varepsilon}^{Q})^2 \qquad (3)$$

when assuming that the units are "liquid" (which is reasonable for $T_c = T_s$, then $C^{(2)}$ is the bulk modulus at T_s) and

$$W(\boldsymbol{\varepsilon}^{Q}) = 1/2\, V_0\, C_{ijkl}\, \varepsilon_{ij}^{Q}\, \varepsilon_{kl}^{Q} \qquad (4)$$

when the units are regarded as solid (which is reasonable for $T_c = T_g$, then C_{ijkl} is the tensor of the elastic constants of the units at T_s). The quantity V_0 denotes the volume of the units. Alternatively to our model, one may argue that the main effect of thermal fluctuations is represented by fluctuations of the number of atoms in the units of the liquid state. Within a continuum approach these "coordination fluctuations" may be again modelled by thermal expansions of the units, with the only exception [4] that the quasi-plastic strains of the different units may no longer be regarded as statistically independent, because the coordination fluctuations arise from the exchange of atoms between the units.

States d–f: After the introduction of the quasi-plastic strains $\varepsilon^Q_{ij}(\mathbf{r})$ the units no longer fit together to a compact material (d). Additional elastic strains $\varepsilon^{el}_{ij}(\mathbf{r})$ are required (e) to restore compatibility of the material (f), and the total strain $\varepsilon_{ij}(\mathbf{r})$ is $\varepsilon_{ij}(\mathbf{r}) = \varepsilon^Q_{ij}(\mathbf{r}) + \varepsilon^{el}_{ij}(\mathbf{r})$. To obtain a compatible strain field $\varepsilon_{ij}(\mathbf{r})$ the elastic strains thereby must be determined in such a way that

$$\mathrm{Ink}\,(\boldsymbol{\varepsilon}^Q(\mathbf{r}) + \boldsymbol{\varepsilon}^{el}(\mathbf{r})) = 0 \tag{5}$$

with $\mathrm{Ink}\,\boldsymbol{\varepsilon} = \nabla \times \boldsymbol{\varepsilon} \times \nabla$. Furthermore, the balance-of-force equation must be fulfilled, which in the absence of external forces reads

$$\mathrm{Div}\,\boldsymbol{\sigma} = 0\,. \tag{6}$$

$\boldsymbol{\sigma}$ denotes the tensor of elastic stresses, which are related to the elastic strains $\boldsymbol{\varepsilon}^{el}$ by

$$\sigma_{ij} = C_{ijkl}\,\varepsilon^{el}_{kl}\,. \tag{7}$$

Here C_{ijkl} is the tensor of elastic constants at room temperature for which we want to calculate the internal stresses. (Note that for the calculation of the atomic level stresses we freeze the quasi-plastic strains $\boldsymbol{\varepsilon}^Q$ the units exhibit in the liquid state at the characteristic temperature T_c, whereas we calculate the elastic strains $\boldsymbol{\varepsilon}^{el}$ required to restore compatibility at room temperature, see above.) The tensor C_{ijkl} is a spatially fluctuating quantity. E.g., for our polycrystalline model the fluctuations $\delta C_{ijkl}(\mathbf{r}) = C_{ijkl}(\mathbf{r}) - \langle C_{ijkl}(\mathbf{r})\rangle$ and the corresponding fluctuations $\delta S_{ijkl}(\mathbf{r})$ of the tensor of the elastic compliances are directly related to the random orientation of the grains [7–10].

Equations (5)–(7) may be solved iteratively by the von Neumann's Green's function perturbation technique (cf. [16, 7–10] with the fluctuations δS_{ijkl} as perturbation parameters, yielding perturbation series for $\sigma_{ij}(\mathbf{r})$ and $\varepsilon_{ij}(\mathbf{r})$. We then can calculate $\langle\varepsilon_{ij}\rangle$ ($\langle\sigma_{ij}\rangle$ is zero due to the theorem of Albenga [1]), the tensors $\langle\sigma_{ij}(\mathbf{r})\,\sigma_{kl}(\mathbf{r})\rangle$ and $\langle\varepsilon_{ij}(\mathbf{r})\,\varepsilon_{kl}(\mathbf{r})\rangle$ and the two invariant combinations of their components, i.e., the hydrostatic pressure $\sqrt{\langle p^2\rangle}$ with $p^2 = 1/9\ \sigma_{ii}\sigma_{jj}$ and the von Mises' shear stress [26] $\sqrt{\langle\tau^2\rangle}$ with $\tau^2 = 1/2\ \sigma_{ij}\sigma_{ji} - 1/6\ \sigma_{ii}\sigma_{jj}$.

3 Results

Magnetostriction. We report only the results for $\langle\varepsilon_{ij}\rangle$ and the effective magnetostriction tensor $\lambda^{\rm eff}_{ijkl}$ defined by $\langle\varepsilon_{ij}\rangle = \lambda^{\rm eff}_{ijkl}\,\gamma_k\,\gamma_l$ from which we can determine the effective magnetostriction constant $\lambda^{\rm eff}_s = 2/3\,(\lambda^{\rm eff}_{1111} - \lambda^{\rm eff}_{1122})$. In this case no length scale enters the theory which characterizes the size of the units. The results are thus valid both for amorphous and for polycrystalline ferromagnets, for which the structural units are given by the crystallites. The local magnetostriction $\langle\varepsilon_{ij}\,\varepsilon_{kl}\rangle$ is discussed in [20]. To obtain physically reasonable results, we insert for the components of the material tensors in the internal coordinate systems of the grains the values for crystalline Co at $T = 300$ K and crystalline Gd at $T = 4$ K.

In zeroth order of the perturbation theory ($\delta S_{ijkl} = 0$) we obtain $\langle\varepsilon_{ij}\rangle = \langle\varepsilon^Q_{ij}\rangle + S_{ijkl}\langle\sigma_{kl}\rangle = \langle\varepsilon^Q_{ij}\rangle$, where we have used the theorem of Albenga [1]. Obviously in this order the effective magnetostriction tensor is identical to the one of an ensemble of independent grains [12]. We have calculated the effective magnetostriction tensor up to second order, and therefore $\lambda^{\rm eff(2)}_s - \lambda^{\rm eff(0)}_s$ yields the deviations from the results of the non-interacting grain model and is a measure for the effects of elastic couplings related to the inhomogeneous character of the magnetostrictive strain field. For the Co-parameters there is only a slight difference between second order ($\lambda^{\rm eff(2)}_s = -6.85 \times 10^{-5}$) and zeroth order ($\lambda^{\rm eff(0)}_s = -6.47 \times 10^{-5}$). In contrast, there is a strong difference for the Gd-parameters ($\lambda^{\rm eff(2)}_s = 7.36 \times 10^{-6}$, $\lambda^{\rm eff(0)}_s = 1.07 \times 10^{-5}$), demonstrating the failure of the non-interacting grain model [12] and the importance of the elastic couplings in this case. Playing around with the material parameters we can also find situations for which the effective magnetostriction constant vanishes although the local magnetostriction tensor and its volume average $\langle\lambda_{ijkl}(\mathbf{r})\rangle$ are finite. In this case the non-zero value of $\langle\lambda_{ijkl}(\mathbf{r})\rangle$ (representing the effective magnetostriction tensor in the approximation of the non-interacting grain model [12]) is compensated by the contribution of elastic couplings. Obviously, zero magnetostriction does not necessarily mean that the local magnetostriction tensor and its volume average vanish. Accordingly, one cannot derive reliable conditions for zero magnetostriction by considering the properties of a single grain only. Altogether, we have shown that a theory neglecting the inhomogeneous character of the magnetostrictive strain field and the concomitant elastic coupling effects between neighbouring units may yield totally wrong results, especially in the case of the technologically appealing near-zero magnetostrictive alloys.

It should be noted that by assuming modified statistical properties of the polycrystalline model the theory is also able to explain the experimentally observed effects of annealing or application of external stress on λ_s^{eff} [7-10].

Atomic level stresses: As an example, we have considered in zeroth order approximation the case of hypothetical amorphous iron, for which the atomistic computer simulations [3] yield $\sqrt{<p^2>} = 0.064$ eV/(Å)^3 and $\sqrt{<\tau^2>} = 0.104$ eV/(Å)^3. Our model of liquid units (eq. (3)) with $T_c = T_s = 1807$ K yields $\sqrt{<p^2>} = 0.0672$ eV/(Å)^3 and $\sqrt{<\tau^2>} = 0.0582$ eV/(Å)^3, whereas for solid units (eq. (4)) with $T_c = T_g = 760$ K we obtain $\sqrt{<p^2>} = 0.0585$ eV/(Å)^3 and $\sqrt{<\tau^2>} = 0.114$ eV/Å)^3, inserting for V_0 the atomic volume of crystalline iron, $V_0 = 11.7$ (Å)^3. Obviously there is an astonishingly good agreement between the microscopic and continuum theoretical calculations, demonstrating that our continuum model of the quenching process of amorphous materials is quite reasonable.

Acknowledgement

The authors are indebted to Professors E. Kröner, H. Kronmüller and T. Egami for helpful discussions.

References

[1] Albenga, G., Atti Acad. Sci. Torino, Cl. fis. mat. nat. **54**, 864 (1918/1919).

[2] Böhm, M.C., Elsässer, C., Fähnle, M. and Brandt, E.H., *Chem. Phys.* **130**, 65 (1989).

[3] Egami, T. and Srolovitz, D., *J. Phys.* **F12**, 2141 (1982).

[4] Egami, T., private communication.

[5] Elsässer, C.,Fähnle, M., Brandt, E.H. and Böhn, H.C., *J. Phys.* **F17**, L301 (1987).

[6] Elsässer, C., Fähnle, M., Brandt, E.H. and Böhm, M.C., *J. Phys.* **F18**, 2463 (1988).

[7] Fähnle, M., Furthmüller, J., Pawellek, R. and Herzer, G., in: *Physics of Magnetic Materials* (Singapore: World Scientific) 1989, p. 228ff.

[8] Fähnle, M., Elsässer, C., Furthmüller, J., Pawellek, R., Brandt, E.H. and Böhm, M.C., *Physica* **B161**, 225 (1989).

[9] Fähnle, M., Furthmüller, J. and Herzer, G., *J. Physique Col.* **C8**, 1329

(1988).
[10] Furthmüller, J., Fähnle, M. and Herzer, G., *J. Phys.* **F16**, L225 (1986); *J. Magn. Magn. Mat.* **69**, 79 (1987); **69**, 89 (1987).
[11] O'Handley, R.C., *J. Appl. Phys.* **62**, R15 (1987).
[12] O'Handley, R.C. and Grant, N.J., Proc. Conf. on Rapidly Quenched Metals, Eds. Steeb, S. and Warlimont, H. (Amserdam: Elsevier) 1985, p. 1125 ff.
[13] Kröner, E., "Kontinuumstheorie der Versetzungen und Eigenspannungen" (Berlin-Göttingen-Heidelberg: Springer-Verlag) 1958.
[14] Kröner, E., in: "Mechanics of Generalized Continua" (Berlin: Springer-Verlag) 1967, p. 330 ff.
[15] Kröner, E., in: "Topics in Applied Continuum Mechanics" (Wien-New York: Springer-Verlag) 1974, p. 22 ff.
[16] Kröner, E. and Koch, H., *SM Archives* **1**, 184 (1976).
[17] Kronmüller, H., Fähnle, M., Domann, M., Grimm, H., Grimm, R. and Gröger, B., *J. Magn. Magn. Mat.* **13**, 53 (1979).
[18] Kronmüller, H., *J. Appl. Phys.* **52**, 1859 (1981).
[19] Kurzyk, J., *J. Magn. Magn. Mat.* **73**, 84 (1988).
[20] Pawellek, R., Furthmüller, J. and Fähnle, M., *J. Magn. Magn. Mat.* **75**, 225 (1988).
[21] Pawellek, R. and Fähnle, M., to be published in *J. Phys.*: Condensed Matter.
[22] Reissner, H., Z. *Angew. Math. Mech.* **11**, 1 (1931).
[23] Suzuki, Y. and Egami, T., *J. Magn. Magn. Mat.* **31-34**, 1549 (1983).
[24] Tsuya, N. and Arai, K.I., *J. Magn. Magn. Mat.* **31-34**, 1594 (1983).
[25] Vitek, V. and Egami, T., *Phys. Stat. Sol.* (b) **144**, 145 (1987).
[26] von Mises, R., Nachr. Ges. Wiss. Göttingen, 582 (1913).

Fähnle, M., Furthmüller, J. and Pawellek, R.,
Institut für Physik
Max-Planck Institut für Metallforschung
Heisenbergstrasse 1
D-7000 Stuttgart 80
F.R.G.

C. HUET

Hierarchies and bounds for size effects in heterogeneous bodies

1 Introduction

The concept of effective properties, or of effective parameters, has been elaborated through the pioneer works by Hill [4, 5], Kröner [12, 13], and others. It relates classically to heterogeneous bodies that can be considered as statistically uniform in the space. Strictly, this requires infinite dimensions, see for instance [13]. In fact, it is generally admitted that the concept of effective properties has still a sense for an elementary volume with dimensions larger than the so-called "representative volume". The latter is generally defined through its expected properties, namely that the effective properties must be independent of the boundary conditions. Then, they can be used in standard structural calculations.

The limitations of this approach have been identified namely by Eimer [3], Mazilu [16] and Kröner [13]. They have shown that, for a body with finite dimensions, we can never eliminate the boundary-layer effects. This leads to non-local theories, like the one developed recently by Kröner [14, 15].

There exist other kinds of limitations to the validity of the effective property concept, like the one defined for elastic solids by the so-called Hill condition, Kröner [13]. In Huet [6, 7], we derived universal conditions that are generalizations of the latter to any kind of materials with any kind of constituents. This was done through the consideration of local ensemble averages, as in the works of Beran [1] for instance, applied to the universal set of balance equations of thermomechanics. But we eliminated the classical ergodic assumption through the use of a spatio-stochastic approach, Huet [8, 9], applied to samples still supposed to be each with dimensions larger than those of the representative volume.

On the other hand, it is of current practice for laboratories to perform tests on samples with smaller dimensions, or made from a material - like wood for instance - that, from origin, exhibits localized defects when used as structural members. In those cases, the "average behaviour" of the samples depends on boundary conditions and the concept of effective properties, does not apply any more. Since such tests look still useful to engineers, it is valuable to examine what kind of information can really be gained from them. The problem does not seem to have been considered, up to now, from this point of view in the literature. Here we present results we recently obtained on this subject and that might be probably considered as first steps in this area.

We consider first the elastic case, for which we provide bounds on the "apparent properties" defined in the sequel. Then, we extend these results to the case of viscoelastic materials through the use of a complex modulus and compliance tensors approach based on results obtained in a previous work [2].

2 Concept of individual and statistical apparent properties

Let us consider any kind of heterogeneous body that we call the initial body D_0. We take it in a shape such that it can be subdivided into small parallelepipeds of smaller sizes (D_α; $\alpha = 1$ to n) that we call the samples. We obtain thus a uniform partition of the initial body. We look for informations that we could gain about the elastic behaviour of the initial body from tests performed separately on the set of samples we have obtained from it.

In general, the results will depend on boundary conditions. We consider here the classical kinematic uniform boundary conditions of the kind ε_0 and the static uniform boundary conditions of the kind σ_0 (ε_0 - KUBC and σ_0-SUBC) of the classical theory of heterogeneous media (see for instance Huet [9]). We know that, in ε_0 - KUBC, the volume average $\langle\varepsilon\rangle$ of the strain is ε_0, and that, in σ_0 - SUBC, the volume average of the stress is σ_0. If the same ε_0 - KUBC or σ_0 - SUBC is applied to the initial body and to each specimen, the initial body and each specimen have one and the same mean strain ε_0, or mean stress σ_0.

For each sample D_α we define, denoting .. the twice contracted tensor product:

- in KUBC, its kinematic apparent elasticity modulus tensor $C_{\varepsilon\alpha}{}^{app}$ and its kinematic compliance tensor $S_{\varepsilon\alpha}{}^{app}$ through

$$\langle\sigma\rangle_\alpha = C_{\varepsilon\alpha}{}^{app} .. \langle\varepsilon\rangle_\alpha = C_{\varepsilon\alpha}{}^{app} .. \varepsilon_0 , \qquad S_{\varepsilon\alpha}{}^{app} = C_{\varepsilon\alpha}{}^{app^{-1}} . \tag{1}$$

- in SUBC, its static apparent elasticity compliance tensor $S_{\sigma\alpha}{}^{app}$ and its static modulus tensor $C_{\sigma\alpha}{}^{app}$ through

$$\langle\varepsilon\rangle_\alpha = S_{\sigma\alpha}{}^{app} .. \langle\sigma\rangle_\alpha = S_{\sigma\alpha}{}^{app} .. \sigma_0 , \qquad C_{\sigma\alpha}{}^{app} = S_{\sigma\alpha}{}^{app^{-1}} . \tag{2}$$

We define the corresponding apparent quantities for the initial body D_0 by making $\alpha = 0$ in eqs. (1) and (2).

Since the Hill condition is satisfied in UBC for any kind of heterogeneity, these definitions are compatible with their energetic counterparts

$$\langle C .. \varepsilon .. \varepsilon\rangle_\alpha = C_{\varepsilon\alpha}{}^{app} .. \langle\varepsilon\rangle_\alpha .. \langle\varepsilon\rangle_\alpha \qquad \text{(KUBC)}, \tag{3}$$

$$\langle S .. \sigma .. \sigma\rangle_\alpha = S_{\sigma\alpha}{}^{app} ... \langle\sigma\rangle_\alpha .. \langle\sigma\rangle_\alpha \qquad \text{(SUBC)}. \tag{4}$$

Let us denote by $\overline{\langle a \rangle_\alpha}$ the stochastic average on the set of samples D_α ($\alpha = 1$ to n) of the spatial average $\langle a \rangle_\alpha$ of the variable a. From applying the stochastic average operator to the first eq. (1) and (2) we may define the statistical apparent compliances and moduli tensors $C_\varepsilon{}^{app}$, $S_\varepsilon{}^{app}$, $C_\sigma{}^{app}$, $S_\sigma{}^{app}$ through

$$C_\varepsilon{}^{app} = \overline{C_{\varepsilon\alpha}{}^{app}}\,,\; S_\varepsilon{}^{app} = C_\varepsilon{}^{app^{-1}}\,, \quad S_\sigma{}^{app} = \overline{S_\sigma{}^{app}}\,,\; C_\sigma{}^{app} = S_\sigma{}^{app^{-1}}\,. \tag{5}$$

This is because we have chosen here the same deterministic boundary conditions, for each specimen.

We then have the following sequence of results.

3 Partition theorem and hierarchies

Let us consider corresponding ε_0 - KUBC applied to D_0 and to the D_α ($\alpha = 1, n$). The union on D_0 of the strain fields obtained separately on each D_α individually submitted to ε_0 - KUBC is an admissible strain field for D_0 also submitted as a whole to ε_0 - KUBC. Thus the classical potential energy minimum theorem can be applied to this case. Let us consider now the dual case of corresponding σ_0 - KUBC for the initial body D_0 and each sample D_α ($\alpha = 1, n$). The union on D_0 of the stress fields obtained separately on each sample separately loaded is an admissible stress-field for D_0 also submitted to σ_0 - SUBC. Thus the classical complementary energy minimum theorem can be applied to this second case. Taking also into account the corresponding inequalities for the reciprocals, this yields finally the following set of inequalities that we call the partition theorem:

$$C_\sigma{}^{app} \le C_{\sigma 0}{}^{app} \le C_{\varepsilon 0}{}^{app} \le C_\varepsilon{}^{app}\,, \quad S_\varepsilon{}^{app} \le S_{\varepsilon 0}{}^{app} \le S_{\sigma 0}{}^{app} \le S_\sigma{}^{app}\,. \tag{6}$$

Let us consider now a uniform partition (denoted by g) of D_0 into "large" samples $D_{\alpha g}$. Let us consider also another partition of the same D_0 into smaller samples that we call $D_{\alpha p}$ being obtained by partitioning each $D_{\alpha g}$. Through iterated application of the partition theorem above, we obtain the following inequalities that sets up a hierarchy between the iterated partitions of a body:

$$C_{\sigma p}{}^{app} \le C_{\sigma g}{}^{app} \le C_{\varepsilon g}{}^{app} \le C_{\varepsilon p}{}^{app}, \quad S_{\varepsilon p}{}^{app} \le S_{\varepsilon g}{}^{app} \le S_{\sigma g}{}^{app} \le S_{\sigma p}{}^{app}\,. \tag{7}$$

Here $C_1 \le C_2$ means $(C_2 - C_1) \,..\, a \,..\, a \ge 0 \; \forall \; a \ne 0$.

4 Limit bounds and separation

Within the limits of applicability of continuum mechanics, the preceding results apply whatever the size of the specimens may be. Let us take a p-type uniform partition for which the dimensions of each sample are smaller than the smaller dimensions of the homogeneous subdomains inside D_0, supposed to be of the non-fractal type (always true for real materials since the atomic dimensions give a limit to fractality). Applying the results of Section 3, we obtain thus the following bounds, for any other coarser partition:

$$\overline{C^{-1}}^{\,-1} \leq C_\sigma{}^{\mathrm{app}} \leq C_\varepsilon{}^{\mathrm{app}} \leq \overline{C}\,, \qquad \overline{S^{-1}}^{\,-1} \leq S_\varepsilon{}^{\mathrm{app}} \leq S_\sigma{}^{\mathrm{app}} \leq \overline{S}\,. \tag{8}$$

Here, we see the striking results that the classical Hill bounds [4] [5] (first-order bounds in the sense of Kröner [13]) are still applicable to the size effects, to which they provide limit bounds. The central inequalities in eqs. (8) follow from a classical result easy to derive from the same variational theorems.

Let us consider now the particular case of a material that is statistically uniform. The effective modulus and compliance tensors C^{eff} and S^{eff} do exist, and are respectively equal to their apparent counterparts when D_0 is larger than the representative volume. Applying the results of Section 3 to any uniform partition of the representative volume into samples that do not have the representative volume yields the following inequalities:

$$C_\sigma{}^{\mathrm{app}} \leq C^{\mathrm{eff}} \leq C_\varepsilon{}^{\mathrm{app}}\,, \qquad S_\varepsilon{}^{\mathrm{app}} \leq S^{\mathrm{eff}} \leq S_\sigma{}^{\mathrm{app}}\,. \tag{9}$$

By combination with the relations (8) above, one sees that, when the kind of boundary conditions are known, the knowledge of the effective moduli or compliances allows one to narrow the bounding interval, since the effective tensor separates the corresponding apparent results obtained on the two kinds of boundary conditions.

The results obtained in Sections 3 and 4 above can be recapitulated in the two following sets of inequalities

$$\overline{C^{-1}}^{\,-1} \leq C_{\sigma p}{}^{\mathrm{app}} \leq C_{\sigma g}{}^{\mathrm{app}} \leq C_{\sigma 0}{}^{\mathrm{app}} \leq C^{\mathrm{eff}} \leq C_{\varepsilon 0}{}^{\mathrm{app}} \leq C_{\varepsilon g}{}^{\mathrm{app}} \leq C_{\varepsilon p}{}^{\mathrm{app}} \leq \overline{C}\,, \tag{10}$$

$$\overline{S^{-1}}^{\,-1} \leq S_{\varepsilon p}{}^{\mathrm{app}} \leq S_{\varepsilon g}{}^{\mathrm{app}} \leq S_{\varepsilon 0}{}^{\mathrm{app}} \leq S^{\mathrm{eff}} \leq S_{\sigma 0}{}^{\mathrm{app}} \leq S_{\sigma g}{}^{\mathrm{app}} \leq S_{\sigma p}{}^{\mathrm{app}} \leq \overline{S}\,. \tag{11}$$

5 Size effects in viscoelastic heterogeneous bodies

In the oral presentation of this paper in Dijon, we dealt with relaxation and creep functions, using new pseudo-convolutive variational theorems based on Brun's results[17] and to be published. Since that time, we found that the corresponding results may be outside the domain validity of these theorems. Then, these results cannot be considered as proved. Here we present results based on another approach.

We consider the effective complex modulus and compliance tensors :

$$C^{*eff}(i\omega) = C^{eff'} + i\,C^{eff''};\; S^{*eff}(i\omega) = S^{eff'} - i\,S^{eff''} \tag{12}$$

where the real parts $C^{eff'}$, $S^{eff'}$ and imaginary parts $C^{eff''}S^{eff''}$ are real positive definite tensors.

It was shown in [3] that lower bounds on the real and imaginary parts of the effective complex modulus and compliance can be found in terms of the effective moduli of auxiliary elastic problems, under the form

$$C'^{\,eff\,el} \leq C^{eff'} \quad ; \quad C''^{\,eff\,el} \leq C^{eff''} \tag{13}$$

$$S'^{\,eff\,el} \leq S^{eff'} \quad ; \quad S''^{\,eff\,el} \leq S^{eff''} \tag{14}$$

In (13), (14), the superscript (eff el) denotes the effective property of a fictitious elastic heterogeneous medium having the same internal geometry as the viscoelastic one, but for which the elasticity modulus or compliance tensor at point x is C′ (x) or C″ (x), or S′ (x), or S″ (x) respectively.

Using (13), (14), the results obtained in Sections 3 and 4 for the elastic case provide information about the size effects for the complex compliances and moduli. Because the elastic compliance of the auxiliary elastic medium with elastic moduli C ′ is not S ′, but $(C')^{-1}$ and the same for the other local quantities, we introduce the following notation:

$$\tilde{S}' = (C')^{-1} \;;\; \tilde{C}' = (S')^{-1} \quad ; \quad \tilde{S}'' = (C'')^{-1} \;;\; \tilde{C}'' = (S'')^{-1} \tag{15}$$

Considering a uniform partition of the auxiliary elastic problem with elastic local modulus C ′, and a uniform partition of the auxiliary elastic problem with elastic local compliance S ′, the inequalities (9) read now, with notations corresponding to (1), (2), (5) through (15) :

$$\tilde{C}_{\sigma}^{'app\,el} \leq C'^{\,eff\,el} \leq C_{\varepsilon}^{'app\,el} \quad ; \quad \tilde{S}_{\varepsilon}^{'app\,el} \leq S'^{\,eff\,el} \leq S_{\sigma}^{'app\,el} \tag{16}$$

From (13), (14) only the lower bounds in (16) can be used, giving :

$$\widetilde{C}_\sigma^{'appel} \leq C^{eff'} \quad ; \quad \widetilde{S}_\varepsilon^{'appel} \leq S^{eff'} \tag{17}$$

with :

$$\widetilde{C}_\sigma^{'appel} = \left(\widetilde{S}_\sigma^{'appel}\right)^{-1} \quad ; \quad \widetilde{S}_\varepsilon^{'appel} = \left(\widetilde{C}_\varepsilon^{'appel}\right)^{-1} \tag{18}$$

(See Section 2), with $\widetilde{S}_\sigma^{'appel}$ and $\widetilde{C}_\varepsilon^{'appel}$ corresponding to $\widetilde{S}'$ and $\widetilde{C}'$ in (15).
In the same way, we obtain also :

$$\widetilde{C}_\sigma^{''appel} \leq C^{eff''} \quad ; \quad \widetilde{S}_\varepsilon^{''appel} \leq S^{eff''} \tag{19}$$

with

$$\widetilde{C}_\sigma^{''appel} = \left(\widetilde{S}_\sigma^{''appel}\right)^{-1} \quad ; \quad \widetilde{S}_\varepsilon^{''appel} = \left(\widetilde{C}_\varepsilon^{''appel}\right)^{-1} \tag{20}$$

In order to calculate upper bounds to $C^{eff'}$, one may start from the following identities, already used in [2] :

$$\left(M^{eff}\right)^{-1} = S^{eff'} + S^{eff''} .. \left(S^{eff'}\right)^{-1} .. S^{eff''} \geq S^{eff'} \tag{21}$$

$$\left(S^{eff'}\right)^{-1} = M^{eff'} + M^{eff''} .. \left(M^{eff'}\right)^{-1} .. M^{eff''} \geq M^{eff'} \tag{22}$$

The results obtained above in this section give then finally, for the real parts $C^{eff'}$ and $S^{eff'}$:

$$\widetilde{C}_\sigma^{'appel} \leq C^{eff'} \leq \widetilde{C}_{\varepsilon\sigma}^{'appel} \leq C_\varepsilon^{'appel} \quad ; \quad \widetilde{S}_\varepsilon^{'appel} \leq S^{eff'} \leq \widetilde{S}_{\sigma\varepsilon}^{'appel} \leq S_\sigma^{'appel} \tag{23}$$

with :

$$\widetilde{C}_{\varepsilon\sigma}^{'appel} = \left(\widetilde{S}_\varepsilon^{'appel} + \widetilde{S}_\varepsilon^{''appel} .. \widetilde{C}_\sigma^{'appel} .. \widetilde{S}_\varepsilon^{''appel}\right)^{-1} \tag{24}$$

$$\widetilde{S}_{\sigma\varepsilon}^{'appel} = \left(\widetilde{C}_\sigma^{'appel} + \widetilde{C}_\sigma^{''appel} .. \widetilde{S}_\varepsilon^{'appel} .. \widetilde{C}_\sigma^{''appel}\right)^{-1} \tag{25}$$

Corresponding upper bounds for the imaginary parts $C^{eff''}$ ans $S^{eff''}$ are obtained by exchange of the $'$ and $''$ superscripts in (23) to (25).This because in (21) and (22) this exchange is valid.

6 Conclusion

The practical applications of the preceding results can be numerous. In particular they may provide a better understanding of the experimental results obtained in the testing laboratories, and lead to new procedures taking more precisely into account the kind of boundary conditions that are, or have to be, applied. We consider that these results can also be useful in the field of numerical simulations like the ones we have, with our co-workers, used in [11] for materials supposed to be statistically uniform. Last, these results provide us with a better feeling of the significance of the classical Hill first-order inequalities, and of the interval that separates them. Note that, for the size effects, they cannot be improved by hypothetical higher-order bounds. Of course, all the results obtained here apply, not only to parallelepipedic bodies and samples, but also to all bodies that can be uniformly partitioned into samples with shapes filling the body. When this is not possible, other corresponding bounds can be obtained by applying the preceding results to two slightly modified initial bodies - a "diminished" one and an "augmented" one - in the sense of Hill's modification theorem that we have used in [11], and that, in the elastic case, still applies here. In addition, it provides us with zeroth order bounds in Kröner's sense [14].

Detailed derivation can be found in [10].

References

[1] Beran, M.J. (1968). *Statistical Continuum Theories.* Interscience, New York.

[2] Chétoui, S., Navi, P., Huet, C., (1986). Researches on the evaluation of macroscopic properties of anisotropic heterogeneous materials with viscoelastic constituents. In : C. Huet, D. Bourgouin, S. Richemond (eds). *Rheology of anisotropic materials.* Cepadeus-Editions, Toulouse, 307-326.

[3] Eimer, C.Z. (1968). The boundary effect in elastic multiphase bodies. *Archiwum Mechaniki: Stosowanej*, 1, **20**, 87-93.

[4] Hill, R. (1952). The elastic behaviour of a crystalline aggregate. *Proc. Phys. Soc., London, Sect. A*, **65**, 349-354.

[5] Hill, R. (1963). Elastic properties of reinforced solids: Some theoretical principles. *J. Mech. Phys. Solids*, **11**, 357-372.

[6] Huet, C. (1981). Remarks on assimilation of a heterogeneous material to a continuous medium (in French). *In*: C. Huet and A. Zaoui (eds.). *Rheological Behaviour and Structures of Materials.* Presses ENPC, Paris, pp. 231-245.

[7] Huet, C. (1982). Universal condition for assimilation of a heterogeneous material to an effective medium. *Mechanics Research Communications*, **9** (3), 165-170.

[8] Huet, C. (1984). On the definition and experimental determination of effective

constitutive equations for heterogeneous materials. *Mechanics Research Communications*, **11** (3), 195-200.

[9] Huet, C. (1988). Some basic principles in the thermorheology of heterogeneous materials. *In*: H. Giesekus and M.F. Hibbert (eds.) *Progress and Trends in Rheology*. Proceeding of the 2nd Conference of European Rheologists, Prague, June 17-20, 1986. Supplement to Rheologica Acta, Vol. 26, pp. 1-8.

[10] Huet, C. (1990). Application of variational concepts to size effects in elastic heterogeneous bodies. In J. Mech Phys. Solids (in press).

[11] Huet, C., Navi, P. and Roelfstra, P.E. (1989). *A homogeneization technique based on Hill's modification theorem.* 6th Symposium on Continuum Models and Discrete Systems, Dijon, June 26-29.

[12] Kröner, E. (1958). Berechnung der elastischen Konstanten des Vielkristalls aus den Konstanten des Einkristalls. *Z. Phys.*, **151**, 504-518.

[13] Kröner, E. (1972). Statistical Continuum Mechanics. Springer-Verlag, Berlin and New York.

[14] Kröner, E. (1977). Bounds for effective elastic moduli of disordered materials. *J. Mech. Phys. Solids*, **25**, 137-155.

[15] Kröner, E. (1981). Linear properties of random media: the systematic theory. *In*: C. Huet, A. Zaoui (eds.), *Rheological Behaviour and Structure of Materials*, Presses ENPC, Paris, 15-40.

[16] Mazilu, P. (1969). Private communication to E. Kröner, Quoted in [13] above, p. 108.

Huet C.
Ecole Polytechnique Fédérale
Département des Matériaux
Chemin de Bellerive 32
CH-1015 Lausanne
SWITZERLAND

C. HUET, P. NAVI AND P.E. ROELFSTRA

A homogenization technique based on Hill's modification theorem

1 Introduction

In many cases, application of classical homogenization techniques to real materials suffers various limitations. For instance, the systematic theory of Kröner [6, 7] based on Neumann series solution of the Lippmann–Schwinger equation fails in the cases of porous media or rigid inclusions, for which the series does not converge any more, and for the materials with a high degree of order, for which a numerous set of correlation functions of increasing orders have to be determined and used. On the other hand, real materials generally considered as periodic materials exhibit random slight deviations from the true periodic case that involve inaccuracy in the evaluation from periodic case techniques like the ones developed by Sanchez–Palencia [16] and others.

Here we show that these various limitations can be overcome to some extent through the use of a method proposed by one of us [2] and that considers appropriate slightly modified materials for which evaluations can be made that are bounds for the real material properties. In a second step, a precise estimation can be derived from these bounds. We present first the principles of the corresponding technique, which is based on a theorem stated by Hill in 1963 [1], and then show examples of application to simple "benchmark problems" on which the properties of the real material can also be calculated, and thus the effectiveness of the method can be verified.

2 Principle of the homogenization technique

In his celebrated paper of 1963 [1], Hill stated and proved a theorem, modestly presented as an auxiliary one, but that proves to be a very powerful one, even if it corresponds directly to common intuition: if the elastic modulus tensor of any heterogeneous material is increased in some region, the effective modulus is also increased. Of course a similar conclusion holds for the compliance tensor, and conversely for a decrease. By increase is understood that the difference between the final tensor, say C^2, and the initial tensor say C^1, is a positive definite tensor, i.e., any quadratic form $(C^2 - C^1)..a..a$ built on these fourth-order tensors is positive-definite, "a" being any symmetric second-order tensor. Thus, in the Voigt matrix form, all the principle determinants must be positive.

Even with the powerful computer codes now available, numerical simulation techniques always need a set of simplifying assumptions about the microstructure of the material that lead to simplified models on which the calculations will be performed.

The basic idea of the proposed technique is to perform these simplifying modifications in ways that the Hill modification theorem can be applied. Doing this on both sides (an increase and a decrease) provides us with two bounds (an upper one and a lower one) on the corresponding effective property of the material. Thus, two computations are performed on two different models of the material obtained through monotonic modifications of the latter. Of course, this can be performed in several fashions, more or less sophisticated depending on the desired precision and the type of real material under consideration.

In the following sections, a few common "benchmark problems" of the type mentioned above will be considered, in order to give an idea of the validity and effectiveness of the method. For complicated materials involving fine details in their structure and a wide granulometry distribution, monotonic modifications will be difficult to realize, and it will be useful to make use of appropriate tools such as mathematical morphology and digital image analysis devices. When the moduli of anisotropic constituents are not comparable, appropriate comparable auxiliary moduli and compliances will be taken.

It should be noted that the principle of the method consists in the consideration of (hopefully) small perturbations from the real material. This is performed instead of considering the material as a perturbation of a homogeneous one as is done in homogenization techniques making use of a homogeneous comparison material. Thus, in our technique, correlation functions of high order are still present in the modified models of the material. In place of (or in addition to) geometric modifications, rheologic modifications can also be considered. For instance, this will be especially useful for anisotropic inclusions oriented at random: isotropic bounds to the inclusion elasticity tensors will facilitate numerical computations and provide bounds to the effective properties. Finally, precise evaluations will be obtained from the bounds through interpolation techniques under continuity assumptions of the dependency of the effective properties about some appropriate modification parameter.

Example of a periodic composite material

As a first simple example allowing direct verification, we consider here a unidirectional periodic fibre composite medium, called M1. A basic unit cell of the medium in the plane perpendicular to the fibre direction is shown in Figure 1(b). The fibres and the matrix are isotropic and are circular in cross-section. The fibre modulus tensor C^{f} is greater than the matrix one C^{m}, meaning here that the same holds for the compressibility and shear moduli, respectively. The values are given in Table 1, in Voigt representation.

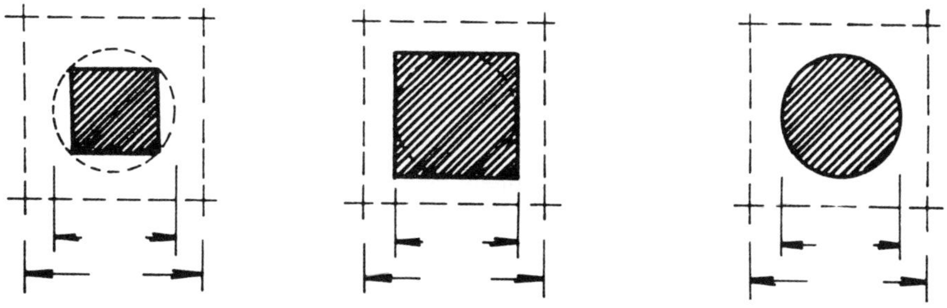

(a) Diminished (b) Real (c) Augmented

Figure 1. Basic cells of the real and monotonically modified materials.

Table 1. Constituents' moduli and calculated transversal effective elastic moduli in GPa ($a = C_{ij}$ or c_f; $\delta C_{ij} = |C_{ij}^1 - C_{ij}^r|$; $r = 2, 3$).

	Fibre	Matrix	Diminished Medium M3	Real Medium M1	Augmented Medium M2	$\frac{a^1 - a^3}{a^1}$ %	$\frac{a^2 - a^1}{a^1}$ %
c_f			0.222	0.349	0.444	27.2	36.4
C_{11}	40	6.4	8.5948	10.1544	11.99	18.1	15.4
C_{22}	40	6.4	8.5948	10.1544	11.99	18.1	15.3
C_{12}	20	3.4	4.0517	4.646	5.001	12.08	7.6
$\sqrt{C_{11} C_{22}}$	40	6.4	8.5948	10.1544	11.99		
δC_{12}			0.5943	0	0.355		
$\sqrt{\delta C_{11} \delta C_{22}}$			1.5596	0	1.8356		
C_{66}	10	1.5	1.9536	2.277	2.588	13.7	24.5

We consider in addition two monotonically modified unidirectional composites M2 and M3 with the basic unit cells illustrated in Figures 1(a) and 1(c) respectively. M2 and M3 are obtained from M1 by transforming the circular fibre into a smallest including square and a largest included square respectively. Thus M2 is the "augmented material" and M3 the "diminished" one. The transverse effective elastic moduli of the three composites are calculated by taking the low frequency limit in the numerical wave dispersion method introduced and developed by Navi [10–12].

The numerical data and results are given in Table 1. As expected, the effective elastic moduli of the medium M1 in the transverse plane is bounded from both sides by the effective moduli of the media M2 and M3. This makes us confident in the fact that a material M1 with fibres of a much more complicated shape with details inaccessible to computation could be evaluated through the same technique.

Better precision can be expected with modifications involving smaller changes in the concentrations. In fact, for a continuous path of slight monotonic transformations with no drastic change in the connectivity going through the real material, the effective properties should be continuous and rather smooth functions of the concentration of the modified constituent. Thus, it can be expected that some rather precise evaluation of the real effective properties can be obtained through linear interpolation between the two bounds in terms of this concentration, since the latter will be known in general. An improved non-linear interpolation would be obtained by calculating two other bounds for two other models that are monotonic transforms for both the real material and the first two modified models. Here, the linear interpolation results in a C_{11} value of 10.537, which equals the real one 10.154 up to an error lower than 3.8% , the other components being still more precisely evaluated in this fashion.

4 Reduction of the case of a quasi-ordered porous material to periodic cases.

Figure 2(b) shows, taken from a transverse microscopic photograph of a piece of wood at the cell level of heterogeneity, the case of a microstructure which is highly porous (more than 50% porosity) and which exhibits a high degree of order. In fact, although not really periodic, the structure is not far from this latter case. For this kind of structure, the assumption of periodicity is generally made, but it is impossible to know how far the results obtained on the periodic model deviate from the real solution.

Figures 2(a) and 2(b) show that, from the initial material of Figure 2(b), one can build, through monotonic transformations, periodic "augmented" and periodic "diminished" models.

The numerical calculations corresponding to this example are still to be performed through Navi's wave dispersion method, which also applies to anisotropic constituents, and to materials with a large amount of voids (see for instance [9, 11]). Interpolation in terms of the concentration in solid constituent is expected to provide a good estimation. Observe that here the material is highly anisotropic, with a low degree of symmetry in the transverse plane.

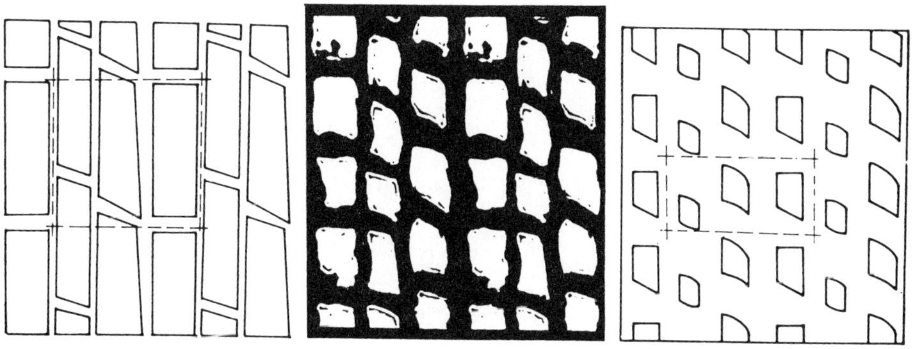

(a) Diminished (b) Real (c) Augmented

Figure 2. Representative element of a quasi-ordered porous medium, and periodic monotonically modified models.

5 Example of a random granular material

An application in the random case has been realized with the help of the "numerical concrete" computer codes developed by Roelfstra [14, 15]. To play the role of the real material, a heterogeneous structure was generated with 10 grains conforming to a given standard granulometry and placed at random in a matrix. Each grain cross-section has the shape of an irregular non-convex ellipse oriented at random. In addition, two other structures, representing the augmented and diminished versions of the initial structure, were created. As modified versions of each grain cross-section, the largest included circle and the smallest including one were taken. This provided an easy automatic mesh generation of about 2500 elements (Figure 3). The Young modulus and Poisson ratio of the aggregate grains and of the matrix were chosen to be (60 GPa, 0.18) and (12 GPa, 0.22), respectively. In order to determine all coefficients of the elasticity matrix for the plane strain condition, three computer analyses in kinematic uniform boundary conditions (see [5]) were carried out for each structure. The results are shown in Table 2.

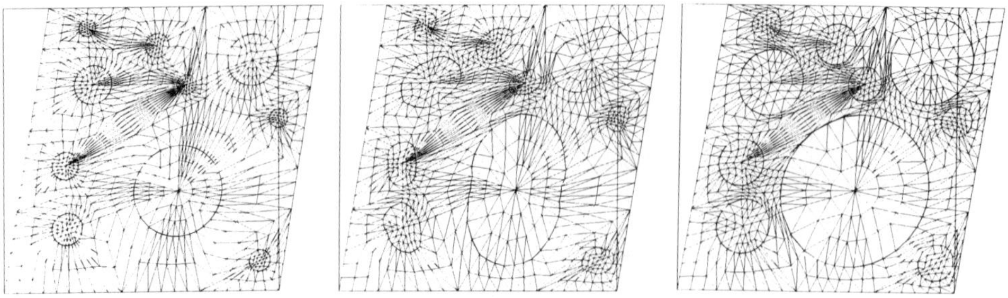

(a) Diminished (b) Real (c) Augmented

Figure 3. Real and monotonically modified realizations of a random material.

It can be verified that the expected inequalities are verified. Linear interpolation in terms of the grain concentration give the real result within an interval of 2%.

Table 2. Computed coefficients of the elasticity matrix for the plane strain condition compared for the three structures ($a = C_{ij}$ or c_f).

	Diminished Material M3	Real Material M1	Augmented Material M2	$\frac{a^1 - a^3}{a^1}$ %	$\frac{a^2 - a^1}{a^1}$ %
c_f	0.155	0.284	0.435	45.4	53.2
C_{11}	16.737	19.300	23.183	13.3	20.1
C_{22}	16.072	19.605	23.565	18.	20.2
C_{12}	4.193	4.613	5.214	9	13.
$\sqrt{C_{11}\, C_{22}}$	16.401	19.452	23.373		
$\delta\, C_{12}$	0.42	0	0.601		
$\sqrt{\delta\, C_{11}\, \delta C_{22}}$	3.009	0	3.921		
C_{66}	5.483	6.222	7.356	11.0	18.2

Note here that the considered models do not have the representative volume, and thus are neither perfectly isotropic, nor quadratic (see [4] for that case).

Conclusion

In the examples given here, the constituents moduli ratios were between 5 and 6. Even for the crude modifications of the initial material performed, it turns out that the accuracy of the bounding evaluation is already quite good. In fact, they do not deviate significantly from the accuracy needed for most engineering purposes and obtained by direct experimentation on materials with a large degree of heterogeneity. Further, the linear interpolation in terms of the concentration provides results that are already very close to the real ones. In fact, this technique possesses at least the same qualities as the self-consistent estimation and is endowed with additional advantages. It will be specially interesting to investigate the efficiency of this bounding and interpolation technique around the percolation threshold concentration, where the material structure exhibits rapid changes corresponding to phase inversion or, at least, connectivity modifications.

An interesting problem, presently under study, is to know if the Hill modification theorem can be extended to thc viscoelastic case. Through this, the homogenization technique presentcd here for the elastic properties could easily, at least in principle, be extended to the viscoelastic effective properties. For practical applications to viscoelastic materials that are periodic (or reducible to the periodic case in the manner presented in Section 5), it could be possible to make use of the extension of Navi's method performed by Naciri and Navi [9]. On the other hand, the practical applications to a particular realization of a random viscoelastic material through a finite element code may involve many numerical difficulties, especially for viscoelastic materials with continuous spectra, or discrete spectra where the number of rays is not very low. A discussion of the significance of effective properties for non purely elastic behaviours can be found in [3].

Let us mention also, that it can be easily shown that Hill's modification theorems still apply to the apparent compliances and moduli that we have introduced in another paper to this Symposium [5] for heterogeneous bodies that are not statistically uniform or have dimensions smaller than the ones of the representative volume. Thus, it will be possible to make use of the technique described here in order to provide numerical evaluation of these apparent properties. In fact, we saw that the random material example presented in Section 5 of this paper belongs to this latter case.

Of course, for standard applications to real materials, the tools of mathematical morphology and image analysis, like the ones based on the works by Matheron [8] using Serra's transforms [17], will be of prime practical importance. In particular, especially useful will be the facts that Matheron-Serra's morphological opening "smooths the contour of the grains, cuts the narrow isthmuses, suppresses the small islands and the sharp capes", while the morphological closing "blocks up the narrow channels, the small lakes and the long thin gulfs" [17]. That erosions and openings are monotonic decreasing transformations will also be useful. It has been shown in [8] and [17] that, if the Serra's structuring set C is morphologically open with respect to the structuring set B (i.e., of the form $A \oplus B$), then the morphological opening of a set X of grains by C is included in the opening of X by B, which is included in X itself. The latter is included in its morphological closing by B, which is included in the closing of X by C. We recognize here the sequence we need in order to perform the boundings and the non-linear interpolations we presented in this paper. All these operations are now implemented in standard digital image analysis systems. These points and other related ones will be developed elsewhere.

Finally, let us mention here that the importance for various problems of Hill's modification (or "strengthening-weakening") theorems was emphasized by Walpole as early as 1970 in [18]. In this paper, of which we had knowledge only after having written the present one, he suggested a bounding technique very similar to the one presented here. This seems to have remained unused, probably because powerful enough calculation tools and methods were lacking at that time. Then, Walpole's suggestion - never quoted elsewhere to our knowledge - seems to have been forgotten. Thus, it is our pleasure to acknowledge it here.

References

[1] Hill, R. (197=63a). Elastic properties of reinforced solids: Some theoretical principles. *J. Mech. Phys. Solids*, **11**, 357-372.

[2] Huet, C. (1988a). Unpublished technical note, ENPC-CERAM, Noisy-Le-Grand.

[3] Huet, C. (1988b). Some basic principles in the thermorheology of heterogeneous materials. *In* : H. Giesekus and M.F. Hibbert (eds.) *Progress and Trends in Rheology.* Proceeding of the 2nd Conference of European Rheologists, Prague, June 17-20, 1986. Supplement to Rheologica Acta, Vol. 26, pp. 1-8.

[4] Huet, C. (1989b). *Hierarchies and bounds for size effects in heterogeneous bodies.* 6th Symposium on Continuum Models and Discrete Systems, Dijon, June 26-29.

[5] Huet, C. (1990). Application of variational concepts to size effects in elastic heterogeneous bodies. *In* : J. Mech. Phys. Solids (in press).

[6] Kröner, E. (1972). *Statistical Continuum Mechanics.* Springer-Verlag, Berlin and New York.

[7] Kröner, E. (1981). Linear properties of random media: the systematic theory. *In*: C. Huet, A. Zaoui (eds.), *Rheological Behaviour and Structure of Materials,* Presses ENPC, Paris 15-40.

[8] Matheron, G. (1967). *Eléments pour une théorie des milieux poreux.* Masson, Paris.

[9] Naciri, T. and Navi, P. (1989). On the harmonic waves propagation in multi-layered viscoelastic media. *In* : Mottershead J.E. (ed.) Proc. Int. Conf. *"Modern Practice in Stress and Vibration Analysis"*, Pergamon, Oxford, 145-151.

[10] Navi, P. (1973). Harmonic wave propagations in composite materials. Ph. D. Thesis, University of California, Los Angeles.

[11] Navi, P. (1986). Application d'une méthode numérique aux aspects micromécaniques de la rhéologie du bois. Actes of the Seminar of the CNRS French Scientific Grouping No. 81 "Rheology of Wood", held in ENITEF, Nogent-sur-Vernisson, France, 65-68.

[12] Navi, P. (1987). Evaluation des constantes élastiques d'un matériau hétérogène par analyse de la dispersion des ondes. Journées Physiques "Les méthodes d'auscultation et d'imagerie", Les Arcs, France, pp. 135-140.

[13] Navi, P. (1988). Three dimensional analysis of the wood microstructural influences on wood elastic properties. *In*: Y. Itani (ed.) *Proc. 1988 Int. Conf. on Timber Engineering, Seattle,* FPRS, Madison, U.S.A., 915-922.

[14] Roelfstra, P.E., Sadouki, H. and Wittmann, F.H. (1985). Le béton numérique. *Mat. & Struct.,* **107**, pp. 327-335.

[15] Roelfstra, P.E. (1989). A numerical approach to investigate the properties of concrete: numerical concrete. Ph.D. Thesis. EPF-Lausanne, Laboratoire des Matériaux de Construction, Suisse.

[16] Sanchez-Palencia, E. (1974). Comportements local et macroscopique d'un type de milieux physiques hétérogènes. *Int. J. Eng. Sci.* **12**, 331-351.

[17] Serra, J. (1982). *Image Analysis and Mathematical Morphology.* Academic Press, London.

[18] Walpole, L.J. (1970). Strengthening effects in elastic solids. *J. Mech. Phys. Solids*, **18**, 343-358, p. 346.

Huet C., Roelfstra P., Navi P.
Ecole Polytechnique Fédérale
Département des Matériaux
Chemin de Bellerive 32
CH-1015 Lausanne
SWITZERLAND

Formerly at:
Ecole Nationale des Ponts
et Chaussées, CERAM
1 avenue Montaigne
F-93167 Noisy-le-Grand Cédex
FRANCE

J.B. LEBLOND AND G. PERRIN

Ductile fracture in a material containing two populations of cavities

1 Introduction

The macroscopic behaviour of a plastic material containing voids of *identical size* is now widely accepted to be well modelled by the yield criterion proposed in 1977 by Gurson [2] and refined in 1981 by Tvergaard [6]. Nevertheless there is some experimental evidence that in many materials, cavities of *different sizes* coexist. It is thus of interest to investigate the influence of the existence of two populations of voids on the macroscopic behaviour.

This question has already been analysed by Perrin and Leblond [5] *before the onset of coalescence*: the Gurson–Tvergaard model with only one internal parameter, namely the total porosity, was shown to account properly for the behaviour of the two-population material. On the other hand, there is experimental evidence that the situation is different when coalescence occurs; Marini et al. [4] suggested for instance that the existence of small cavities scattered between the large ones could considerably modify the final coalescence of the latter. The subject of the present work is to investigate these effects of interaction between two populations of voids of different sizes *after the onset of coalescence.*

The present work is based on the study of the same problem as in [5], namely a hollow sphere made of rigid plastic porous material (accounting for one large void surrounded by many small cavities) and subjected to hydrostatic tension. The matrix is treated as a homogeneous material obeying the Gurson–Tvergaard criterion.

In [5], only the stresses were evaluated. The calculation of the displacement is presented here; it leads to an evaluation of the cavity growth rate and of the porosity, as functions of position and time, and this allows for the investigation of coalescence phenomena.

The paper is organized as follows. Section 2 introduces a simplified, more tractable version of the Gurson–Tvergaard model. Using this model, we calculate the cavity growth rate as a function of position at time $t = 0$, when the porosity is still uniform. It is shown that the presence of a large cavity greatly enhances the growth rate of small cavities in its immediate neighbourhood. It is thus possible for the porosity of the second population (of small cavities) to reach the critical value for coalescence before that of the first population (of large cavities), even if it is smaller initially. In that case the fracture process should consist of three stages:

(i) The small cavities located near the big ones grow more quickly than the latter.
(ii) Close to the large cavities, the porosity of the second population reaches the

critical value for coalescence, and a shell of ruined matter loosens from the porous matrix around each large void; this phenomenon accelerates the growth of the latter.

(iii) The material is eventually ruined when the large cavities coalesce.

Stages (i) and (ii) of this mechanism are analysed in Section 3, through the numerical computation of the porosity for $t > 0$, when it is no more uniform. Section 4 develops a macroscopic model with few internal parameters which describes these two stages in an approximate way. Finally Section 5 studies the changes brought to the above scenario by the use of the complete Gurson–Tvergaard model.

2 Initial growth rate of the cavities

We consider a porous metal containing two populations of cavities of very different sizes, subjected to hydrostatic tension. An elementary volume of this material is schematized by a hollow sphere of internal radius a, external radius b, made of porous material, subjected to a hydrostatic tensile stress Σ (Figure 1). The initial porosity of the matrix is uniform and denoted f_2^0. The initial porosity due to the big central hole, denoted by f_1^0, is defined in such a manner that the total porosity of the sphere be simply the sum of f_1^0 and f_2^0; a straightforward calculation leads then to $f_1^0 = \frac{a^3}{b^3}(1 - f_2^0)$. The large hole is accounted for through the boundary condition $\sigma_{rr}(r = a) = 0$, and the small ones by assuming the matrix to obey the following criterion:

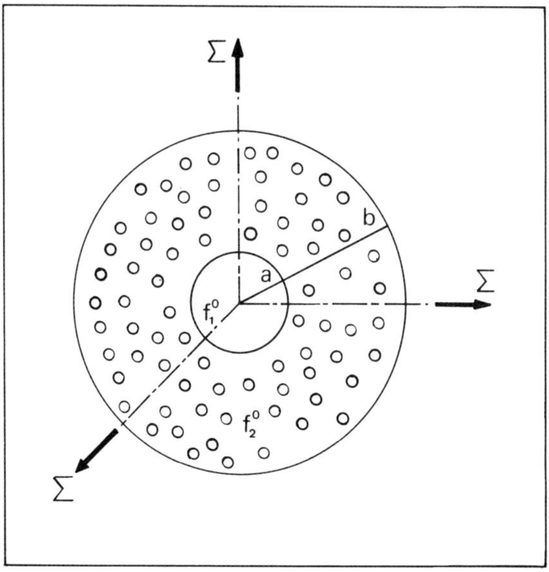

Figure 1

$$\Phi(\sigma_{eq}, \sigma_m, f_2^0) = \left(\frac{\sigma_{eq}}{\sigma_0}\right)^2 + qf_2^0 \exp\left(\frac{3}{2}\frac{\sigma_m}{\sigma_0}\right) - 1 = 0, \quad q = \frac{4}{e}, \tag{1}$$

where σ_{eq}, σ_m, σ_0 are the local Von Mises, hydrostatic and yield stresses, and q the Tvergaard parameter; the value $q = 4/e$ derived in [5] by a *self-consistent* method will be used here. This criterion is an acceptable approximation for that of Gurson and Tvergaard (see eq. (16) below) for low porosities and high hydrostatic stresses.

Let us introduce the following reduced variables:

$$u = \frac{\sigma_{rr}}{\sigma_0}, \qquad v = \frac{\sigma_{\theta\theta} - \sigma_{rr}}{\sigma_0} \tag{2}$$

(where r, θ, φ denote spherical coordinates). Moreover, let $\mathcal{U} \equiv \mathcal{U}\ \mathbf{e}_r$ denote the velocity field and $\mathbf{D} = \frac{1}{2}(\nabla\mathcal{U} + {}^t\nabla\mathcal{U})$ the strain rate. The equations of the problem are as follows:

Equilibrium and boundary conditions:

$$\frac{du}{dr} = \frac{2v}{r}; \quad u(a) = 0\,; \quad u(b) = \frac{\Sigma}{\sigma_0}. \tag{3}$$

Yield criterion:

$$\forall\, r \in [a, b] : \; v^2 + qf_2^0 \exp\left(\frac{3}{2}\,u + v\right) - 1 = 0. \tag{4}$$

Flow rule:

$$\frac{D_m}{D_{eq}} = \frac{\dfrac{1}{3}\dfrac{\partial\Phi}{\partial\sigma_m}}{\dfrac{\partial\Phi}{\partial\sigma_{eq}}} \Rightarrow \frac{\dfrac{1}{3}\left(\dfrac{\partial\mathcal{U}}{\partial r} + \dfrac{2\mathcal{U}}{r}\right)}{\dfrac{2}{3}\left(\dfrac{\mathcal{U}}{r} - \dfrac{\partial\mathcal{U}}{\partial r}\right)} = qf_2^0\,\frac{\exp\left(\dfrac{3}{2}u + v\right)}{4\,v}. \tag{5}$$

Growth rate of the small cavities:

$$\frac{\dot\rho}{\rho} = \frac{\dot f_2}{3f_2^0(1 - f_2^0)} = \frac{D_m}{f_2^0} = \frac{1}{f_2^0}\,\frac{1}{3}\left(\frac{\partial\mathcal{U}}{\partial r} + \frac{2\mathcal{U}}{r}\right). \tag{6}$$

In (6), ρ denotes the radius of a small cavity located at the distance r from the centre.

The method to solve this system can be summarized as follows. First get the stresses from (3) and (4); then observe from (3), (4) and (5) that the product (σ_{eq} · matter flux) $\propto v\, r^2 u$ is independent of r and get the velocity field; eventually use (6) to get the growth rate of small cavities at time $t = 0$ at the distance r from the centre:

$$\frac{\dot{\rho}}{\rho} = \frac{\dot{a}}{a} \frac{1}{f_2^0} \frac{1 - v(a)}{1 + v(a)} \left(\frac{1}{2} + \frac{\mu + 3}{2\sqrt{\mu^2 + 6\mu + 1}} \right),$$

where

$$\mu = \frac{1 - v(a)}{v(a)\,(1 + v(a))} \cdot \frac{r^3}{a^3}.$$

In this expression, $v(a)$ can be obtained by solving the equation

$$v^2(a) + q f_2^0\, e^{v(a)} - 1 = 0$$

which results from eqs. (3_2) and (4).

It can be shown that for small porosities, this simplifies into

$$\frac{\dot{\rho}}{\rho} = \frac{\dot{a}}{a} \left(\frac{1}{2} + \frac{\mu + 3}{2\sqrt{\mu^2 + 6\mu + 1}} \right), \quad \mu = f_2^0 \frac{r^3}{a^3}. \tag{7}$$

Expression (7) shows that the ratio $\dfrac{\dot{\rho}/\rho}{\dot{a}/a}$ is a decreasing function of r; it tends towards 2 for $r \to a$ and towards unity for $r \to \infty$. There is thus a notable influence of the large cavity on the growth rate of the small ones; this influence is maximal close to the central hole, and vanishes at large distances.

It can be shown that the property $\dfrac{\dot{\rho}/\rho}{\dot{a}/a}(r = a) = 2$ does not depend on whether $f_2(r)$ is uniform (as assumed in this section) or not; hence it remains true for $t \geq 0$. This implies that at every instant

$$\frac{\rho(a(t), t)}{\rho_0} = \left(\frac{a(t)}{a_0} \right)^2 \Rightarrow f_2(a(t), t) = f_2^0 \left(\frac{f_1(t)}{f_1^0} \right)^2. \tag{8}$$

This equation shows clearly that the porosity of the small voids located near the large one can become considerably greater than the porosity of the latter, even if it is smaller initially.

3 Evolution of the porosity for $t > 0$ - Numerical simulation of the mechanism of damage

Because of the effect just evidenced, the fracture process in ductile materials containing two populations of cavities should follow the scenario sketched in the Introduction. The aim of this section is to study stages (i) and (ii) of this scenario, for $t > 0$. The stresses will be supposed to vanish as soon as f_2 reaches the critical value for coalescence f_2^c ; this means that the loss of stress bearing capacity upon coalescence is instantaneous and complete.

Solving the equations of the problem analytically is no longer possible for two reasons: first, for $t > 0$, $f_2(r)$ is no longer uniform; second, from the onset of coalescence at time t_c onwards, the ruined matter must be progressively eliminated. We hence resort to a numerical solution in which both variables r and t are discretized. The mesh moves with time in order to take large transformation effects into account. Damage is accounted for by shifting the boundary condition $\sigma_{rr} = 0$ across the ruined zone from $r = a$ to $r = a^*$, as shown in Figure 2.

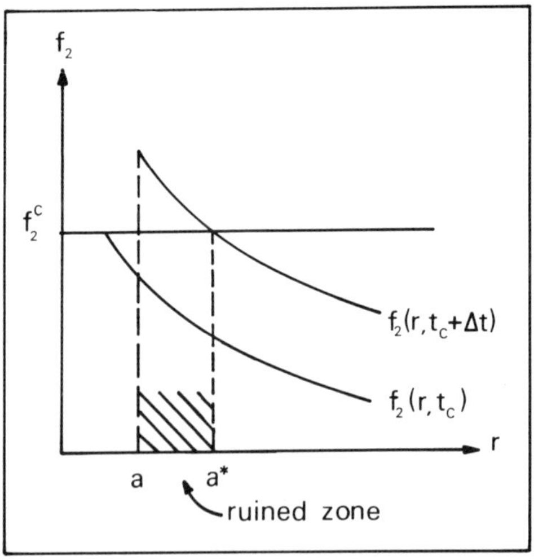

Figure 2

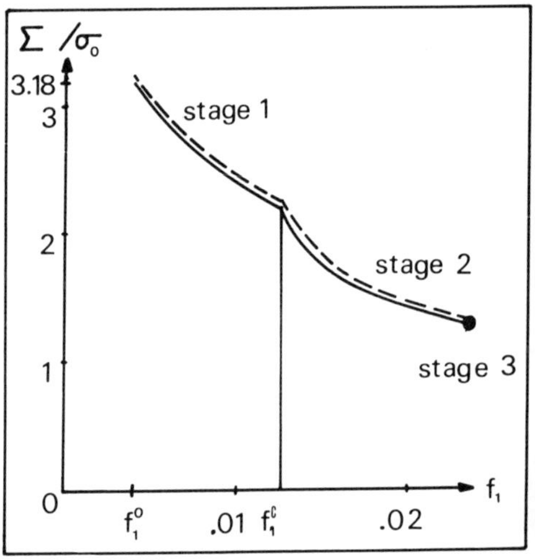

Figure 3

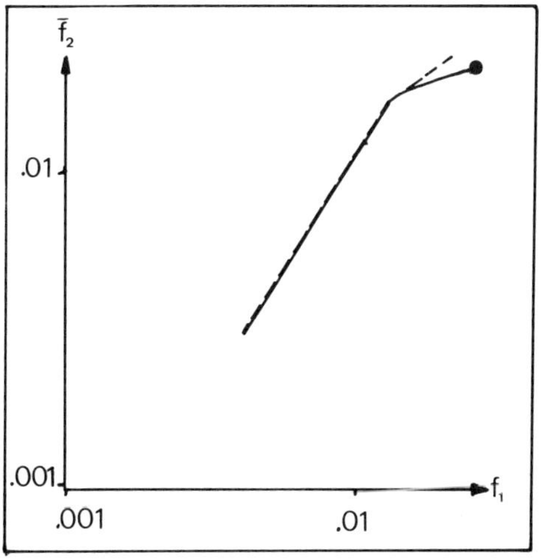

Figure 4

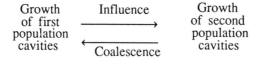

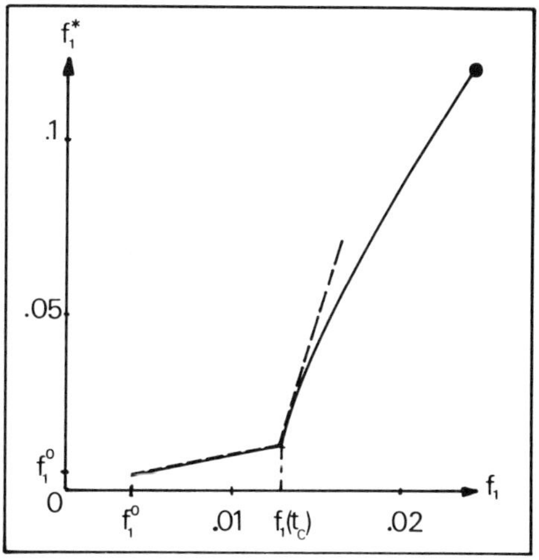

Figure 5

Figures 3, 4, 5 present the results (in full lines) of computations carried out for some reasonable values of f_1^0, f_2^0, f_2^c (see [3]): $f_1^0 = 0.004$, $f_2^0 = 0.003$ and $f_2^c = 0.03$. The evolution of the hydrostatic stress Σ (Figure 3) clearly evidences the change from stage (i) to stage (ii). Figure 4 displays the mean value $\bar{f}_2$ of the porosity of the second population over the sound part of the matrix, as a function of the porosity f_1 of the central void and ruined zone, in log-log scale. The slope is greater than unity during stage (i), because the growth rate of the small cavities is enhanced by the presence of the large one; it is smaller than unity during stage (ii) because the growth rate of the large hole is then increased by the coalescence of small cavities in its vicinity. Finally, Figure 5 shows the evolution of the external radius a^* of the ruined zone (connected with the parameter f_1^* by eq. (9) below); a^* is seen to increase very fast as soon as stage (ii) begins.

4 A two-population model

The aim of this section is to develop a macroscopic model for two-population ductile materials that should account for both interaction effects

Growth of first population cavities	Influence ⟶ ⟵ Coalescence	Growth of second population cavities

The model consists of a yield criterion and some evolution laws for the internal parameters, namely the porosities of the two populations. Since it is obviously impossible to take the whole distribution of the porosity of the small voids into account, these will be accounted for only through their *average* porosity.

Let us first define the criterion. Before the onset of coalescence (stage (i)), the yield locus is well described by the Gurson–Tvergaard criterion with porosity $f = f_1 + \bar{f}_2$, where f_1 is the porosity of the big voids and $\bar{f}_2$ the average value of that of the small voids (see [5]). After the onset of coalescence (stage (ii)), the criterion is the same as if the ruined zone were void (since $\sigma_{rr}(a^*) = 0$); it is thus well described by the Gurson–Tvergaard criterion with porosity $f = f_1^* + \bar{f}_2$, where $\bar{f}_2$ is the average porosity of the second population over the residual sound volume and f_1^* is defined by

$$f_1^* = \frac{a^{*3}}{b^3}(1 - \bar{f}_2) \tag{9}$$

(note that f_1^* is different from the real porosity f_1 of the big cavities and ruined regions, since the latter are only *partially* void). Putting $f_1^* \equiv f_1$ for $t \leq t_c$, the criterion then reads

$$\left(\frac{\Sigma_{eq}}{\sigma_0}\right)^2 + \mathrm{p}\exp\left(\frac{3}{2}\frac{\Sigma_m}{\sigma_0}\right) - 1 = 0, \text{ where } p = \frac{4}{e}\left(f_1^*(f_1) + f_2(f_1)\right). \tag{10}$$

Let us now define the evolution laws of the three internal parameters f_1, f_1^* and $\bar{f}_2$. The normality rule yields the macroscopic strain rate, the hydrostatic part of which is connected to the total porosity rate by

$$\dot{f} = \frac{d}{dt}(f_1 + \bar{f}_2) = 3\left(1 - f_1 - \bar{f}_2\right)D_m . \tag{11}$$

This equation is obviously insufficient to define the evolution of f_1, f_1^* and $\bar{f}_2$; there remains (A) to share $\dot{f}$ between $\dot{f}_1$ and $\dot{\bar{f}}_2$, and (B) to define f_1^* as a function of f_1.

A. Sharing $\dot{f}$ between $\dot{f}_1$ and $\dot{\bar{f}}_2$

The numerical calculations of Section 3 show $\log \bar{f}_2$ to be quite close to a piece-wise linear function of $\log f_1$. Thus the values of $\dfrac{\dot{\bar{f}}_2 / \bar{f}_2}{\dot{f}_1 / f_1}$ during stages (i) and (ii) are almost constant. An analytical calculation then yields

$$\frac{\dot{\bar{f}}_2 / \bar{f}_2}{\dot{f}_1 / f_1} \simeq \alpha = \left.\frac{d \log \bar{f}_2}{d \log f_1}\right|_{t=0} = \frac{1}{2}\left(1 - \gamma + \sqrt{1 + 6\gamma + \gamma^2}\right), \quad \gamma = \frac{f_1^0}{f_2^0} \quad \text{for } t < t_c$$

$$\simeq \left.\frac{d \log \bar{f}_2}{d \log f_1}\right|_{t=t_c} = \frac{2\alpha\left(f_2^c - \sqrt{f_2^0 f_2^c}\right) - f_1(t_c)\left(\dfrac{f_2^c}{\bar{f}_2(t_c)} - 1\right)}{3f_2^c - \bar{f}_2(t_c) - 2\sqrt{f_2^0 f_2^c}} \quad \text{for } t > t_c \tag{12}$$

up to first order with respect to the porosities.

B. *Defining f_1^* as a function of f_1*

f_1^* will be approximated by a piece-wise linear function of f_1 :

$$f_1^* = f_1 \text{ for } t < t_c, \quad f_1^* = f_1(t_c) + \delta \cdot (f_1 - f_1(t_c)) \text{ for } t > t_c. \tag{13}$$

An analytical calculation then yields the value of the slope δ

$$\delta = \left.\frac{d f_1^*}{d f_1}\right|_{t=t_c} = \frac{1}{3f_2^c - \bar{f}_2(t_c) - 2\sqrt{f_2^0 f_2^c}} \tag{14}$$

In eqs. (12)–(14), the critical time t_c is defined by the equation $f_2(a(t_c), t_c)) = f_2^c$, which reads by eq. (8)

$$f_2^0 \left(\frac{f_1(t_c)}{f_1^0}\right)^2 = f_2^c. \tag{15}$$

Equations (10)–(15) define the model completely. Figures 3, 4, 5 display the comparison between its predictions (dashed lines) and the results obtained numerically by the procedure sketched in Section 3 (full lines).

5. Changes brought by the use of the complete Gurson-Tvergaard criterion

The matrix is now supposed to obey the *complete* Gurson–Tvergaard criterion

$$\Phi(\sigma_{eq}, \sigma_m, f_2^0) = \left(\frac{\sigma_{eq}}{\sigma_0}\right)^2 + 2q f_2^0 \cosh\left(\frac{3}{2}\frac{\sigma_m}{\sigma_0}\right) - 1 - q^2 \left(f_2^0\right)^2 = 0, \quad q = 4/e, \tag{16}$$

instead of its simplified version (1).

An analytical calculation analogous to that of Section 2 (but much more involved) yields

$$\frac{\dot\rho}{\rho} = \frac{\dot a}{a}\left(\frac{1}{2} + \frac{\mu + 3}{2\sqrt{\mu^2 + 6\mu + 1}} - \frac{2}{e^2}\frac{a^6}{r^6}\right), \quad \mu = f_2^0 \frac{r^3}{a^3}$$

at time $t = 0$, up to the first order with respect to f_2^0. The only difference with respect to (7) is the last term. It is important only for $r \le 2a$, and its influence reduces to a shift of the maximal value of $\dfrac{\dot\rho/\rho}{\dot a/a}$ from $r = a$ to $r = \dfrac{a}{9\sqrt{e^2 f_2^0}}$.

Numerical calculations analogous to those reported in Section 3 show that during the subsequent growth of the cavities, the position of the maximum of $f_2(r)$ is almost a material point. At time $t = t_c$, the small cavities coalesce thus at this point, *right inside* the matrix. Hence a large shell of sound matter should loosen from the matrix at this instant. This prediction might explain some experimental observations of Batisse ([1], Plate IX A, Figure 1).

References

[1] Batisse, R., Contribution à la modélisation de la rupture ductile des aciers, Thesis, Université de Technologie de Compiègne, France, 1988.

[2] Gurson, A.L., Continuum theory of ductile rupture by void nucleation and growth - Part I : Yield criteria and flow rules for porous ductile media, *J. Eng. Mat. Tech.,* **99**, 2, 1977.

[3] Koplik, J., and Needleman, A., Void growth and coalescence in porous plastic

solids, *Int. J. Solids Structures,* **24**, 835, 1988.

[4] Marini, B., Mudry, F., and Pineau, A., Experimental study of cavity growth in ductile rupture, *Eng. Fract. Mech.*, **22**, 989, 1985.

[5] Perrin, G., and Leblond, J.B., Analytical study of a hollow sphere made of porous plastic material and subjected to hydrostatic tension - Application to some problems in ductile fracture of metals, to be published in *Int. J. Plast.*

[6] Tvergaard, V., Influence of voids on shear band instabilities under plane strain conditions, *Int. J. Fract.,* **17**, 389, 1981.

Leblond J.-B., and Perrin G.
Laboratoire de Modélisation en Mécanique
Université Pierre et Marie Curie
Tour 66, 4 place Jussieu
F-75252 Paris Cédex 05
FRANCE

L. LIMAT

Fractal models of micropolar elastic percolation

1 Introduction

This work deals with the mechanical properties of a two-dimensional lattice, in which sites A_i are rigid bodies connected by elastic ligaments of random rigidity. When these ligaments are symmetrical upon both the bond axes and the bond mediatives, the general expression of elastic energy consistent with rotational invariance is given by [14]:

$$2U = \sum_{<ij>} \left\{ k_{||}^{ij} (\mathbf{u}_{ij})_{||}^2 + k_{\perp}^{ij} [(\mathbf{u}_{ij})_{\perp} - (1/2)(\theta_i + \theta_j)(\hat{z} \times \mathbf{R}_{ij})]^2 + k_{\theta\theta}^{ij} (\theta_{ij})^2 \right\} \quad (1)$$

In this equation, $\mathbf{u}_{ij} = \mathbf{u}_i - \mathbf{u}_j$ and $\theta_{ij} = \theta_i - \theta_j$ designate, respectively, the relative displacement and the relative rotation of two neighbouring sites A_i and A_j, of initial separation $\mathbf{R}_{ij} = \mathbf{R}_i - \mathbf{R}_j$. $\hat{z}$ is the unit vector normal to the plane, and the k^{ij} are three bond rigidities with ratios which depend on the particular shape of the ligaments. This model has recently been applied to the description of granular media [12, 20], but U can also be understood as the elastic energy of a lattice of elastic beams rigidly joined at nodes [1, 14, 18]. In particular, the dynamic properties of non-random lattices of this kind have been studied earlier, by Askar and Cakmak [4] with a less general form of equation (1). They showed that this "oriented" lattice was described at the continuum level by the micropolar theory of elasticity [11]. In this non-symmetrical theory, the energy density involves four generalized Lamé coefficients

$$2\mathrm{d}U/\mathrm{d}S = \lambda u_{\alpha\alpha} u_{\beta\beta} + 2\mu u_{\alpha\beta} u_{\alpha\beta} + \mu' [(\nabla \times \mathbf{u})_z - \theta]^2 + \alpha(\nabla\theta)^2 \quad (2)$$

with $u_{\alpha\beta} = (1/2)(\partial_\alpha u_\beta + \partial_\beta u_\alpha)$.

In this work, we will be interested in percolation-type lattices [7], and more precisely in the two limiting cases:

- "dilute problem": a fraction $1-p$ of the ligaments are suppressed,
- "reinforced problem": a fraction p of the ligaments are rigid.

The studied question is as follows: is it possible to define two equivalent electrical problems, in which the rigid domains are treated as perfect conductors, and the elastic ones are resistive. It is known that in the vicinity of the "percolation threshold" p_c, the average conductivity of the system exhibits a critical behaviour [16]: $\sigma \sim (p-p_c)^t$ in the dilute case, and $\sigma \sim (p_c-p)^{-s}$ in the reinforced one ($t = s = 1.30$ in two dimensions). Is there any relation between this behaviour and that of the elastic moduli in our problem? Such a study is motivated by the possible applicability of percolation theory to random mechanical systems [9] such as: the sol and gel phases observed in polymerization processes [2], the aerogels [22], and some kinds of sintered materials [10].

In fact, the available approaches of this question are essentially numerical [3, 5, 12], or based on qualitative descriptions of chain elasticity [13, 18]. We summarize here some results obtained with a different approach [14, 15] based on the conception of fractal models. It is known that below a particular length scale, namely the correlation length, that diverges as $\xi \sim |p-p_c|^{-\nu}$ ($\nu = 4/3$ in two dimensions), the system exhibits a self-similar geometry [7]. This one can be roughly modelled by fractal Koch curves [8, 17] such as that suggested in Figure 1. We have developed a method that allows exact calculations of elasticity on such objects, and that we are now going to present.

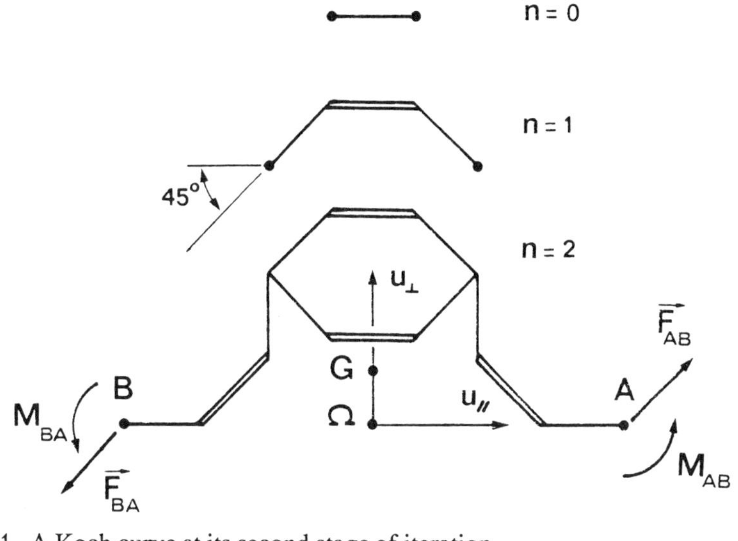

Figure 1. A Koch curve at its second stage of iteration.

2 Elasticity of a fractal curve

As suggested in Figure 1, a fractal curve is not necessarily symmetrical upon the AB axis. Moreover, the symmetry upon the AB mediative can also disappear in the case of a completely asymmetrical motive. In this case, the three rigidities involved in (1) must be completed by non-diagonal coupling terms

$$2U_{AB} = \mathbf{v}_{AB} \cdot \mathcal{K} \cdot \mathbf{v}_{AB} + 2\theta_{AB}\,\mathbf{K}_\theta \cdot \mathbf{v}_{AB} + K_{\theta\theta}\,(\theta_{AB})^2 \tag{3}$$

with: $\mathbf{v}_{AB} = \mathbf{u}_{AB} - (1/2)\,(\theta_A + \theta_B)\,(\hat{z} \times \mathbf{R}_{AB})$.

This expression, summarized under the form $2U_{AB} = {}^3\mathbf{v}_{AB} \cdot {}^3\mathcal{K}{:}{}^3\mathbf{v}_{AB}$, is the equivalent of a dissipation law in a resistor: $2U_{AB} = G_{AB}(V_{AB})^2$. An equivalent of Ohm's law can be obtained with the stress decomposition suggested in Figure 2

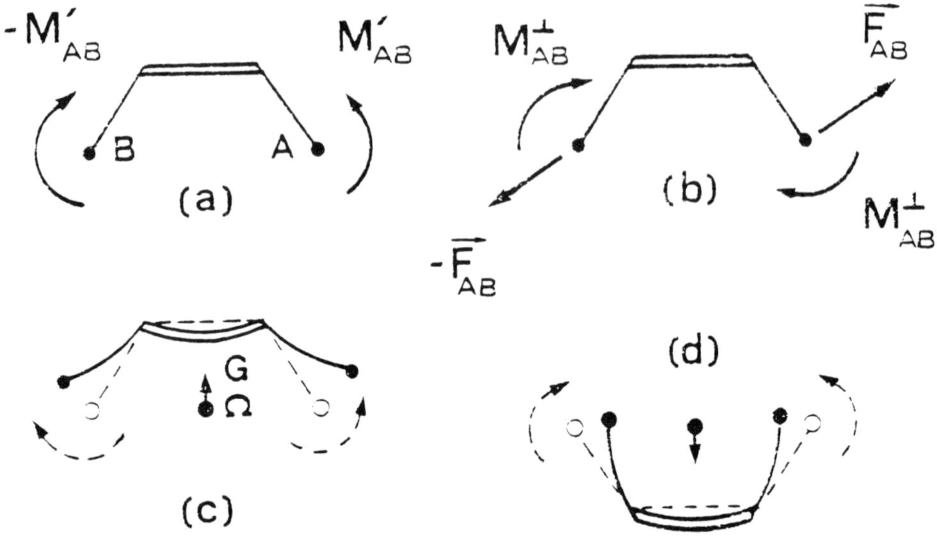

Figure 2. The external stresses can be viewed as the superposition of a bending (a) and of a symmetrical stretching (b). (c), (d): The eccentricity ΩG couples these two modes.

$$\begin{bmatrix} \mathbf{v}_{AB} \\ \theta_{AB} \end{bmatrix} = \begin{bmatrix} \mathcal{S} & \mathbf{S}_\theta \\ \mathbf{S}_\theta & S_{\theta\theta} \end{bmatrix} \begin{bmatrix} F_{AB} \\ M'_{AB} \end{bmatrix} . \tag{4}$$

Here the ${}^3\mathcal{S}$ compliance matrix is just obtained by inverting the 3×3 stiffness matrix ${}^3\mathcal{K}$. A complete set of formulae has been developed that allows one to solve any series or parallel association of elastic elements. For simplicity, we give here only the result for a series association of N identical and symmetrical bonds, i.e., for a simple chain [14]

$$S_{\theta\theta} = N s_{\theta\theta}, \qquad \mathbf{S}_\theta = -N s_{\theta\theta} \hat{z} \times \langle \mathbf{r} \rangle, \tag{5}$$

$$\mathcal{S} = N\langle s \rangle + N s_{\theta\theta} \langle (\hat{z} \times \mathbf{r}) \otimes (\hat{z} \times \mathbf{r}) \rangle ,$$

where the average is performed over the N links, $\mathbf{r}$ being their position relative to the point Ω defined in Figure 1. These equations are to be compared with that giving the electrical resistance: $R = Nr'$, r' being the resistance of a link. We observe that, in the $S_{\theta\theta}$ equation, the rotational degrees of freedom seem to play the role of the electrical potential. In the case of a fractal curve, $S_{\theta\theta}$ and R will thus behave in a rather similar way: $R \sim L^{\zeta_0}$ and $S_{\theta\theta} \sim L^{\zeta}$, with $\zeta \cong \zeta_0$. Note, however, that the equality is rigorous only for an unlooped curve, and in this case $\zeta = \zeta_0 = d_F$, d_F being the fractal dimension of the chain. The behaviour of the other rigidities is modified by an r-dependence: $S_\theta \sim L^{\zeta+1}$ and $S \sim L^{\xi+2}$. This dependence comes from the non-conservative nature of the bending moments when a force is applied to the chain. The critical behaviour of elastic moduli in the dilute problem can be deduced from these scaling laws by using the "node-link-blob" description [7]: the percolating lattice is treated as a superlattice in which macrobonds are fractal structures similar to that of Figure 1, and for which $L = \xi$. This finally gives

$$\lambda \sim \mu \sim (p - p_c)^f \text{ and } \alpha \sim (p - p_c)^{t'} \tag{6}$$

with $f = t' + 2\nu$ and $t' \cong t$. We obtain that the rotational modulus α seems to behave as the conductivity, but not the other moduli where the exponent is larger. It is, however, noteworthy that the equality $t' = t$ would be correct only for unlooped macrolinks. Thus we must now improve this description by taking the loops into account.

3 Loops and eccentricity

In eqs. (4), the physical meaning of the coupling term S_θ now becomes more clear: as suggested in Figure 2 the vector $\mathbf{\Omega G}$, namely the "eccentricity" of a chain, introduces a coupling effect between displacements and rotations [14]. In the previous section, $S_{\theta\theta}$ was found to be additive in series. It is clear that in a parallel association, the eccentricity effect should tend to make the loops more rigid, and thus should modify the $S_{\theta\theta}$ parallel association rule. This effect can be illustrated by the example of Figure 1. In this case, an exact calculation gives the following recursion relationships between two length scales L and $L' = L(1 + \sqrt{2})$

$$\begin{aligned} S'_{\theta\theta} &= (5/2)S_{\theta\theta} - (S_{\theta\|})^2/2S_\|, \\ S'_{\theta\|} &= -\sqrt{2}\, S_{\theta\|} + (3/2\sqrt{2})\, LS_{\theta\theta} - (L/2\sqrt{2})\, (S^2_{\theta\|}/S_\|), \\ S'_\| &= (3/2)S_\| + S_\perp - LS_{\theta\|} + L^2 S_{\theta\theta}/2 - S^2_{\theta\|}[(1/2S_{\theta\theta}) - L^2/4S_\|], \\ S'_\perp &= S_\| + (3/2)S_\perp - (1 + \sqrt{2})LS_{\theta\|} + (3/4 + 1/\sqrt{2})L^2 S_{\theta\theta}\,. \end{aligned} \tag{7}$$

These equations, especially the first one, are to be compared with that of the electrical resistance: $R' = (5/2)R$. This simpler recursion relationship leads to the scaling law: $R \sim L^{\zeta_0}$ with $\zeta_0 = \text{Ln}(5/2)/\text{Ln}(1+2\sqrt{2}) \cong 1.04$. When one iterates eqs. (7), the system tends toward a fixed point associated to particular values of the ratios $S_{\perp}/S_{\|}$, $S_{\|}/L^2 S_{\theta\theta}$, $S_{\theta\|}/L S_{\theta\theta}$. Asymptotically, this gives $S_{\theta\theta} \sim L^{\zeta}$, $S_{\theta\|} \sim L^{\zeta+1}$ and $S_{\|} \sim S_{\perp} \sim L^{\zeta+2}$ where $\zeta \cong 0.903$ is appreciably smaller than ζ_0. Clearly, the eccentricity effect will modify the critical behaviour and should introduce a correction in the t' exponent

$$t' = t - \Delta \quad \text{and} \quad f = t + 2\nu - \Delta \, . \tag{8}$$

Various calculations on different fractals [14,15], lead to the estimates $\Delta \cong 0.2$, $t' \cong 1.1$ and $f \cong 3.8$, to be compared with $t = 1.30$ and $t + 2\nu = 3.96$ [16].

4 The reinforced problem

The reinforced problem can be treated in a very similar way [15]. We have suggested in Figure 3(a) a possible model of the dilute problem, and in Figure 3(b) the "dual" object defined as follows: each elastic bond is replaced by a perpendicular elastic bond, and each empty bond is replaced by a perpendicular rigid bond. This new structure can be considered as a model of the "reinforced problem".

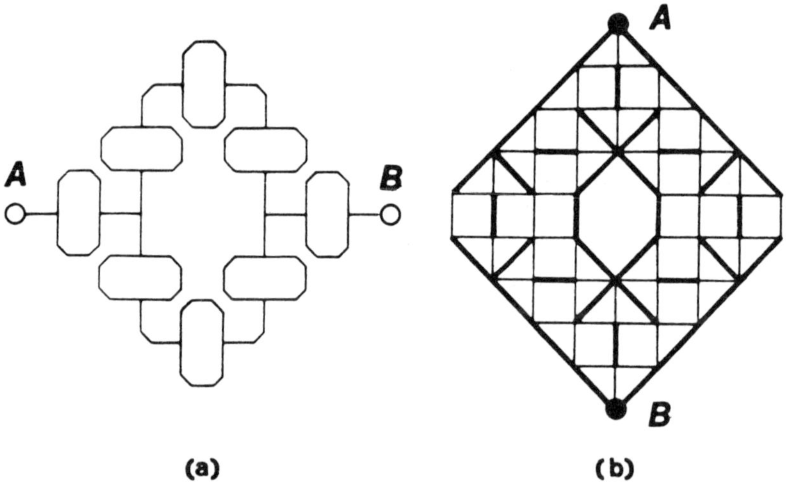

Figure 3. Another possible representation of the dilute problem (a), and (b) the dual representation of the reinforced problem. The oblique junctions avoid overlaps without modifying the results.

The developed method also allows the exact calculation of structures of this kind. For simplicity, we give only a brief summary of the obtained results. In the reinforced problem, the translational degrees of freedom become the equivalent of the electrical potential, and it is now the bending rigidity $K_{\theta\theta}$ that becomes r-dependent. This can roughly be understood as follows: in a translational displacement of the upper rigid boundary A of Figure 3(b), each bridge connecting this boundary to the lower one B is submitted to the same strain. This situation is very similar to that encountered in the electrical problem, in which each bridge is submitted to the same voltage drop. On the contrary, a rotation of A will induce a non-uniform strain in the bridge, similar to the non-uniform bending moment obtained in a stressed chain. We finally obtain: $\mathcal{K} \sim L^{\zeta'}$, $K_\theta \sim L^{\zeta'+1}$ and $K_{\theta\theta} \sim L^{\zeta'+2}$, where ζ' is a new exponent. In the percolation problem this gives:

$$\lambda \sim \mu \sim \mu' \sim (p_c - p)^{-s'} \quad \text{and} \quad \alpha \sim (p_c - p)^{-f'} \tag{9}$$

with $f' = s' + 2\nu$ and $s' = s - \Delta'$. $\cdot\Delta'$ is a new correction that comes from the rotation of the rigid subislands in transverse displacements of two rigid boundaries. This correction can be rigorously related to Δ by the inequality $\Delta' < \Delta$, and thus we have $t' < s' < t = s$ in two dimensions. The calculations performed on various dual Koch curves [15] lead to the estimate $s' \cong 1.2$, that clearly satisfies the above inequalities.

5 Critical Poisson ratios

Besides the critical exponents, other quantities of interest [5, 6] are the values reached by the ratio K/μ of bulk to shear moduli when $p \to p_c$. This quantity can be related to the ratios $S_\perp/S_\parallel$ and $K_\parallel/K_\perp$ of the studied curves [15–16]. These fractals lead to Poisson ratios, respectively, of order $\eta^+ \cong 0.4$ and $\eta^- \cong 0.05$ in the dilute and reinforced cases. The behaviour of η^- can be qualitatively related to that of the anisotropy $(\sigma_{xx}/\sigma_{yy} - 1)$ in the anisotropic conductivity problem. This explains the rather low value obtained.

6 Conclusion

In summary, the elasticity behaviour can be qualitatively related to that of the conductivity. However the vectorial nature of the problem, in the case of a rotationally invariant system, prevents quantitative relationships. New geometrical parameters are involved, as for instance the "eccentricity", that may lead to significant corrections. The comparison of the obtained results with the numerical treatment of the granular model [12, 20] is rather satisfactory, especially for the correction Δ'. A correction of this kind has also been found in the case of the central-force model [19], but this problem is not necessarily related with the present one. We must also mention that the most precise simulations [3, 5, 21] have been performed with a third model [13] based on Kirkwood's molecular Hamiltonian (angular bond-to-bond elasticity). These computations give values of the Poisson ratio in complete agreement with those of Section 5, but seem to imply that $\Delta \cong \Delta' \cong 0$. We believe - but this is not proved - that this result is due to the occurrence of very strong corrections to scaling involved by the vectorial nature of the problem. More precisely, eqs. (7) constitute a four-parameter recursion process. The correct solution is in fact a sum of various power laws, involving combinations of four different exponents. If this suggestion is correct, the fractal approach developed above was the only available possibility to point out Δ and Δ'. We have here a particular problem in which the concept of fractal is not only a descriptive tool, but is also a predictive one, that may give important and new indications.

Acknowledgements

This work has benefitted from discussions with C. Allain, M. Cloitre, J.C. Charmet and G.A. Maugin.

References

[1] Allain, C., Charmet, J.C., Clément, M. and Limat, L., *Phys. Rev.* **B32**, 7552 (1985).
[2] Allain, C. and Salomé, L., *Macromol.* **20**, 2597 (1987).
[3] Arbabi, S. and Sahimi, M., *Phys. Rev.* **B38**, 7173 (1988).
[4] Askar, A. and Càkmak, A.S., *Int. J. Eng. Sci.* **6**, 583 (1968).
[5] Bergman, D.J., *Phys. Rev.* **B33**, 2073 (1986).
[6] Bergman, D.J. and Duering, E., *Phys. Rev.* **B34**, 8199 (1986).
[7] Coniglio, A., *J. Phys.A* **15**, 3829 (1985).
[8] De Arcangelis, L., Redner, S., Coniglio, A., *Phys. Rev.* **B31**, 4725 (1985).
[9] De Gennes, P.G., *J. Phys. (Paris)* **45**, 1939 (1984).
[10] Deptuck, D., Harrison, J. and Zawadzki, P., *Phys.Rev.Lett.* **54**, 913 (1985).
[11] Eringen, A.C., in "Fracture" edited by H. Liebowitz (Academic, New York) Vol. II, 621. See also Nowacki, W., "Theory of Micropolar Elasticity", (Springer-Verlag, Udine 1972).
[12] Feng, S., *Phys. Rev.* **B32**, 510 (1985).
[13] Kantor, Y. and Webman, I., *Phys.Rev.Lett.* **52**, 1891 (1984).
[14] Limat, L., *Phys.Rev.* **B37**, 672 (1988); **38**, 512 (1988).
[15] Limat, L., *Phys.Rev.* **B38**, 7219 (1988); **40**, 3253 (1989).
[16] Lobb, C.J. and Frank, D.J., *Phys.Rev.* **B30**, 4090 (1984). See also Hermann, H.J., Derrida, B. and Vannimenus, J., p. 4079 and Zabolitzki, J.G., p. 4076.
[17] Mandelbrot, B.B. and Given, J.G., *Phys.Rev.Lett.* **52**, 1853 (1984).
[18] Roux, S., *J.Phys.* **A19**, L351 (1986).
[19] Sahimi, M. and Goddart, J., *Phys.Rev.* **B32**, 1869 (1985).
[20] Schwartz, L.M., Feng, S., Thrope, M.F. and Sen, P.N., *Phys.Rev.* **B32**, 4607

(1985).
[21] Zabolitzki, J.G., Bergman, D.J. and Stauffer, D., *J.Stat. Phys.* **44**, 211 (1986).
[22] Recent data can be found in *Rev.Phys.Appl.Coll.* C4, T.24 (1989).

Limat L.
L.P.M.M.H., URS CNR5 857
L.P.C.T., URA CNRS 1382
Ecole Supérieure de Physique et
de Chimie Industrielles
10 rue Vauquelin
F-75231 Paris Cédex 05
FRANCE

A. LITEWKA

Continuum model for creep damage of metals

1 Introduction

Existing phenomenological models of the creep damage and creep rupture of solids are generally the extrapolation of the classical rheology. However, because of a complicated character of a tertiary creep, this approach leads to semi-empirical formulae which can be utilized after determining the numerous constants. The satisfactory agreement of theoretical predictions and experimental data is then obtained only for some selected materials and usually this is possible due to a carefully performed curve-fitting procedure. Such theories, usually based on a limited number of experimental data resulted in the opinion that for an aluminium alloy, creep rupture is governed by the second invariant of the stress deviator, whereas for copper the maximum principlal stress is most significant [2, 4, 5, 15].

The aim of this study is to formulate the creep rupture criterion based on the assumption that the tertiary creep and creep rupture are the result of the stiffness and strength reduction of the material due to crack and void growth [3, 9]. The current state of the deteriorated material microstructure is described in the theory proposed by the second rank damage tensor, and the appropriate constitutive equations are derived by employing the tensor function representations. The creep rupture criterion proposed consists of the damage evolution equation and the failure criterion for material with oriented deteriorated internal structure. To illustrate the applicability of this theory, the multi-axial creep rupture experimental data for various metals obtained by Johnson, Henderson and Mathur [4, 5], Finnie and Abo el Ata [1] and Murakami and Sanomura [14] were used.

2 Creep damage constitutive equations

Creep rupture description proposed in this work is based on the concept of the representation of mechanical behaviour of the material in the presence of changing internal structure as presented by Onat and Leckie [15]. In particular, it was assumed that the creep rupture occurs when the failure criterion for material with deteriorated microstructuve is satisfied. This criterion can be represented in the form of a scalar-valued function

$$f(\boldsymbol{\sigma}, \mathbf{S}_n, T) = 0\,, \tag{1}$$

where $\boldsymbol{\sigma}$ is a stress tensor, T is the temperature and $\mathbf{S}_n$ is a set of n time-dependent tensorial variables accounting for the current state of a damaged material. The number and character of the variables $\mathbf{S}_n$ depends on the geometry and symmetries of the deteriorated material structure. The changing internal structure is described by the evolution equation

$$\dot{\mathbf{S}}_n = F(\sigma, \mathbf{S}_n, T), \tag{2}$$

where $\dot{\mathbf{S}}_n$ stands for the time derivative of an appropriate damage variable.

If we confine our considerations to perfectly elastic-plastic materials and to isothermal processes the creep rupture can be easily explained by means of a graphical interpretation shown in Figure 1. For perfectly elastic-plastic material the failure criterion (1) corresponds to yield loci represented in Figure 1. The actual loading applied to the solid, described by the stress tensor σ_{ij}, does not exceed the strength of

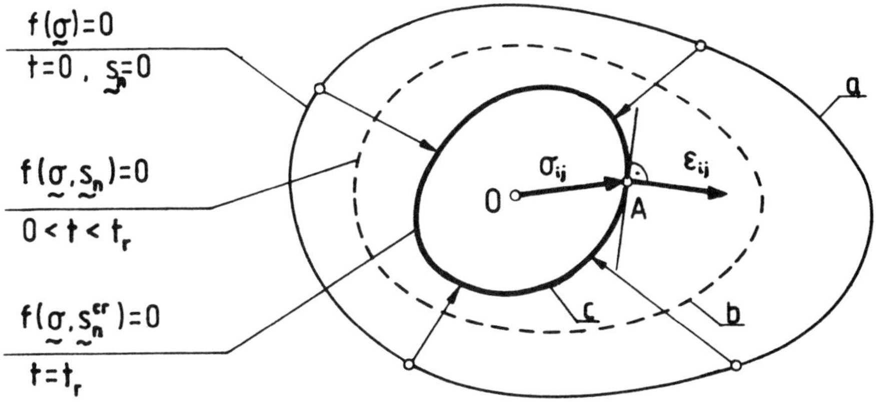

Figure 1. Yield surfaces for damaged solid.

an undamaged material. It is seen from Figure 1 where point A is inside of the yield surface for the undamaged material (curve a). However, from the beginning of the creep process the microstructural damage and the reduction of the overall material strength occurs. This growth of the damage is described by the damage evolution equation (2) and the relevant strength reduction is expressed by the yield criterion (1) represented in Figure 1, for time $0 \leq t < t_r$ by the curve b. In a limit case when the yield surface is reduced to such a size that it touches the point A (curve c) the failure criterion for damaged material is satisfied and creep rupture occurs. The amount of microstructural defects at rupture $\mathbf{S}_n^{cr}$ is referred to as a critical damage [6, 7, 16]. Creep rupture is accompanied by the increasing creep strain rate and this is associated with the plastic flow shown in Figure 1 as a vector $\dot{\varepsilon}_{ij}^{p}$. However, the problem of strains in creep will not be discussed here and we shall confine our attention to the analysis of the creep rupture criterion only.

To obtain the effective and accurate description of the creep rupture process it is necessary to select the appropriate damage variables $\mathbf{S}_n$ and to find the explicit form of the eqs. (1) and (2). This was done by the author in his previous papers [7, 9, 10, 12] and here only the final form of the equations will be shown.

The damage evolution equation (2) as proposed in [10] has a form

$$\dot{\mathbf{\Omega}} = C\Phi_e^2\, \boldsymbol{\sigma}^* , \tag{3}$$

where C is a temeprature dependent material constant, $\dot{\mathbf{\Omega}}$ is the time derivative of the damage tensor $\mathbf{\Omega}$ defined by Vakulenko and Kachanov [17] and Murakami and Ohno [13], and $\boldsymbol{\sigma}^*$ is a modified stress tensor expressed in terms of the principal values of the stress tensor $\boldsymbol{\sigma}$. It means that in the tensor $\boldsymbol{\sigma}^*$ the compressive principal stress components of $\boldsymbol{\sigma}$ are replaced by zeros, whereas tensile principal stress components are left unchanged. The scalar multiplier Φ_e is an elastic strain energy density of the damaged solid expressed [9, 10] in the form

$$\Phi_e = \frac{1-2\nu}{6E}\,\mathrm{tr}^2\,\boldsymbol{\sigma} + \frac{1+\nu}{2E}\,\mathrm{tr}\,\mathbf{s}^2 + \frac{D_1}{2(1+D_1)E}\,\mathrm{tr}\,\boldsymbol{\sigma}^2\,\mathbf{D} , \tag{4}$$

where E and ν are Young's modulus and Poisson's ratio of the undamaged material at the actual temeprature and $\mathbf{s}$ is the stress deviator. The damage effect tensor $\mathbf{D}$ described in [8,9] has principal values related to those of the damage tensor $\mathbf{\Omega}$ through

$$D_i = \Omega_i\,/\,(1-\Omega_i),\quad i = 1,2,3 . \tag{5}$$

The failure criterion for damaged solids (1) was proposed as a scalar funciton [9, 10]

$$C_1\,\mathrm{tr}^2\,\boldsymbol{\sigma} + C_2\,\mathrm{tr}\,\mathbf{s}^2 + C_3\,\mathrm{tr}\,\boldsymbol{\sigma}^2 D - \sigma_u^2 = 0 , \tag{6}$$

where C_1, C_2 and C_3 are temeprature and damage dependent material constants and σ_u is the temperature dependent ultimate strength of the undamaged material. According to [7, 9] the constants C_1, C_2, C_3 can be calculated from the set of equations

$$\begin{aligned}
&(1-\Omega_1)^2 C_1 + \frac{2}{3}(1-\Omega_1)^2 C_2 + (1-\Omega_1)\,\Omega_1\,C_3 = 1,\\
&(1-\Omega_2^2\, C_1 + \frac{2}{3}(1-\Omega_2)^2 C_2 + (1-\Omega_2)\,\Omega_2\,C_3 = 1,\\
&4(1-\Omega_1)^2 C_1 + \frac{2}{3}(1-\Omega_1)^2 C_2 + (1-\Omega_1)\,\Omega_1\,C_3 = 1 .
\end{aligned} \tag{7}$$

The equations (3), (6) and (7) represent the creep rupture criterion which enables one to calculate the time to rupture for an arbitrary multiaxial state of stress. To this end only standard material constants E, ν and σ_u for undamaged material along with the only constant C responsible for the damage evolution are needed.

Creep rupture criterion

The constitutive equations for creep rupture (3) and (6) can be readily used in numerical calculations. The details of such calculations including the integration of the damage evolution equation can be found elsewhere [7, 10, 11, 12] that is why only the final numerical results concerning a copper at 523 K and aluminium alloy at 473 K are shown here. These two materials were chosen to show that, contrary to existing theories [2, 4, 5, 15], creep rupture for these metals is not governed by simple criteria like that of maximum principal stress or Huber–Mises equivalent stress. Rather a combination of them should be taken into account and that is why both of these factors are included in the creep rupture criterion proposed in this work. The influence of the maximum principal stress σ_1 is accounted for in eq. (3) by means of the modified stress tensor $\boldsymbol{\sigma}^*$ and Huber–Mises equivalent stress is included in tr $\mathbf{s}^2$ in (4) and (6). As is seen from (4) and (6) the effect of the first stress invariant is also accounted for in the creep rupture criterion proposed.

The dependence of the rupture time on the stress level for aluminium alloy and copper is shown for biaxial states of stress in Figures 2 and 3. The theoretical curves for various cases of biaxiality expressed in terms of the ratio of the principal stresses $m = \sigma_2/\sigma_1$ were opbtained from eqs. (3), (4), (6) and (7). To this end the following material constants were used

Material	Temperature K	σ_u MPa	ν	$k = C/4E^2$ $\text{MPa}^{-5} \cdot \text{h}^{-1}$
Copper	523	145	0.35	$1.90 \cdot 10^{-12}$
Alluminium alloy	473	145	0.3	$1.11 \cdot 10^{-14}$

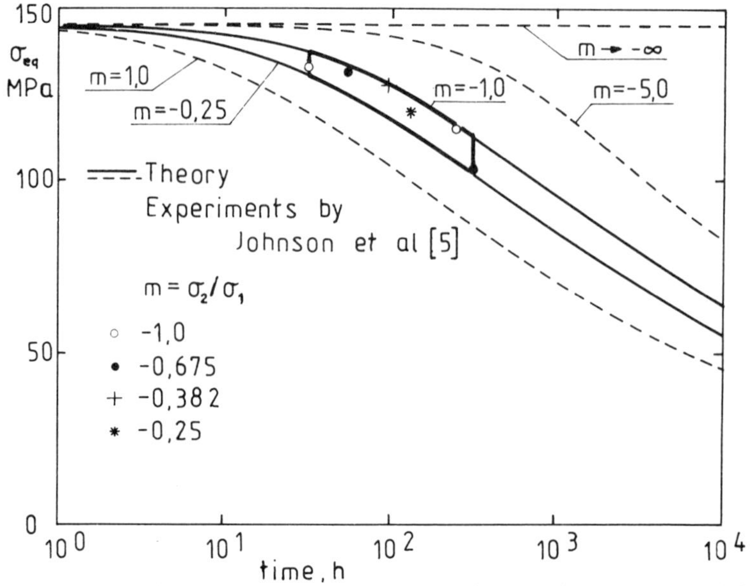

Figure 2. Equivalent stress versus rupture time for aluminium alloy at 473 K.

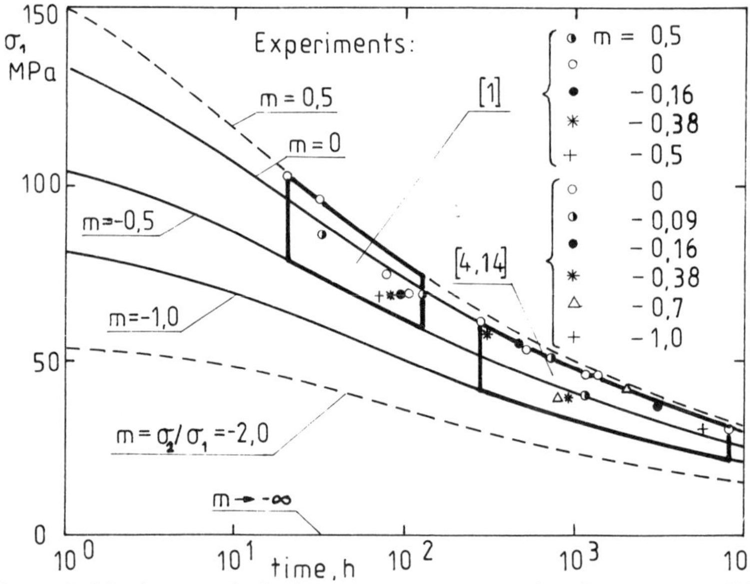

Figure 3. Maximum principal stress versus rupture time for copper at 523 K.

It is seen from Figure 2 that in experiments for aluminium alloy by Johnson et al, [5] the number of specimens tested was too small and the range of biaxiality $(-1.0 < m < -0.25)$ was too narrow to assess if the equivalent stress is most significant. Analysis of the data presented in Figure 3 reveals that the maximum principal stress could hardly be considered as most significant when the creep rupture of copper is considered. Experimental points obtained by Johnson et al. [4] and

Murakami and Sanomura [14] are located in the region of the graphs in Figure 3 where the criterion of maximum principal stress could be assumed only as rough approximation of the material behaviour. Finnie and Abo el Ata [1] carried out similar experiments for copper at the same temperature of 523 K but for shorter time to rupture and different range of biaxiality. The diagrams shown in Figure 3 explain why Finnie and Abo el Ata [1] arrived at the conclusion that rupture life of copper depends on both effective stress and maximum principal stress.

4 Conclusions

A stress level dependent creep rupture criterion was used to explain which factors are significant in the theoretical description of the creep damage and life time of solids. It was found that the creep rupture of metals is governed by the combination of simple failure criteria and that is why the combined effect of maximum principal stress, hydrostatic stress as well as effective stress should be taken into account. The constitutive equations proposed in this work were used to show that this rule also applies to such metals like copper and aluminium alloy where usually simple failure criteria were assumed as a sufficient approximation of the material behaviour observed in experiments.

Acknowledgements

This work was finacially supported by the grant C.P.B.P. 02.01, t. 1. 10.

References

[1] Finnie, I. and Abo el Ata, M.M., Creep and creep rupture of copper tubes under multiaxial stress, in: *Advances in Creep Design*, Eds. : A.J. Smith, A.M. Nicolson, John Wiley, New York, 1971, pp. 329-352.

[2] Hayhurst, D.R., Creep rupture under multiaxial states of stress, *J. Mech. Phys. Solids*, **20**, pp. 381-390, 1972.

[3] Hult, J., Stiffness and strength of damaged material, *Z. Angew. Math. Mech.*, **68**, pp. T31-T39, 1988.

[4] Johnson, A.E., Henderson, J. and Mathur, V.D., Combined stress fracture of commerical copper at 250 deg. cent., *The Engineer*, **24**, pp. 261-265, 1956.

[5] Johnson, A.E., Henderson, J. and Mathur, V.D., Complex stress creep fracture of an aluminium alloy, *Aircraft Eng.*, June, pp. 2-11, 1960.

[6] Lemaitre, J., How to use damage mechanics, *Nucl. Eng. Design*, **80**, pp.233-245, 1984.

[7] Litewka, A. and Hult, J., One parameter CDM model for creep rupture prediction, *European J. Mech.*, A/Solids, **8**, 1990

[8] Litewka, A., Effective material constants for orthotropically damaged elastic

solid, *Arch. Mech.*, **37**, pp. 631-642, 1985.

[9] Litewka, A., On stiffness and strength reduciton of solids due to crack development, *Eng. Fract. Mech.*, **25**, pp. 637-643, 1986.

[10] Litewka, A., Analytical and experimental study of fracture of damaging solids, in *Yielding, Damage and Failure of Anisotropic Solids*, IUTAM/ICM Symp. 1987, Ed. J.P. Boehler, Mechanical Energinnering Pulbications, London, 1989, pp. 653-663.

[11] Litewka, A., *Creep Rupture criterion for Metals*, IIIrd Symp. Creep Problems of Metals, Białystok 1989, pp. 295-302.

[12] Litewka, A., Creep rupture of metals under multi-axial state of stress, *Arch. Mech.*, **41**, 1989, pp. 3-23.

[13] Murakami, S. and Ohno, N., A continuum theory of creep and creep damage, in: *Creep in Structures*, Eds. : A.R.S. Ponter, D.R. Hayhurst, Springer-Verlag, Berlin, 1981, pp. 422-444.

[14] Murakami, S. and Sanomura, Y., Creep and creep damage of copper under multiaxial state of stress, in: *Plasticity Today*, Eds. : A. Sawczuk, G. Bianchi, Elsevier, London, 1985, pp. 535-551.

[15] Onat, E.T. and Leckie, F.A., Representation of mechanical behavior in the presence of changing internal structure, *J. Appl. Mech.*, **55**, pp. 1-10, 1988.

[16] Piatti, G., Bernasconi, G. and Cozarelli, F.A., *Damage Equations for Creep Rupture in Steels*, Trans. 5th Int. Conf. SMiRT, Berlin 1979, North-Holland, Amsterdam, 1979, vol. L, L11/4, pp. 1-9.

[17] Vakulenko, A.A. and Kachanov, M.L., Continuum theory of medium with cracks, *Izv. Akad. Nauk S.S.S.R., M.T.T.,,* **4**, pp. 159-166, 1971 (in Russian).

Litewka A.
Technical University of Poznań
Institute of Building Structures
ul. Piotrowo 5
PL-60-965 Poznań
POLAND

K.Z. MARKOV

Analysis of random particulate media via factorial functional expansions

1 Introduction

The aim of this work is to demonstrate how the so-called factorial functional expansions can be applied to the solution of a wide class of transport problems in random particulate media. The solution is understood hereafter in a statistical sense, i.e., one should evaluate all multipoint correlation functions for the needed random fields (temperature, defect concentration, displacement, etc.) making use of the given statistical description of the medium. In particular one would be thus able to interrelate rigorously the effective properties and the microstructural description. For simplicity we deal with a rigid random dispersion of non-overlapping spheres of radius a. We first recall (Section 2) the definition and the basic virial property of the factorial series [7]. In Section 3 the procedure of identification for the kernels is briefly discussed. In Section 4 we consider several examples (heat propagation, steady-state diffusion and an elastostatic counterpart of the sedimentation problem), and find the needed random fields to the order c^2, where c is the volume fraction of the spheres.

2 Factorial functional series

Let $\mathbf{x}_j$ be the random set of sphere's centres. After [9] we introduce the random density function $\psi(\mathbf{x}) = \sum_j \delta(\mathbf{x}-\mathbf{x}_j)$. The field $\psi(\mathbf{x})$ is uniquely defined by the set $\mathbf{x}_j$. In particular, the multipoint moments of $\psi(\mathbf{x})$ can be expressed by means of the probability densities f_k that define the set $\mathbf{x}_j$ [9]

$$\langle\psi(\mathbf{y})\rangle = f_1(\mathbf{y}),$$
$$\langle\psi(\mathbf{y}_1)\psi(\mathbf{y}_2)\rangle = f_1(\mathbf{y}_1)\delta(\mathbf{y}_1-\mathbf{y}_2) + f_2(\mathbf{y}_1, \mathbf{y}_2), \dots, \tag{2.1}$$

where $\langle\ \rangle$ denotes ensemble averaging. Hereafter we assume the set $\mathbf{x}_j$ statistically homogeneous and isotropic; then, in particular, $f_1(\mathbf{y}) = n$, where $n = c/V_a$ is its number density and $V_a = 4/3\,\pi a^3$; also $f_2(\mathbf{y}_1, \mathbf{y}_2) = n^2 g(\mathbf{y}_1 - \mathbf{y}_2)$, where $g(\mathbf{y})$ is the radial distribution function for the set $\mathbf{x}_j$.

Let $u(\mathbf{x})$ be a certain random field in the dispersion that appears in a given problem. Provided the external impacts are determinstic, $u(\mathbf{x})$ is uniquely defined by the field $\psi(\mathbf{x})$. This allows us to consider $\psi(\mathbf{x})$ as the "input" that generates the "output" $u(\mathbf{x})$. Following the general idea of system theory, we develop $u(\mathbf{x})$ as the functional series

$$u(\mathbf{x}) = K_0(\mathbf{x}) + \int K_1(\mathbf{x}-\mathbf{y})\, \psi(\mathbf{y})\, d^3\mathbf{y}$$
$$+ \int\int K_2(\mathbf{x}-\mathbf{y}_1, \mathbf{x}-\mathbf{y}_2)\, \psi(\mathbf{y}_1)\, \psi(\mathbf{y}_2)\, d^3\mathbf{y}_1\, d^3\mathbf{y}_2 + \dots , \tag{2.2}$$

with certain non-random kernels $K_0, K_1, \dots$. (Hereafter, if the integration domain is not explicitly indicated, the integrals are taken over the whole $\mathbb{R}^3$.) It is then natural to truncate (2.2) after the pth tuple term in order to obtain certain approximations, $u^{(p)}(\mathbf{x})$, for $u(\mathbf{x})$. Such a truncation immediately brings forth two basic questions:

(i) How to rearrange the terms of the series so that the trunctions $u^{(p)}(\mathbf{x})$ "converge" to the field $u(\mathbf{x})$

$$u^{(p)}(\mathbf{x}) \xrightarrow[p \to \infty]{} u(\mathbf{x}). \tag{2.3}$$

(ii) In what sense the convergence in (2.3) is to be understood.

In [7] the following answer to these questions is given. Let us consider a family of dispersions with different number densities n, produced by means of a certain manufacturing process. Then the probability densities f_k depend on n as a parameter. Suppose that

$$f_k = f_k(\mathbf{y}_1, \dots, \mathbf{y}_k; n) = n^k f_{kk}(\mathbf{y}_1, \dots, \mathbf{y}_k) + o(n^k), \tag{2.4}$$

i.e., $f_k \sim n^k$, $k = 1,2,\dots$, which is a usual assumption. We introduce, after [2, 7], the so-called factorial fields:

$$\Delta_\psi^{(0)} = 1, \quad \Delta_\psi^{(1)}(\mathbf{y}) = \psi(\mathbf{y}), \dots ,$$
$$\Delta_\psi^{(k)}(\mathbf{y}_1, \dots, \mathbf{y}_k) = \psi(\mathbf{y}_1)\, [\psi(\mathbf{y}_2) - \delta(\mathbf{y}_2 - \mathbf{y}_1)] \tag{2.5}$$
$$\dots [\psi(\mathbf{y}_k) - \delta(\mathbf{y}_k - \mathbf{y}_1) - \dots - \delta(\mathbf{y}_k - \mathbf{y}_{k-1})],$$

$k = 2,3,\dots$. The name factorial may be explained to a certain extent by the formula $\Delta_\psi^{(k)}(\mathbf{y}_1, \dots, \mathbf{y}_k) = \psi(\mathbf{y}_1) \dots \psi(\mathbf{y}_k)$ if $\mathbf{y}_i \neq \mathbf{y}_j$ and vanishes otherwise, see [7] for a simple proof.

Let us rearrange the terms of the series (2.2) replacing the products $\psi(\mathbf{y}_1) \dots \psi(\mathbf{y}_k)$ by the factorial fields (2.5)

$$u(\mathbf{x}) = T_0(\mathbf{x}) + \int T_1(\mathbf{x}-\mathbf{y})\,\Delta_\psi^{(1)}(\mathbf{y})\,d^3\mathbf{y}$$
$$+ \int\int T_2(\mathbf{x}-\mathbf{y}_1, \mathbf{x}-\mathbf{y}_2)\,\Delta_\psi^{(2)}(\mathbf{y}_1,\mathbf{y}_2)\,d^3\mathbf{y}_1\,d^3\mathbf{y}_2 + \dots . \tag{2.6}$$

Series of the type (2.6) are called in [7] factorial. The basic result of [7] states that for the class of dispersions that comply with the assumption (2.4) the series (2.6) is virial. This means that the convergence in (2.4) is virial in the sense that all correlation functions for $u^{(p)}(\mathbf{x})$ and $u(\mathbf{x})$ asymptotically coincide to the order c^p at $c \to 0$, $p \geq 1$.

3 Identification of the kernels

The identification of the kernels T_i in the factorial series (2.6) can be performed by means of the following procedure, see [7] for more details and references. Suppose we want to determine the field $u(\mathbf{x})$ to the order n^p, i.e., c^p, only, $p = 1,2,\dots$. Then the truncation $u^{(p)}(\mathbf{x})$ is solely needed and thus the kernels $T_0, T_1, \dots, T_p$ are to be specified. Consider the equation that governs $u(\mathbf{x})$, insert there $u^{(p)}(\mathbf{x})$ instead of $u(\mathbf{x})$, multiply by 1, $\Delta_\psi^{(1)}(0), \dots, \Delta_\psi^{(p)}(0,\mathbf{z}_1,\dots,\mathbf{z}_{p-1})$ and average the results. Making use of the formulae for the averages of the respective products of the factorials [7] (they are straightforward consequences of (2.1)), and truncating them to the same order c^p, we get a system of $p+1$ equations for the needed kernels T_0 to T_p. The specific form of this system depends on the problem under study. However, two important features, common for all such systems, could be traced out. First, they can be split into sets of equations that are solved successively, if the needed kernels are expanded in powers of n as follows:

$$T_i = T_{i\,0} + nT_{i\,1} + \dots + n^k T_{i\,k},\quad k = p - i,\ i = 0,1,\dots,p . \tag{3.1}$$

Second, the leading coefficients $T_{i\,0}$ in (3.1) appear to be connected with the respective single, double, etc., sphere interaction fields. This fact implies that the pth term of the factorial series which is accountable for the n^p-contribution to the field $u(\mathbf{x})$ results, loosely speaking, from interactions of groups of l spheres up to $l = p$. In this sense the series (2.6) could be viewed as a cluster (or group) expansion for $u(\mathbf{x})$ because the latter is broken up in a sum of consecutive terms that result from interactions within successively larger groups of spheres. Such cluster expansions have been widely used in the theory of random dispersions, see [3,4] et al. In these works, however, the cluster expansions concern the effective properties only and not the full statistical description of the random fields under study, and they are introduced on a certain heuristic base. In our approach the cluster interpretation emerges naturally as a certain by-product of the virial property of the factorial series (2.6).

4 Some applications of the factorial series

Here we shall illustrate, skipping all technical details, the performance of the factorial series on several transport problems for random dispersions. The case $p = 2$ will only be treated, i.e., the statistical description of the needed random fields, correct to the order c^2, will be found.

A. Heat propagation problem.

The basic equations read

$$\nabla \bullet (\kappa(\mathbf{x})\nabla\theta(\mathbf{x})) = 0, \; \langle \nabla\theta(\mathbf{x}) \rangle = \mathbf{G}, \tag{4.1}$$

where $\theta(\mathbf{x})$ is the temperature field, $\mathbf{G}$ is the given macroscopical temperature gradient and $\kappa(\mathbf{x})$ is the conductivity field with the representation

$$\kappa(\mathbf{x}) = \kappa_m + [\,\kappa\,]\int h(\mathbf{x}-\mathbf{y})\Delta_\psi^{(1)}(\mathbf{y})\,d^3\mathbf{y}, \; [\,\kappa\,] = \kappa_f - \kappa_m, \tag{4.2}$$

κ_f and κ_m are the conductivity of the spheres and matrix, respectively, $h(\mathbf{x}) = 1$ if $|\mathbf{x}| \leq a$ and vanishes otherwise.

The temperature field in the dispersion is sought as

$$\begin{aligned}\theta(\mathbf{x}) &= \mathbf{G}\bullet\mathbf{x} + \int T_1(\mathbf{x}-\mathbf{y})\,D_\psi^{(1)}(\mathbf{y})\,d^3\mathbf{y} \\ &+ \iint T_1(\mathbf{x}-\mathbf{y}_1, \mathbf{x}-\mathbf{y}_2)\,D_\psi^{(2)}(\mathbf{y}_1, \mathbf{y}_2)d^3\mathbf{y}_1\,d^3\mathbf{y}_2,\end{aligned} \tag{4.3}$$

where

$$\begin{aligned}D_\psi^{(1)}(\mathbf{y}) &= \Delta_\psi^{(1)}(\mathbf{y}) - n = \psi'(\mathbf{y}), \\ D_\psi^{(2)}(\mathbf{y}_1, \mathbf{y}_2) &= \Delta_\psi^{(2)}(\mathbf{y}_1, \mathbf{y}_2) - n\,g_0(\mathbf{y}_{21})\,[\,\psi'(\mathbf{y}_1) + \psi'(\mathbf{y}_2)\,] - n^2 g_0(\mathbf{y}_{21})\end{aligned} \tag{4.4}$$

are the centred c^2-orthogonal fields, introduced in [6, 7]; here $\mathbf{y}_{21} = \mathbf{y}_2 - \mathbf{y}_1$ and $g_0(\mathbf{y}) = f_{22}(\mathbf{y}) = g(\mathbf{y}) + O(n)$, cf. (2.4). Using the method, described in Section 3, one derives the system for the kernels T_1 and T_2 and finds its solution in the form

$$T_1(\mathbf{y}) = T_1(\mathbf{y};\, n) = T_{10}(\mathbf{y}) + nT_{11}(\mathbf{y}),$$
$$T_2(\mathbf{y}_1, \mathbf{y}_2) = T_1(\mathbf{y}_1, \mathbf{y}_2;\, n) = T_{20}(\mathbf{y}_1, \mathbf{y}_2), \tag{4.5}$$

where $T^{(1)}(\mathbf{x})$ and $T^{(2)}(\mathbf{x};\, \mathbf{z})$ are the disturbances to the temperature field in an unbounded matrix, introduced by one and two (located at 0 and at $\mathbf{z}, |\mathbf{z}| > 2a$) spherical inclusions, respectively, when the temperature gradient at infinity is $\mathbf{G}$. The coefficient $T_{11}(\mathbf{x})$ can also be expressed by means of the fields $T^{(1)}(\mathbf{x})$ and $T^{(2)}(\mathbf{x};\, \mathbf{z})$. In this way one could calculate to the order c^2 all statistical characteristics of the temperature field and, in particular, the effective conductivity κ^*. The so-obtained formula for κ^* coincides with this of Jeffrey's [4] but the mode of integration of the conditionally convergent integrals is here unambiguously defined and no need of renormalization arises, see [6, 8].

B. Diffusion problem.

Consider the steady-state diffusion equation

$$\Delta\varphi(\mathbf{x}) - k^2(\mathbf{x})\varphi(\mathbf{x}) + K = 0 \tag{4.7}$$

that governs, at the expense of some simplifying assumptions, the concentration $\varphi(\mathbf{x})$ of a diffusing species, generated at the rate K, in an unbounded lossy dispersion. The sink strength field $k^2(\mathbf{x})$ of the latter has the form (4.2) so that it equals k_{f}^2 and k_{m}^2 depending on whether $\mathbf{x}$ lies in the spheres or in the matrix, see [10] for references and more details. The c^2-solution of eq. (4.7) has the form

$$\varphi(\mathbf{x}) = T_0 + \int T_1(\mathbf{x} - \mathbf{y})\Delta_\psi^{(1)}(\mathbf{y})\, d^3\mathbf{y}$$
$$+ \int\!\!\int T_2(\mathbf{x} - \mathbf{y}_1, \mathbf{x} - \mathbf{y}_2)\, \Delta_\psi^{(2)}(\mathbf{y}_1, \mathbf{y}_2)\, d^3\mathbf{y}_1\, d^3\mathbf{y}_2\,. \tag{4.8}$$

The virial coefficients of the kernels, needed in the c^2-analysis, are

$$T_0 = T_0(n) = T_{00} + nT_{01} + n^2T_{02},$$
$$T_1(\mathbf{y}) = T_1(\mathbf{y};n) = T_{10}(\mathbf{y}) + nT_{11}(\mathbf{y}), \tag{4.9}$$
$$T_2(\mathbf{y}_1, \mathbf{y}_2) = T_2(\mathbf{y}_1, \mathbf{y}_2;\, n) = T_{20}(\mathbf{y}_1, \mathbf{y}_2).$$

Having solved the respective system for the kernels in the form (4.9), one eventually obtains [7]

$$T_{00} = K/k_m^2,\ T_{01} = 0,$$

$$T_{02} = -\frac{[k^2]}{k_m^2}\iint h(\mathbf{y}_1) g_0(\mathbf{y}_2 - \mathbf{y}_1) T_{10}(\mathbf{y}_2) d^3\mathbf{y}_1\, d^3\mathbf{y}_2$$

$$-\iint [1 + 2\frac{[k^2]}{k_m^2} h(\mathbf{y}_1)]\, g_0(\mathbf{y}_2 - \mathbf{y}_1) T_{20}(\mathbf{y}_2)\, d^3\mathbf{y}_1\, d^3\mathbf{y}_2, \qquad (4.10)$$

$$T_{10}(\mathbf{x}) = T_{00} H^{(1)}(\mathbf{x}),\ T_{11}(\mathbf{x}) = 0,\ T_{20}(\mathbf{x} - \mathbf{z}, \mathbf{x}) = T_{00} H_{20}(\mathbf{x} - \mathbf{z}, \mathbf{x}),$$

$$2H_{20}(\mathbf{x} - \mathbf{z}, \mathbf{x}) = H^{(2)}(\mathbf{x}; \mathbf{z}) - H^{(1)}(\mathbf{x}) - H^{(1)}(\mathbf{x} - \mathbf{z}),$$

where $H^{(1)}(\mathbf{x})$ and $H^{(2)}(\mathbf{x}; \mathbf{z})$ are the solutions of the one- and two-sphere problems, governed here by the equations

$$\Delta H^{(1)}(\mathbf{x}) - [\, k_m^2 + [\, k^2\,]\, h(\mathbf{x})\,]\, H^{(1)}(\mathbf{x}) - [\, k^2]\, h(\mathbf{x}) = 0,$$

$$\Delta H^{(2)}(\mathbf{x}; \mathbf{z}) - [\, k_m^2 + [\, k^2\,](h(\mathbf{x}) + h(\mathbf{x} - \mathbf{z}))] H^{(2)}(\mathbf{x}; \mathbf{z}) \qquad (4.11)$$

$$-[\, k^2\,](h(\mathbf{x}) + h(\mathbf{x} - \mathbf{z})) = 0.$$

C. Elastostatic Robin's problem.

The governing equations are

$$L[u] = \mu\Delta \mathrm{u} + (\lambda + \mu)\nabla\nabla\bullet\mathrm{u} = \mathbf{0},\ \ \mathrm{u}(\mathbf{x}) \xrightarrow[x \to \infty]{} 0,$$

$$\qquad (4.12)$$

$$\mathrm{u}(\mathbf{x}) I_f(\mathbf{x}) = \mathrm{U}_0 I_f(\mathbf{x}),\ \ I_f(\mathbf{x}) = \int h(\mathbf{x} - \mathbf{y}) \Delta_{\psi}^{(1)}(\mathbf{y})\, d^3\mathbf{y},$$

so that $I_f(\mathbf{x})$ is the indicator function of the region occupied by the spheres. Thus $\mathrm{u}(\mathbf{x})$ is the displacement field in the matrix, assumed elastic, when the rigid spheres undergo constant displacement U_0. The problem consists in evaluating the mean force F_0 which acts on each sphere. If the Poisson ratio $\nu = 0.5$, we may regard U_0 as the tantamount to the well-known sedimentaiton problem, see [1] for more details and references.

The c^2-solution of (4.12) has the form

$$\mathbf{u}(\mathbf{x}) = \int \mathbf{T}_1(\mathbf{x} - \mathbf{y})\Delta_\psi^{(1)}(\mathbf{y})\, d^3\mathbf{y}$$

$$+\int\int \mathbf{T}_2(\mathbf{x} - \mathbf{y}_1, \mathbf{x} - \mathbf{y}_2)\Delta_\psi^{(2)}(\mathbf{y}_1, \mathbf{y}_2)\, d^3\mathbf{y}_1\, d^3\mathbf{y}_2, \tag{4.13}$$

The c^2-expansion of the kernels has the form (4.5). Eventually one obtains

$$\mathbf{T}_{10}(\mathbf{x}) = \mathbf{T}^{(1)}(\mathbf{x}),$$

$$2\mathbf{T}_{20}(\mathbf{x} - \mathbf{z}, \mathbf{x}) = \mathbf{T}^{(2)}(\mathbf{x}; \mathbf{z}) - \mathbf{T}_{10}(\mathbf{x}) - \mathbf{T}_{10}(\mathbf{x} - \mathbf{z}), \tag{4.14}$$

where $\mathbf{T}^{(1)}(\mathbf{x})$ and $\mathbf{T}^{(2)}(\mathbf{x}; \mathbf{z})$ are the solution of the Robin's problems for one and two-spheres governed, respectively, by the equations [5]:

$$L\,[\mathbf{T}^{(1)}] = 0, \ [\,\mathbf{T}^{(1)}(\mathbf{x}) - \mathbf{U}_0]\, h(\mathbf{x}) = 0,$$

$$L_{\mathbf{x}}\,[\,\mathbf{T}^{(2)}(\mathbf{x}; \mathbf{z})\,] = 0, \ [\,\mathbf{T}^{(2)}(\mathbf{x}; \mathbf{z}) - \mathbf{U}_0]\,[\,h(\mathbf{x}) + h(\mathbf{x} - \mathbf{z})\,] = 0. \tag{4.15}$$

In turn

$$\mathbf{F}_0 = (A_0 + A_1 c)\mathbf{U}_0 + o(c), \ A_0 = \frac{24\,\pi\,\mu(1 - \nu)}{5 - 6\,\nu}\, a,$$

$$A_1 = \frac{2}{V_a}\int d^3\mathbf{z}\, g_0(\mathbf{z}) \int_{\Omega_0} \mathbf{S}_{20}(\mathbf{x} - \mathbf{z}, \mathbf{x})\circ\mathbf{n}\, d\Omega_a\,, \tag{4.16}$$

where $\mathbf{S}_{20} = \lambda\,\mathbf{I}\nabla\bullet\mathbf{T}_{20} + \mu(\nabla\mathbf{T}_{20} + \mathbf{T}_{20}\nabla)$ and $\mathbf{n}$ is the outward unit normal to the sphere $\Omega_a = \{\mathbf{x} |\, |\mathbf{x}| = a\}$. The integrand in the right-hand side of $(4.16)_2$ is absolutely integrable having the order $|\mathbf{z}|^{-4}$ at $|\mathbf{z}| >> 1$. Thus, unlike the approach of Batchelor [1] et al., there is no need of renormalization. Details, concerning the above solution, will be given elsewhere.

5 Conclusions

The foregoing examples show that the factorial series are a well-adapted tool in studies of random particulate media because they allow us to obtain explicit stochastic solutions with controlled accuracy in powers of the volume fractions c. In this way one can, in particular, calculate unambiguously and without any renormalization the effective properties of the media. This fact confirms the author's view [8] that these properties can be rigorously evaluated, in general, only if they are extracted from the full stochastic solutions of the respective random problems.

References

[1] Batchelor, G.K., Sedimentation in a dilute suspension of spheres. *J. Fluid Mechanics*, **52**, pp. 245-268, 1972.

[2] Christov, C.I., A further development of the concept of random density function with application .... *C.R. Acad. Bulg. Sci.*, **31**(1), pp. 35-38, 1985.

[3] Felderhof, B.V., G.W. Ford and E.G.D. Cohen. Two-particle cluster integral in the expansion of the dielectric constant. *J. Stat. Phys.*, **28**, pp. 649-672, 1982.

[4] Jeffrey, D.J., Conduction through a random suspension of spheres, *Proc. Roy. Soc. London*, **A335**, pp. 355-367, 1973.

[5] Lurie, A.I. *Theory of Elasticity*, Nauka, Moscow, 1970, Ch.V, §3.3. (In Russian.)

[6] Markov, K.Z., A virial stochastic expansion for a random suspension of spheres. In: *Continuum Models of Discrete Systems*, Ed. A.J.M. Spencer, Balkema, Rotterdam, 1987, pp. 97-105.

[7] Markov, K.Z., On the factorial functional series and their application to random media. *SIAM J. Appl. Math.*, **50**, 1990 (in press).

[8] Markov, K.Z., On the heat propagation problem for random dispersions of spheres, *Math. Balkanica* (New Series), **3**, pp. 399-417, 1989.

[9] Stratonovich, R.L., *Topics in theory of random noise*, Vol. 1, Gordon and Breach, New York, 1963.

[10] Talbot, D.R.S. and J. Willis, The overall sink strength of an inhomogeneous lossy medium, p.I and p.II. *Mech. Materials*, **3**, pp. 171-181, 183-191, 1984.

Markov K.
University of Sofia
Faculty of Mathematics and Informatics
5 Boulevard A. Ivanov,
1126 Sofia
BULGARIA

P. MAZILU

Green's operator for determination of the effective properties in random media

1 Introduction

A random heterogeneous medium is characterized by the correlations related to the mechanical or field properties of the materials. Here we shall focus our attention on the heterogeneous elasticity. Most of the methods developed in heterogeneous elasticity are applicable for other types of heterogeneous problems. Let $L(\mathbf{x})$ be a random Hooke's tensor defined by the correlations

$$\langle L(\mathbf{x}_1) \ldots L(\mathbf{x}_n)\rangle, \ \mathbf{x}_1,\mathbf{x}_2,\ldots,\mathbf{x}_n \in E, \ n \in N, \tag{1.1}$$

where E denotes the whole tri-dimensional Euclidean space and the brackets are volume or an ensemble average. Because of the assumed randomness the following decorrelations hold at great distances

$$\lim_{\mathbf{x}_n \to \infty} \langle L(\mathbf{x}_1)L(\mathbf{x}_2) \ldots L(\mathbf{x}_n)\rangle = \langle L(\mathbf{x}_1) \ldots L(\mathbf{x}_{n-1})\rangle \langle L\rangle . \tag{2.1}$$

A special class of random heterogeneous materials is formed by the statistically homogeneous materials. For these materials the correlations (1.1) are invariant to any shifting $\mathbf{h}$ of the points $\mathbf{x}_1,\mathbf{x}_2,\ldots,\mathbf{x}_n$.

$$\langle L(\mathbf{x}_1 + \mathbf{h})L(\mathbf{x}_2 + \mathbf{h}) \ldots L(\mathbf{x}_n + \mathbf{h})\rangle = \langle L(\mathbf{x}_1)L(\mathbf{x}_2) \ldots L(\mathbf{x}_n)\rangle . \tag{1.3}$$

In the study of statistically homogeneous elasticity an important role is played by the reference medium. This medium is a homogeneous comparison material defined by a constant Hooke's tensor $\tilde{L}$ = const. In the following we use isotropic reference media defined by

$$\tilde{L} = [\lambda\delta_{ij}\delta_{kl} + \mu(\delta_{ik}\delta_{jl} + \delta_{il}\delta_{jk})]_{i,j,k,l\ =1,2,3} . \tag{1.4}$$

Let $\mathbf{G}(\mathbf{x},\mathbf{x}') = [G_{ij}(\mathbf{x},\mathbf{x}')]_{i,j=1,2,3}$ be the Green's matrix corresponding to $\tilde{L}$ defined by (1.4) and to a domain D. The components of this matrix satisfy the boundary value problem

$$\mu \frac{\partial^2 G_{ij}}{\partial x_k \partial x_k} + (\lambda + \mu) \frac{\partial^2 G_{ik}}{\partial x_k \partial x_j} = 0, \quad j = 1,2,3, \quad \mathbf{x} \in D - \mathbf{x}',$$

$$G_{ij}(\mathbf{x},\mathbf{x}') = 0 \quad \textit{for } \mathbf{x} \in \partial D, \tag{1.5}$$

and have singularities of the form

$$G_{ij}(\mathbf{x},\mathbf{x}') = \frac{\lambda + 3\mu}{8\pi\mu(\lambda + 2\mu)} \frac{\delta_{ij}}{|\mathbf{x} - \mathbf{x}'|} + \frac{\lambda + \mu}{8\pi\mu(\lambda + 2\mu)} \frac{(x_i - x_i')(x_j - x_j')}{|\mathbf{x} - \mathbf{x}'|^3} +$$

$$+ \text{regular} \equiv \nu_{ij}(\mathbf{x},\mathbf{x}') + G^0_{ij}(\mathbf{x},\mathbf{x}') . \tag{1.6}$$

In the next section, by means of this Green's matrix a peculiar operator will be constructed. This operator represents a fundamental tool in the solving of any non-homogeneous elastic problem.

2 Definition of modified Green's operators

Let us denote by $\mathcal{T}$ the set of tensor functions defined on the domain D

$$\mathcal{T} = \{\mathbf{f} = (f_{ij}) / f_{ij} : D \to R, f_{ij} \in C^1(D)\}, \quad i, j = 1,2,3 \tag{2.1}$$

and by $\mathcal{T}_s$ the subset of $\mathcal{T}$ formed on symmetrical tensor functions

$$\mathcal{T}_s = \{\mathbf{f} \in \mathcal{T} / f_{ij} = f_{ji}\}, \quad i, j = 1,2,3. \tag{2.2}$$

By definition the modified Green's operators are the operator Λ and Γ defined on $\mathcal{T}$ and $\mathcal{T}_s$ by the relations

$$(\Lambda \mathbf{f})_{ij}(\mathbf{x}) = g_{ipqj} f_{pq}(\mathbf{x}) + \int_D \frac{\partial^2 G_{ip}(\mathbf{x},\mathbf{x}')}{\partial x_j \partial x_q'} f_{pq}(\mathbf{x}') dv', \; f \in \mathcal{T}, \tag{2.3}$$

$$g_{ipqj} = \int_{|\mathbf{x}'| = 1} |\mathbf{x}'|^2 \frac{\partial \nu_{ip}}{\partial x_q}(0,\mathbf{x}') \cos(r, x_j') d\omega, \tag{2.4}$$

$$(\Gamma \mathbf{f})_{ij} = \frac{1}{2}[(\Lambda \mathbf{f})_{ij} + (\Lambda \mathbf{f})_{ji}], \; \mathbf{f} \in \mathcal{T}_s, \tag{2.5}$$

respectively. Let us consider the following boundary value problem

$$\mu \frac{\partial^2 u_i}{\partial x_k \partial x_k} + (\lambda + \mu) \frac{\partial^2 u_k}{\partial x_k \partial x_i} = \frac{\partial f_{iq}}{\partial x_q}, \tag{2.6}$$

$$u_i|_{\partial D} = 0 .$$

It was proved (see [3]) that the gradient of displacement $\nabla \mathbf{u}$ and the strain ε of (2.6) have the following expressions

$$\nabla \mathbf{u} = \Lambda \mathbf{f}, \tag{2.7}$$

$$\varepsilon = \frac{1}{2}(\nabla \mathbf{u} + \nabla \mathbf{u}^T) = \Gamma \mathbf{f}. \tag{2.8}$$

For the proof of (2.7) and (2.8) the Green's representation

$$u_i(\mathbf{x}) = -\int_D G_{ip}(\mathbf{x},\mathbf{x}') \frac{\partial f_{pq}}{\partial x'_q}(\mathbf{x}')dv' \tag{2.9}$$

is recast first into the equivalent form

$$u_i = \int_D \frac{\partial G_{ip}(\mathbf{x},\mathbf{x}')}{\partial x'_q} f_{pq}(\mathbf{x}')dv' \tag{2.10}$$

From (2.10), by applying the form of differentiation for integrals with weak singularities (see [5]), one obtains (2.7) and (2.8). The operator Γ defined by (2.5) presents a local part γ given by

$$\gamma_{ipqj} = \frac{1}{2}(g_{ipqj} + g_{jpqi})$$

with $\mathbf{g}$ given by (2.4). A direct evaluation of the integrals (2.4) gives

$$\gamma_{ipqj} = \frac{1}{6\mu}(\delta_{ik}\delta_{jl} + \delta_{il}\delta_{jk}) - \frac{\lambda + \mu}{15\mu(\lambda + 2\mu)}(\delta_{ik}\delta_{jl} + \delta_{kj}\delta_{il} + \delta_{ji}\delta_{kl}) . \tag{2.11}$$

This value of γ was first published, without any proof, by Eimer in 1971 (see [2]).

3 Deterministic and statistic solutions of heterogeneous elasticity

Let $L(\mathbf{x})$ be a piece-wise continuous Hooke's tensor defined in a bounded domain D.

The linear non-homogeneous elastic problem reads

$$(L_{ij}^{hk}(\mathbf{x})u_{h\,k})_{,k} + F_i = 0, \quad \mathbf{x} \in D, \; i = 1, 2, 3, \tag{3.1}$$

$$\mathbf{u}\,|_{\partial D} = h(\mathbf{x}), \quad \mathbf{x} \in \partial D, \tag{3.2}$$

with the continuity conditions

$$[\mathbf{u}] = 0 \quad [L_{ij}^{hk}(\mathbf{x})u_{h\,k}]\,n_j(\mathbf{x}) = 0, \quad \mathbf{x} \in D\,, i = 1, 2, 3. \tag{3.3}$$

This boundary value problem corresponds to a discrete non-homogeneity. Because any piece-wise continuous function can be approximated by smooth functions, it will be assumed that $L(\mathbf{x})$ belongs to $C^1(D)$. By this the continuity conditions (3.3) are automatically fulfilled. For the solving of the boundary value problem (3.1)-(3.2) a successive approximation method is used. Let us split $L(\mathbf{x})$ into

$$L(\mathbf{x}) = \tilde{L} - \mathring{L}(\mathbf{x}) \tag{3.4}$$

and let us introduce a family $L_\eta(\mathbf{x})$ of Hooke's tensors defined by

$$L_\eta(\mathbf{x}) = \tilde{L} - \eta\,\mathring{L}(\mathbf{x}), \quad \eta \in R\,. \tag{3.5}$$

It was proved [3] that for sufficiently small $|\,\eta\,|$ the strains and stress corresponding to (3.1)-(3.2) with $L_\eta(x)$ as Hooke's tensor reads

$$\varepsilon = \varepsilon_0 + \eta\Gamma\mathring{L}\,\varepsilon^0 + \eta^2\Gamma\mathring{L}\,(\Gamma\mathring{L}\,\varepsilon^0) + \ldots + \eta^n\Gamma(\ldots\,\Gamma\mathring{L}\,\varepsilon^0) + \ldots\,, \tag{3.6}$$

$$\sigma = L\varepsilon^0 + \eta L\,\Gamma\mathring{L}\,\varepsilon^0 + \eta^2 L(\Gamma\mathring{L}\,(\Gamma\mathring{L}\,\varepsilon^0)) + \ldots + \eta^n L\Gamma\mathring{L}\,(\ldots\,\Gamma\;\varepsilon^0) + \ldots\,, \tag{3.7}$$

where ε_0 denotes the strains in the reference medium. As was proved in [4] for sufficiently small $|\,\eta\,|$ these series converge absolutely with respect to the norm $\mathcal{L}_2(D)$, moreover analytical prolongation of (3.6) till $\eta = 1$ is always possible. Now let us assume that $L(\mathbf{x})$ is a realization of the random Hooke's tensor defined by the correlations (1.1). Then it is possible to prove (see [3]) that the ensemble averages $\langle\varepsilon\rangle$ and $\langle\sigma\rangle$ of the stress and strain can be expressed by means of the correlations (1.1) as follows:

$$\langle\varepsilon_1\rangle = \varepsilon_{01} + \eta\Gamma_{12}\langle\mathring{L}_2\rangle\,\varepsilon_{02} + \eta^2\Gamma_{12}\Gamma_{23}\langle\mathring{L}_2\,\mathring{L}_3\rangle\,\varepsilon_{03} + \dots$$

$$+\eta^n\,\Gamma_{12}\dots\Gamma_{n-1,n}\,\langle\mathring{L}_2\dots\mathring{L}_n\rangle\varepsilon_{0n} + \dots = M_\eta[\varepsilon_0]\,, \tag{3.8}$$

$$\langle\sigma_1\rangle = \langle L\rangle\,\varepsilon_{01} + \eta\Gamma_{12}\langle L_1L_2\rangle_{02} + \eta^2\Gamma_{12}\Gamma_{23}\langle\mathring{L}_{12}\mathring{L}_3\rangle\,\varepsilon_{03} + \dots$$

$$+\eta^n\,\Gamma_{12}\dots\Gamma_{n-1,n}\,\langle\mathring{L}_1\mathring{L}_2\dots\mathring{L}_n\rangle\,\varepsilon_{0n} + \dots = N_\eta[\varepsilon_0]\,.$$

If one dentoes by M and N the operators M_η, N_η corresponding to $\eta = 1$, then for the considered heterogeneous material defined by (1.1) holds

$$\langle\varepsilon\rangle = M\,[\varepsilon_0]\,,$$

$$\langle\sigma\rangle = N\,[\varepsilon_0]\,. \tag{3.9}$$

From (3.8) follows the macroscopical stress-strain relation

$$\langle\sigma\rangle = NM^{-1}\,[\langle\varepsilon\rangle]\,. \tag{3.10}$$

Generally the stress-strain relation (3.9) is a non-local constitutive law. This law becomes local if the boundary ∂D will be thrown at infinity and the domain D fills the whole space E. If one considers a sequence of domains D_n whose boundaries ∂D_n tends to infinity then it can be proved that the operator Γ tends to $\overset{\infty}{\Gamma}$ defined by

$$\overset{\infty}{\Gamma}\,\mathbf{f}(\mathbf{x}) = \lim_{n\to\infty}\,(\Gamma_n\mathbf{f})(\mathbf{x}) = \gamma(\mathbf{f}(\mathbf{x}) - \mathbf{f}_\infty) + \int_E \frac{\varphi_v(\mathbf{x},\mathbf{x}')}{r^3}(\mathbf{f}(\mathbf{x}) - \mathbf{f}_\infty)dv\,. \tag{3.11}$$

The mean strain and stress corresponding to the whole space are obtained directly by setting in (3.7) $\overset{\infty}{\Gamma}$ instead of Γ. In this case the constitutive law (3.9) will become local and will be defined by the effective Hooke's tensor

$$L^{\mathrm{eff}} = NM^{-1}\,. \tag{3.12}$$

4 Associated integral equation

The strain ε solution of the boundary value problem (3.1)-(3.3) satisfies the integral equation

$$\varepsilon - \Gamma \overset{\circ}{L} \varepsilon = \varepsilon_0 . \tag{4.1}$$

Indeed the equilibrium equation (3.1) can be written in the equivalent form

$$(\tilde{L}_{ij}^{hk} \overset{(n)}{u}_{h,k})_{,j} = (\overset{\circ}{L}_{ij}^{hk} \overset{0}{u}_{h,k})_{,j} , \tag{4.2}$$

where $\mathbf{u}_0$ denotes the displacement in the reference medium. By taking

$$f_{ij} = \overset{\circ}{L}_{ij}^{kl} u_{k,l}^0 \tag{4.3}$$

one regains the boundary value problem (2.6). The eq. (4.1) is a direct consequence of the form (2.8). If one denotes by $(I + \Gamma \overset{\circ}{L})^{-1}$ the operator defined by

$$(I - \eta\Gamma \overset{\circ}{L})^{-1} = \varepsilon_0 + \eta\Gamma \overset{\circ}{L} \varepsilon_0 + \eta^2\Gamma \overset{\circ}{L} (\Gamma \overset{\circ}{L} \varepsilon) + \ldots + \eta^n\Gamma \overset{\circ}{L} (\ldots \Gamma \overset{\circ}{L} \varepsilon) + \ldots \tag{4.4}$$

then

$$\begin{aligned} \varepsilon &= (I - \Gamma \overset{\circ}{L})^{-1} \varepsilon_0 , \\ \sigma &= L(I - \Gamma \overset{\circ}{L})\varepsilon_0 , \end{aligned} \tag{4.5}$$

and

$$\begin{aligned} \langle\varepsilon\rangle &= \langle(I - \eta\Gamma \overset{\circ}{L})^{-1}\rangle \varepsilon_0 , \\ \langle\sigma\rangle &= \langle L(I - \eta\Gamma \overset{\circ}{L})^{-1}\rangle \varepsilon_0 , \end{aligned} \tag{4.6}$$

whence follows

$$\langle\sigma\rangle = \langle L(I - \eta\Gamma \overset{\circ}{L})^{-1}\rangle \langle(I - \eta\Gamma \overset{\circ}{L})^{-1}\rangle^{-1} \langle\varepsilon\rangle . \tag{4.7}$$

The forms (4.5)-(4.7) in which one replaces the operator Γ by $\overset{\infty}{\Gamma}$ supply the results for the whole space.

Remark. The forms (4.6) are not completely identical with (3.8). As we have seen the fundamental forms (3.8) are expressed only by mens of correlations (1.1). For convenience, however, we shall further attribute to (4.6) the same significances as (3.8).

5 Self-embedded heterogeneity and self-consistent schema

A special type of boundary condition different to Dirichlet's conditions (3.2) is obtained if one assumes the domain D containing the heterogeneity $L(\mathbf{x})$ embedded into a matrix having $\tilde{L}$ as Hooke's tensor. The Green's operator corresponding to this boundary value problem reduces in this case into local and singular parts

$$(\Gamma\mathbf{f})(\mathbf{x}) = \gamma\mathbf{f} + \int_D \frac{\Phi_\nu(\mathbf{x}, \mathbf{x}')}{r^3}\mathbf{f}(\mathbf{x}')dv \,. \tag{5.1}$$

If the boundary ∂D is thrown to infinity, i.e., the heterogeneity fills the whole space, the Green's operator will be

$$\mathring{\Gamma}\,\mathbf{f} = \lim_{\partial D \to \infty} \Gamma\mathbf{f} \tag{5.2}$$

with this operator the mean values of strain and stress are

$$\begin{aligned} \langle\varepsilon\rangle &= (I + \mathring{\Gamma}\langle\mathring{L}\rangle + \mathring{\Gamma}_{12}\mathring{\Gamma}_{23}\langle\mathring{L}_2\mathring{L}_3\rangle + \ldots)\varepsilon_0 \equiv \langle(I - \mathring{\Gamma}\mathring{L})^{-1}\rangle\varepsilon_0 \,, \\ \langle\sigma\rangle &= (\langle L\rangle + \mathring{\Gamma}_{12}\langle L_1\mathring{L}_2\rangle + \mathring{\Gamma}_{12}\mathring{\Gamma}_{23}\langle L_1\mathring{L}_2\mathring{L}_3\rangle + \ldots)\varepsilon_0 \\ &\equiv \langle L(I - \mathring{\Gamma}\mathring{L})^{-1}\rangle\varepsilon_0 \,. \end{aligned} \tag{5.3}$$

The effective Hooke's tensor will be

$$L^{\text{eff}} = \langle L(I - \mathring{\Gamma}\mathring{L})^{-1}\rangle\,\langle(I - \mathring{\Gamma}\mathring{L})^{-1}\rangle^{-1} \,. \tag{5.4}$$

A particular reference medium corresponds just to L^{eff}. This case will be called self-embedded heterogeneity. Then (5.4) represents an implicit equation. If one assumes the ergodic hypothesis and the ensemble mean value is replaced by volume mean value over a representative volume element (RVE), then the same equation (5.4) will follow as if one considers a single RVE embedded in the matrix L^{eff}.

Remark. The above method of self-embedded heterogeneity is valid equally for the periodic heterogeneity provided the condition that a cell will be taken instead of the RVE. In the particular case of a heterogeneity formed by grains the white noise hypothesis allows a remarkable simplification of the computations. According to this hypothesis the strains and stress values in a given grain remain unchanged if the heterogeneous materials standing at a certain distance of the grain are replaced by a homogeneous one having L^{eff} as Hooke's tensor. A limit case of the applicability of the white noise hypothesis consists of considering a single grain embedded into L^{eff}–matrix. This limit case, called self-consistent method, can be applied with success for heterogeneities which are formed from indefinitely fractured homothetical grains. The self-consistent method becomes simpler if the grains are spheres. In this case the Green's operator $\overset{\circ}{\Gamma}$ reduces to its local part γ. The general problem of the approximation of the Green's operator by its local part will be considered in the following section.

6 Approximative Green operators

The error made by neglecting the non-local part depends on the chosen reference medium. An optimal choice of $\tilde{L}$ can lead to a minimal error. In order to give an evaluation of the error made, let us write the integral equation (4.1) in the equivalent form

$$\varepsilon - \gamma \overset{\circ}{L} \varepsilon + (\gamma - \overset{\infty}{\Gamma})\overset{\circ}{L} \varepsilon = \varepsilon_0 . \tag{6.1}$$

By neglecting the non-local part this equation will be replaced by

$$\varepsilon - \gamma \overset{\circ}{L} \varepsilon = \varepsilon. \tag{6.2}$$

The error made will be as great as the neglected part

$$e = (\gamma - \overset{\infty}{\Gamma})\overset{\circ}{L} \varepsilon = (\gamma - \overset{\infty}{\Gamma})(\tilde{L}\varepsilon - \sigma) . \tag{6.3}$$

Let us consider the norm

$$\| L \|^2 = \Sigma_{i,j,k,l} \int_{RVE} | L_{ij}^{kl} |^2 \, d\mathbf{v} \tag{6.4}$$

then from (6.3) follows

$$\begin{aligned} \| e \| &= \| (\gamma - \overset{\infty}{\Gamma})(\overset{\circ}{L}\varepsilon - \sigma) \| = \| (\gamma - \overset{\infty}{\Gamma}) [(\tilde{L}\varepsilon - \sigma) - (L\varepsilon - \sigma)] \| \\ &= \| (\gamma - \overset{\infty}{\Gamma}) (\tilde{L} - L)\varepsilon \| \le K | \varepsilon | \, \| \tilde{L} - L \| \end{aligned} \tag{6.5}$$

where K denotes the norm of $(\gamma - \Gamma)$ and $|\varepsilon|$ is the maximal absolute value of the strain components. The following lemma holds:

Lemma. $\| \tilde{L} - L \|$ attains its minimum for $\tilde{L} = \langle L \rangle$.

Proof.

$$\| \tilde{L}-L \|_2 = (\tilde{L}-L, \tilde{L}-L) = (\tilde{L} - \langle L\rangle - L', \tilde{L} - \langle L\rangle - L') = \| \tilde{L} - \langle L\rangle \|^2 + \| L' \|^2 .$$

This lemma recommends the Voigt's medium $\tilde{L} = \langle L\rangle$ as an appropriate reference medium. For the evaluation of scattering, Reuss' reference medium $\tilde{L} = \langle L^{-1}\rangle^{-1}$ can be considered. In Table 1 the calculated and the measured (see [1]) elastic moduli are compared.

Table 1. Comparison between calculated and measured moduli

	E_{calc}/MP_a		$\frac{E_{meas}}{MP_a}$	ν_{calc}		ν_{meas}
	$\langle L\rangle$	$\langle L^{-1}\rangle^{-1}$		$\langle L\rangle$	$\langle L^{-1}\rangle^{-1}$	
Ag	82.85	81.90	77	0.36	0.36	0.38
Al	70.03	70.03	70	0.34	0.34	0.38
Au	79.79	78.81	84	0.42	0.42	0.42
Cu	129.64	127.50	126	0.34	0.34	0.35
Ni	226.22	224.88	217	0.29	0.30	0.31
Pb	25.33	26.29	18.2	0.41	0.49	0.45
Th	75.37	74.11	77	0.28	0.28	0.30
K	2.67	3.06	3.5	0.36	0.34	0.31
Li	11.69	16.71	11.62	0.33	0.26	0.08
Na	6.01	7.29	9.1	0.35	0.35	0.32
Nb	105.49	105.26	105	0.39	0.39	0.38
Ge	131.68	131.60	77	0.21	0.21	-
Si	162.91	162.85	112	0.22	0.22	-
W	410.16	410.16	350	0.28	0.28	0.17

7 Conclusions

A short review of the properties of the modified Green's operators was performed. The possible applications to the self-embedded heterogeneity and to the self-consistent method are discussed. The last section is devoted to the approximation of the Green's operator by its local part.

References

[1] Cotrell, A.H., *The mechanical properties of matters*, J. Wiley & Sons, Chicester, New York 1964.

[2] Eimer, Cz., The viscoelasticity of multi-phase media, *Arch. Mech. Stosow.*, **23**, (3-15), 1971.

[3] Mazilu, P., On the theory of linear elasticity in statistically homogeneous media (I), *Rev. Roum. Math. Pures et Appl.*, **23** (261-272), 1972.

[4] Mazilu, P., On the theory of liner elasticity in statistically homonogeneous media (II), *Arch. of Mechanics*, **28** (517-529), 1976.

[5] Michlin, S.G., *Multidimensional singular integrals and integral equations*, Pergamon Press, Oxford, New York, 1965.

Mazilu P.
Institut für Umformteknik
TH-Darmstadt
Petersen Strasse 30
D-6100 Darmstadt
F.R.G.

M.V. MIĆUNOVIĆ

Materials with thermoplastic memory

1 Introduction

The principal objective of this work is to investigate a relation between inelastic materials with internal variables and inelastic materials with memory. If an evolution equation for plastic strain rate is given for the first class of materials, then its integration inevitably leads to the description represented by integrals whose kernels are responsible for a plastic memory. Here the opposite and more difficult way is suggested: to see how functionals appearing in a description of inelastic materials with memory may be transformed into the corresponding evolution equations.

2 Geometrical prelimianries

As a prerequisite, a correct geometric description of an inelastic deformation process analysed is needed. Consider a crystalline body in a real configuration (ψ_t) with dislocations and an inhomogeneous temperature field $\theta(X, t)$ (where t stands for time and X for the considered particle of the body) subject to surface tractions. Corresponding to (ψ_t) there exists, usually, an initial reference configuration (κ_0) with (differently distributed) dislocations at a homogeneous temperature θ_0 without surface tractions. Due to these defects such a configuration is not stress free but contains an equilibrated residual stress (sometimes named as "back-stress"). It is generally accepted that the linear mapping function $\mathbb{F}(\cdot, t)$: $(\kappa_0) \rightarrow (\psi_t)$ is a compatible second-order *total deformation gradient tensor*. Here t as scalar parameter allows for a time developing family of deformed configurations (ψ_t). In the papers dealing with continuum representations of dislocation distributions (ψ_t)-configuration is imagined to be cut into small elements denoted by (ν_t), these being subsequently brought to (θ_0) free of neighbours. The deformation tensor $\Delta_E(\cdot, t)$: $(\nu_t) \rightarrow (\psi_t)$ obtained in such a way is incompatible and should be called the *thermoelastic distortion tensor* whereas (ν_t)-elements are commonly named as *natural state local reference configurations* (cf. e.g. [8, 17, 11]). Moreover, often a plastic distortion tensor $\Pi(\cdot, t)$: $(\kappa_\pi) \rightarrow (\psi_t)$ is defined, where (κ_π) is a global ideal crystal having the same intrinsic crystalline structure as (ν_t)-elements themselves. However, such a distortion is not unique since there are many indistinguishable configurations (κ_π) with various shapes but the same intrinsic strucutre. The difficulty is bypassed by the following definition of the *plastic distortion tensor*

$$\Delta_P(\cdot, t) := \Delta_E(\cdot, t)^{-1}\, \mathbb{F}(\cdot, t), \tag{1}$$

where $\mathbb{F}(\cdot, t)$ is found by comparison of material fibres in (κ_0) and (ψ_t), whereas $\Delta_E(\cdot, t)$ is determined by crystallographic vectors in (ν_t) and (ψ_t). Multiplying (1) from the left-hand side by $\Delta_E(\cdot, t)$ we reach Kröner's decomposition rule which is often wrongly named as Lee's decomposition formula. The above two definitions of plastic distortion are easily connected by $\Pi(\cdot, t) = \Delta_P(\cdot, t)\Delta_0(\cdot)$. It is worthy of note that curl $\Delta_E(\cdot, t)^{-1} \neq \mathbb{O}$ and this *incompatibility* is commonly connected to an asymmetric second rank tensor of dislocation density.

3 Materials with thermoplastic memory

Let $\mathcal{R}$ denote the set of all real numbers, $\mathcal{R}^+$ the set of all positive numbers, $\mathcal{L}$ the set of all second-order tensors, and $\mathcal{L}^+$ its subset whose elements have positive determinant. Dropping (for simplicity) the dependence of distortion tensors on the particle considered in the sequel and introducing notations

$$\theta^t(s) := \theta(t) - \theta(t - s),\ \ \mathbb{P}^t(s) := \Delta_P(t) - \Delta_P(t - s),$$

where $s \in [0,\infty)$, $\theta^t : \mathcal{R}^+ \to \mathcal{R}$ is the relative temperature history and $\mathbb{P}^t : \mathcal{R}^+ \to \mathcal{L}$ is the relative plastic deformation history, we formulate a constitutive equation for Cauchy stress in the following way:

$$\mathbb{T}(t) = \mathfrak{T}(\Delta_E(t), \Delta_P(t), \theta(t), \theta^t, \mathbb{P}^t) =: \mathfrak{T}(\gamma, \theta^t, \mathbb{P}^t) =: \mathfrak{T}(\mathbb{Y}^t). \tag{2}$$

Here $\theta(t) \in \mathcal{R}^+$ and $\mathbb{T}(t)$ is the actual Cauchy stress at the considered time t. Denoting the collection of thermoplastic histories by $I := \{\mathbb{P}^t, \theta^t\}$ the constitutive functional may be presented by the mapping

$$\mathfrak{T} : S_d \to \mathcal{L}, \tag{3}$$

where $S_d := \mathcal{L}^+ \times \mathcal{L}^+ \times \mathcal{R}^+ \times I$ is the space of thermostrain histories and $\mathbb{Y}^t \in S_d$. For the postulated functional the following properties are assumed:

$$\mathfrak{T}(\mathbb{I}(1 + \alpha\Delta\theta),\ \Delta_P(t), \theta(t), \mathbb{P}^t, \theta^t) = \mathbb{O}\,, \qquad (\alpha \in \mathcal{R}^+), \tag{P1}$$

$$\mathfrak{T}(\mathbb{Y}^t) = \mathbb{T}^*(\gamma) \tag{P2}$$

$$\mathbb{Q}(t)\mathbb{T}(t)\mathbb{Q}(t)^T = \mathfrak{T}(\mathbb{Q}(t)\Delta_E(t), \Delta_P(t), \theta(t), \mathbb{P}^t, \theta^t) \tag{P3}$$

$$\text{with } \mathbb{Q}(t)\mathbb{Q}(t)^{\mathrm{T}} = \mathbb{I}, \det \mathbb{Q}(t) = 1,$$

$$\mathbb{T}(t) = \mathcal{T}(\Delta_{\mathrm{E}}(t)\mathbb{H}, \mathbb{H}^{-1}\Delta_{\mathrm{P}}(t)\mathbb{H}_{\kappa}, \mathbb{H}^{-1}\mathbb{P}^{t}\mathbb{H}_{\kappa}, \theta(t), \theta^{t}) \tag{P4}$$

$$\text{with } \mathbb{H} \in \mathcal{G}, \mathbb{H}_{\mathrm{K}} \in \mathcal{G}_{\mathrm{K}} = \Delta_0 \mathcal{G} \Delta_0^{-1} .$$

The first of these claims that in the case when pure elastic distortion is unit tensor (corresponding to vanishing pure elastic strain) stress disappears, whereas $\Delta\theta \equiv \theta - \theta_0$ and α is the coefficient of thermal expansion. The property (P2) describes a rapid unloading at fixed $\theta(t)$ with elastic moduli dependent on actual temperature as well as on actual value of the plastic distortion tensor generalizing in such a way the corresponding P3 property of [16] where such a dependence was neglected. The next property is the principle of material frame indifference, i.e., objectivity. Here this is understood in the way that an observer of (ψ_t)-configuration may rotate, while observers of (ν_t) and (κ_0) (or, equivalently, (κ_π)) configurations are fixed and connected either to intrinsic crystal structure or to the fixed shape of (κ_0). Such an objectivity is different from [16] where it was accepted that the same (rotating) observer was responsible for both (ψ_t) and (ν_t)-configurations. Finally, the last of the above listed properties establishes material symmetry whose group with respect to (ν_t)-configuration is denoted by $\mathcal{G}$ and that of (κ_0)-configuration by $\mathcal{G}_{\mathrm{K}}$. Their relationship easily follows from [18] p. 77. Sometimes a difference between these groups is called plastic strain induced anisotropy. It is worthy of note that for an inhomogeneous Δ_0 the group $\mathcal{G}_{\mathrm{K}}$ varies from particle to particle of the body.

Let us denote the material (time) derivative of the plastic distortion tensor by $\dot{\Delta}_{\mathrm{P}}(t)$ and the related plastic "velocity gradient" by $\mathbb{L}_{\mathrm{P}}(t) := \dot{\Delta}_{\mathrm{P}}(t)\Delta_{\mathrm{P}}(t)^{-1}$. Then the corresponding plastic deformation path length $p(t)$ may be defined as:

$$p(t) = \int_{c}^{t} \dot{p}(\tau)d\tau \quad \text{where } \dot{p}(t)^2 := \|\mathbb{L}_{\mathrm{P}}(t)\|^2 \equiv \mathrm{tr}\,\{\, \mathbb{L}_{\mathrm{P}}(t)\mathbb{L}_{\mathrm{P}}(t)^{\mathrm{T}} \ . \tag{4}$$

In the so-called endochronic theories of plasticity (cf. e.g. [4]) intrinsic time is introduced as an increasing function of $p(t)$. If intrinsic time is identified with $p(t)$ itself, then $p(t) \in [0, p_f]$, p_f being the *intrinsic life time*, i.e., the intrinsic time at rupture. Obviously, $p(t)$ may be thus imagined to be an effective age mesure (i.e., *intrinsic elapsed time*) of material of the considered body.

Consider now a space dual to $\mathcal{S}_d$ namely a space $\mathcal{S}_s$ with the elements

$$\mathbb{X}^t := (\mathbb{T}(t), \Delta_{\mathrm{P}}(t), \theta(t), \mathbb{P}^t, \theta^t) \in \mathcal{S}_s \tag{5}$$

called the space of thermostress histories. Let us define a function

$$f(t) = \mathfrak{F}(\mathbb{X}^t) \quad \text{or} \quad \mathfrak{F} : \mathcal{S}_s \to \mathcal{R} \tag{P5}$$

which is named the yield function whose sign is the indiator of the nature of deformation process considered. Namely, if we define sets

$$\mathcal{U} := \{\mathbb{X}^t \in \mathcal{S}_s \mid \mathfrak{F}(\mathbb{X}^t) < 0, \dot{p}(t) = 0\}, \quad \mathcal{V} := \{\mathbb{X}^t \in \mathcal{S}_s \mid \mathfrak{F}(\mathbb{X}^t) > 0, \dot{p}(t) > 0\},$$

$$\mathcal{B} := \{\mathbb{X}^t \in \mathcal{S}_s \mid \mathfrak{F}(\mathbb{X}^t) = 0, \dot{p}(t) = 0\},$$

then $\mathcal{U}$ is the elastic region, $\mathcal{B}$ the corresponding elastic–plastic frontier (i.e., *yield surface*) and $\mathcal{V}$ elastic–viscoplastic region. For simplicity their obvious dependence on time t which is usually called *hardening* (when $\mathcal{B}$ expands) or *softening* (when $\mathcal{B}$ shrinks) is suppressed. Similar sets to $\mathcal{U}$ and $\mathcal{B}$ but without temperature influence taken into account were considered in [10].

The property (P3) allows an equivalent formulation of (2). By the polar decomposition theorem applied to the elastic distortion, i.e., $\Delta_E(t) = \mathbb{R}_E(t)\mathbb{U}_E(t)$ where $\mathbb{C}_E(t) := \Delta_E(t)^T \Delta_E(t) = \mathbb{U}_E(t)^2$ and choosing a rotation tensor to be $\mathbb{Q}(t) = \mathbb{R}_E(t)^T$ the functional (2) may be transformed into

$$\mathbb{S}(t) = \mathfrak{S}(\mathbb{C}_E(t), \Delta_P(t), \theta(t), \mathbb{P}^t, \theta^t) =: \mathfrak{S}(\gamma_1, \mathbb{P}^t, \theta^t)\,. \tag{6}$$

Here $\mathbb{S}(t) = (\det \mathbb{C}_E(t))^{1/2} \Delta_E(t)^{-1} \mathbb{T}(t)\Delta_E(t)^{-T}$ is the second Piola–Kirchhoff stress tensor. The following remarks may be interesting.

(R1) *In most plasticity theories plastic deformation history,* $\mathbb{P}^t$, *is replaced by the intrinsic elapsed time* $p(t)$. *Such a replacement delimits considerably the variety of considered phenomena.*

(R2) *For cyclic processes the so-called discrete memory* [13, 14] *is often represented by maximum previous plastic strain length which accounts for yield surface expansion during cyclic inelastic processes.*

(R3) *Another approach to yield surface expansion was given in* [6] *where the maximum previous stress intensity* $\max \|\mathbb{T}(\tau)\|$, $(\tau \in (-\infty, t])$, *was adopted to be the diameter of the yield surface with softening being neglected.*

(R4) *As an additional postulate* $\mathbb{U}_P(t) = \mathfrak{U}(\mathbb{F}^t)$ *(with* $\mathbb{F}^t : \mathcal{R}^+ \to \mathcal{L}^+$ *being the history of the total deformation gradient tensor and* $\mathbb{U}_P(t)^2 := \Delta_P(t)^T\Delta_P(t)$*) was assumed in* [7].

For the time being let us restrict our attention to processes with negligible temperature history dropping in such a way θ^t from the arguments appearing in (5) and (6). Moreover, if an *obliviator function* $h : [0,\infty) \to [0,1]$ is introduced (cf. [3, 18]), then the collection of all plastic deformation histories with the following inner product

$$<\mathbb{P}_1^t, \mathbb{P}_2^t> := \int_0^\infty h(s)^2 \operatorname{tr}\{\mathbb{P}_1^t(s)\mathbb{P}_2^t(s)^{\mathrm{T}}\}ds \tag{7}$$

forms a Hilbert space $\mathcal{H}$ with the norm $\| \mathbb{P}^t \|_h^2 := < \mathbb{P}^t, \mathbb{P}^t >$. On the other hand let us introduce

$$\mathbb{S}(\gamma_1, \mathbb{P}^t) = \mathbb{S}(\gamma_1, \mathbb{O}^t) + \mathfrak{s}(\mathbb{P}^t) \equiv \mathbb{S}^\# + \mathfrak{s}(\mathbb{P}^t) \tag{8}$$

with $\mathbb{O}^t(s) = \mathbb{O}$, and assume that $\mathfrak{s}$ is Fréchet-differentiable at this zero relative plastic deformation history [3], i.e.

$$\forall\, \varepsilon > 0, \exists\, \eta > 0 \text{ such that } \forall \mathbb{P}_1^t, \mathbb{P}_2^t \in \mathcal{H}, \| \mathbb{P}_1^t - \mathbb{P}_2^t \|_h < \eta \Rightarrow \| \mathfrak{s}(\mathbb{P}_1^t) - \mathfrak{s}(\mathbb{P}_2^t) \| < \varepsilon \tag{F1}$$

and, moreover [3]

$$\mathfrak{s}(\mathbb{P}^t) = \delta\mathfrak{s}(\mathbb{P}^t) + \tau(\mathbb{P}^t), \text{ with } \lim_{\| \mathbb{P}^t \|_h \to 0} \| \mathbb{P}^t \|_h^{-1} \tau(\mathbb{P}^t) \to 0. \tag{F2}$$

The linear functional $\delta\mathfrak{s} : \mathcal{H} \to \mathcal{L}$ appearing in the property (F2) is the *first variation* or *Fréchet differential* of $\mathfrak{s}$ at the zero relative plastic deformation history. From the theorem of the theory of Hilbert spaces, every bounded continuous linear functional may be written as an inner product. Such products may be not only scalars but second rank tensors as well. Therefore [3, 18]

$$\delta\mathfrak{s}(\mathbb{P}^t) = \int_0^\infty \mathcal{K}(\gamma_1, s)\, [\, \mathbb{P}^t(s)]\, ds \tag{F3}$$

with the kernel being a fourth rank tensor and, as beforehand, $\gamma_1 := (\mathbb{C}_{\mathrm{E}}(t), \Delta_{\mathrm{P}}(t), \theta(t))$. In this way by means of (F1–F3) the representation (8) reduces into

$$\mathbb{S}(t) \approx \mathbb{S}^\#(\gamma_1) + \int_0^\infty \mathcal{K}(\gamma_1, s)\, [\, \mathbb{P}^t(s)]\, ds\,. \tag{9}$$

The above formula is said to describe a *linear viscoplastic material of integral type*. Obviously, a similar procedure could lead to higher-order materials of integral type.

(R5) *The above procedure is extended to the more general case of thermoplastic histories by introducing an extended ten-dimensional Hilbert space with a norm based on dimensionless plastic and on dimensionless temperature history (with corresponding orthogonal subspaces of S_d). This allows us to write*

$$\mathbb{S}(t) = \mathbb{S}^{\#}(\gamma_1) + \int_0^{\infty} \mathcal{K}(\gamma_1,s)\,[\,\mathbb{P}^t(s)]\,\mathrm{d}s + \int_0^{\infty} \mathbb{K}(\gamma_1,s)\;\theta^t(s)\,\mathrm{d}s \tag{10}$$

extending the representation (9). *Here the new kernel is a second rank symmetric tensor* $\mathbb{K}(\gamma_1,s)$.

(R6) *An alternative approach is to take the discrete memory influence neglecting* $\mathbb{K}(\gamma_1, s)$ *and extending* γ_1 *to include*

$$\theta_m := \max_{\tau} \theta(\tau) \text{ as well as } \delta_{pm} := \max_{\tau} \| \mathbb{U}_P(\tau) - \mathbb{I} \|$$

with is especially suitable for cyclic inelastic processes (cf. [13]).

(R7) *Introducing the mapping* $p : (-\infty, t] \rightarrow [0,p(t)]$ *responsible for the intrinsic elapsed time into* (9) *and* (10) *these become general endochronic constitutive equations (cf., e.g.* [4]). *The inverse transformation, however, is not possible if a sequence of elastic and plastic deformation processes occurs.*

Suppose, now, that the considered plastic motion is slow and $\mathfrak{s}$ is k times Fréchet-differentiable. Under such conditions its asymptotic approximation (cf., Theorem 2 in [2]) may be written in the following way:

$$\mathbb{S}(t) = \mathbb{S}^{\#}(\gamma_1) + \sum_{I_k} \mathcal{P}_{m_1 \ldots m_k}(\gamma_1)\;\overset{(m_1)}{\theta}(t)[\overset{(m_2)}{\Delta_P}(t) ,\ldots, \overset{(m_k)}{\Delta_P}(t)\,] + o(\alpha^k)\,, \tag{11}$$

where I_k is the maximal set of ordered indices $m_1 ,\ldots, m_k$ whose sum is never larger than k and

$$\overset{(k)}{\Delta_P}(t) = \frac{\mathrm{d}^k}{\mathrm{d}\tau^k} \Delta_P(t-\tau)\,|_{\tau=0}, \quad \overset{(k)}{\theta}(t) = \frac{\mathrm{d}^k}{\mathrm{d}\tau^k} \theta(t-\tau)\,|_{\tau=0}\,,$$

whereas the residual term satisfies the condition

$$\lim_{\alpha \to 0} \alpha^k o(\alpha^k) = 0 .$$

A material characterized by (11) may be called an *elastic-visco-plastic material of plastic differential type and complexity* k. Its following special classes deserve special attention

$$\mathbb{S}(t) = \mathbb{S}^{\#}(\gamma_1) + \mathcal{P}_1(\gamma_1)\,[\dot{\Delta}_P(t)] + \mathbf{p}(\gamma_1)\,\dot{\theta}(t) \tag{12}$$

for $k = 1$ and

$$\mathbb{S}(t) = \mathbb{S}^{\#}(\gamma_1) + \mathcal{P}_1(\gamma_1)\,[\dot{\Delta}_P(t)] + \mathbf{p}(\gamma_1)\,\dot{\theta}(t) + \mathcal{P}_{11}(\gamma_1)\,[\dot{\Delta}_P(t), \dot{\Delta}_P(t)] \tag{13}$$

$$+ \mathcal{P}_2(\gamma_1)\,[\ddot{\Delta}_P(t)] + \mathcal{M}_1(\gamma_1)\,[\dot{\Delta}_P(t)]\,\dot{\theta}(t) + \mathbf{p}_{11}(\gamma_1)\,\dot{\theta}(t)^2 + \mathbf{p}_2(\gamma_1)\,\ddot{\theta}(t)$$

for $k = 2$. In (11)-(13) $\mathbb{S}^{\#}(\gamma_1)$ is termed *static stress* while $\mathbb{S} - \mathbb{S}^{\#}$ (for simplicity, difference between functions and their values is neglected here) is usually named *overstress tensor.* A danger of losing contact with real materials forces us, however, to check carefully the meaning of all the coefficients in (13) having support in the experimental evidence and thermodynamical considerations (cf., e.g. [15, 1, 9]).

In Müller's extended thermodynamics of thermoelastic bodies [15] (which corresponds to $\dot{\Delta}_P(t) = 0$ and $\ddot{\Delta}(t) = 0$) the elastic stress (i.e., stress during elastic deformation) is assumed to depend not only on γ_1 but on $\dot{\theta}(t)$ as well. This suggests dropping of $\mathbf{p}_2(\gamma_1)$ from (13). It seems logical to neglect $\mathcal{P}_2(\gamma_1)$ as well. Then (13) may be expressed by means of

$$\mathbb{S}(t) = \mathbb{S}^{\#}(\gamma_1, \dot{\theta}(t)) + \mathcal{P}(\gamma_1, \dot{\theta}(t))\,[\dot{\Delta}_P(t)] + \mathcal{P}_{11}(\gamma_1)\,[\dot{\Delta}_P(t), \dot{\Delta}_P(t)\,], \tag{14}$$

where $\mathcal{P}(\gamma_1, \dot{\theta}(t))$ is the fourth rank *tensor of plastic viscosity coefficients* and $\mathcal{P}_{11}(\gamma_1)$ the sixth rank tensor of plastic viscosity coefficients.

(R8) *The assumption of fading memory requires that* $\lim s^r h(s) = 0$ *when* $s \to \infty$. *In the terminology of* [5] *a material has total memory if it remembers everything (without above property) and finite* ω*-memory if* $h(s) = 0$, $s \in [\omega, \infty)$. *If the state of the particle is understood as the set of all equivalent histories leading to the same response at the actual time* t, *then* [5] *material is equipped with absolute memory if states could be only singletons. In our opinion, the notion of infinitesimal memory may be of* *interest for further exploration especially in connection with plastic materials of differential type.*

(R9) *Instead of the elastic distortion the total deformation tensor* $\mathbb{F}(t)$ *might be used in* (2) *as well. However, for such a replacement left and right Fréchet-differentials of the obtained stress functions are different for loading and unloading processes and this complicates the alternative formulation of* (F1-F3) *properties.*

4 Concluding remarks

In this paper the principle of determinism based on temperature as well as on plastic deformation history is formulated. In the case of fading (as well as a finite) memory, materials of integral and differential type are obtained showing a natural transition from a description based on functionals to a description characterized by discrete memory and evolution equations. As a future research direction, following directly from the paper, a more detailed consideration on infinitesimal plastic memory might be interesting.

Acknowledgement

Discussions with Professor H. Zorski and Dr. W. Kosinski on the subject are highly appreciated.

References

[1] Albertini, C., Montagnani, M. and Mićunović, M., Transactions of SMIRT-10, Los Angeles (1989).

[2] Coleman, B.D. and Noll, W., *Arch. Rational Mech. Anal.*, **6**, 355 (1960)

[3] Coleman, B.D. and Noll, W., *Rev. Mod. Phys.*, **33**, 239 (1961).

[4] Fazio, C.J., Proceedings of MECAMAT, ed. G. Cailletaud et al, Besançon : Mecamat, 185 (1988).

[5] Frischmuth, K. and Kosinski, W., *Arch. Mech.* (in press).

[6] Gurtin, M.E., *ASME J. Appl. Mech.*, **50**, 894 (1983).

[7] Holsapple, K.A., *Acta Mechanica*, **17**, 277 (1973).

[8] Kröner, E., *Arch. Rational Mech. Anal.*, **4**, 273 (1970).

[9] Maruszewski, B. and Mićunović, M., *Int. J. Engng. Sci.*, **27**, 955 (1989).

[10] Matsumoto, E., *Quart. J. Mech. Appl. Math.*, **35** (2), 197 (1982).

[11] Mićunović, M., *Bull. Acad. Polon. Sci., Ser. Sci. Techn.*, **22**, 663 (1974).

[12] Mićunović, M., Proceedings of the 1st National Congress on Mechanics, Athens- HSTAM, 364 (1986).

[13] Mićunović, M., Transactions of SMIRT-9, L, Rotterdam : Balkema, 195 (1987).
[14] Mroz, Z., Shrivastava, H.R. and Dubey, R.N., *Acta Mechanica*, **17**, 277 (1973).
[15] Müller, I., *Arch. Rational Mech. Anal.*, **41**, 319 (1971).
[16] Owen, D.R., *Arch. Rational Mech. Anal.*, **37**, 85 (1970).
[17] Stojanović, R., *Physica Status Solidi*, **2**, 566 (1962).
[18] Truesdell, C. and Noll, W., *The Non-Linear Field Theories of Mechanics*, Handbuch der Physik, Vol. III/3, ed. S. Flügge, Berlin, Springer-Verlag (1965).

Mićunović M.
Faculty of Mechanical Engineering
Univesity Svetozar Markovic
Sestre Janjic 6
YU-34000 Kragujevac
YUGOSLAVIA

J.P. NOWACKI

Dislocation dynamics in magneto-thermo-elastic continua

1 Introduction

In recent years, due to the technological applications of paramagnetic materials working in the presence of a high magnetic field in the domain of magnetic levitation, magnetic forming, fusion reactors, etc, the mechanical behaviour of such materials has received considerable attention from researchers (see, e.g. [1]).

The reversible generation of heat, thermal and elastic fields, caused by moving disloctions in an anisotropic solid with thermal conductivity has been investigated in [2]. After formulating the system of basic governing equations, the problem was solved completely by the method of Green's functions. For an infinite straight dislocation, the integrations were carried out by the aid of integral exponential functions.

The mechanical and electrical fields produced by moving dislocations in a paramagnetic solid with electrical conduction in the presence of a large magnetic field was investigated in [3]. The problem has been solved with the aid of Green's functions and the expressions for the fields produced by the infinite straight dislocation, which is moving at subsonic velocity, were given.

In recent years the general theory of moving dislocations and disclinations in thermo-piezoelectric continua was introduced [4]. This approach is based on the modelling of defects by the initial plastic distortion. The final expressions for thermo-piezoelectric fields were obtained in terms of the defect density and defect-flux tensor with the aid of Green's functions and dynamic Green's potential. Recently Minagawa and Shintani [5, 6] studied the electric and mechanical fields of moving dislocations by the method of continuously distributed dislocations. The formal solution in the most general form is exhibited by means of convolution integral in space-time. Next the piezoelectric field of a uniformly moving infinite straight dislocation is considered and the integrals are calculated completely. Several examples have been evaluated by numerical computation.

In the last two years a series of papers, concerned with the description of a piezoelectric field of dislocation, in the frame of the theory of dielectrics with polarization gradient [7], has been published by Nowacki and Hsieh [8] and Nowacki [9, 10]. The theory of dislocation, based on the generalization of Weingarten's theorem, as well as the theory based on Burgers' condition has been given. The distortion and electric fields due to dislocations have been calculated and the analogy and difference between the linear and surface concept of dislocation have been discussed.

2 Basic equations

The starting point of our studies will be the Maxwell relations for the electromagnetic theory of moving media. In MKS units we have

$$\operatorname{curl} \mathbf{H} = \dot{\mathbf{D}} + \mathbf{J}, \qquad \operatorname{div} \mathbf{D} = \rho_e, \tag{2.1}$$

$$\operatorname{curl} \dot{\mathbf{E}} = \mathbf{B}, \qquad \operatorname{div} \mathbf{B} = 0, \tag{2.2}$$

where $\mathbf{D}, \mathbf{E}, \mathbf{H}, \mathbf{B}$ designate the vectors of electric displacement, electric field, magnetic field and magnetic induction, respectively, while $\mathbf{J}$, ρ_e, and $\mathbf{v}$ stand for vector of electric current density, density of electric charge and velocity vector.

The following constitutive relations hold

$$\mathbf{D} = \varepsilon\, \mathbf{E}, \qquad \mathbf{B} = \mu_e \mathbf{H}\,.$$

If we introduce the potentials $\mathbf{A}$ and φ by the relations

$$\mathbf{E} = \operatorname{grad} \varphi + \dot{\mathbf{A}}\,, \qquad \mathbf{B} = \operatorname{curl} \mathbf{A}\,, \tag{2.3}$$

then Maxwell's equations (2.2) are satisfied identically.

Next, we shall derive the equations of motion for paramagnetic solids using Hamilton's principle. By variation of the action integral with respect to potentials φ and $\mathbf{A}$ as in [13] or by use of Lagrange multipliers as in [12] we can obtain the two remaining Maxwell equations (2.1). Varying the action integral with respect to the displacement $\mathbf{u}$, we obtain the equations of motion.

Finally, we will use the balance of energy and the Clausius-Duhem inequality in the following form:

$$\begin{aligned}\frac{\mathrm{d}}{\mathrm{d}t}\int [\tfrac{1}{2}\rho\mathbf{v}\cdot\mathbf{v} + \rho U + U_e]\, \mathrm{d}V &= \int \mathbf{X}\cdot\mathbf{v}\, \mathrm{d}V + \int \rho r\, \mathrm{d}V + \\ &+ \int [\,\mathbf{p}\cdot\mathbf{v} - \mathbf{q}\cdot\mathbf{n} - (\mathbf{E}\times\mathbf{H})\cdot\mathbf{n} + U_e\, \mathbf{v}\cdot\mathbf{n}\,]\, \mathrm{d}A\,,\end{aligned} \tag{2.4}$$

$$\rho\dot{S} + \left(\frac{q_i}{T}\right)_{,i} + \frac{\rho r}{T} \geq 0 \tag{2.5}$$

where U_e is the electromagnetic energy related to unit volume, U is the internal energy, r denotes the intensity of heat sources, $\mathbf{q}$ stands for the vector of heat flow. $\mathbf{X}$ is the body force, $\mathbf{p}$ represents the generalized stress vector associated with the thermoelastic and electromagnetic action. The expression $(\mathbf{E}\times\mathbf{H})\cdot\mathbf{n}$ represents the transmission of electromagnetic energy into a medium through the surface of the body, while $U_e\,\mathbf{v}\cdot\mathbf{n}$ describes the flow of electromagnetic energy produced by the body movement through the external electromagnetic fields.

The electromagnetic energy is expressed in the form:

$$U_e = \frac{1}{2}(\varepsilon_0 \mathbf{E}\cdot\mathbf{E} + \mu_0 \mathbf{H}\cdot\mathbf{H}). \tag{2.6}$$

Following the procedure described in [11], after linearization and assuming that the electric field is quasi-static, and in the absence of the heat sources, one obtains the following system of basic relations:

Equation of motion

$$\sigma_{ij,j} + (\mathbf{J}\times\mathbf{B})_i = \rho \ddot{u}_i\,. \tag{2.7}$$

Ohm's law

$$J_i = \sigma(E_i + \varepsilon_{ijk}\, v_j\, B_k^0 - \kappa\,\theta_{,i}). \tag{2.8}$$

Fourier's law

$$q_i = \lambda_{ji}\,\theta_{,j} + \kappa\, T_0\, J_i\,. \tag{2.9}$$

Field equations

$$\mathbf{E} = -\,\nabla\,\varphi, \quad \nabla\cdot\mathbf{J} = 0. \tag{2.10}$$

Constitutive relations

$$\sigma_{ij} = c_{ijkl}\,\beta_{kl} - \gamma_{ij}\theta\,, \tag{2.11}$$

$$S = \gamma_{ij}\,\beta_{ij} + c_\varepsilon \frac{\theta}{T_0}.$$

Residual entropy balance

$$(\mathbf{E} + \mathbf{v}\times\mathbf{B})\times\mathbf{J} + \nabla\cdot\dot{\mathbf{q}} + \rho S\, T = 0. \tag{2.12}$$

We consider here the medium embedded in the constant, primary magnetic field of an intensity H^0.

3 Mechanical, thermal and electric fields

In order to find the fields produced by moving dislocations, the method of continuously distributed dislocations is used. Let an infinite paramagnetic body contain the initial distortions β^{P}_{ij}. If the elastic distortions are denoted by β_{ij}, the total distortion may be written in the form:

$$\beta^{\mathrm{T}}_{ij} = \beta_{ij} + \beta^{\mathrm{P}}_{ij} \tag{3.1}$$

and it can be also expressed in terms of displacement

$$\beta^{\mathrm{T}}_{ij} = \nabla_i u_j . \tag{3.2}$$

Substituting (3.1) and (3.2) into the field equations (2.7)-(2.11) one arrives at the system of basic equations

$$\begin{gathered}
(c_{ijpq}\nabla_q \nabla_j - \rho\nabla_t \nabla_t \delta_{ip} + \sigma\mu^2 \varepsilon_{ijq}\,\varepsilon_{jpl}\, H_1 H_q \nabla_t) u_p - (\gamma_{ij} + \\
\sigma\kappa\mu\varepsilon_{pjq} H_q)\nabla_j\theta - \sigma\mu\varepsilon_{ipq} H_q \nabla_p\varphi = c_{ijkl}\, \beta^{\mathrm{P}}_{kl,j} , \\
(\gamma_{pj} + \sigma\kappa\mu\varepsilon_{pjq} H_q)\nabla_j\nabla_t u_p - T_0^{-1}\,(\lambda_{ip}\,\nabla_i\nabla_p - \kappa^2\sigma\, T_0\nabla_i\nabla_i - \\
c_\varepsilon \nabla_t)\theta + \kappa\hat{\sigma}\nabla_i\,\nabla_i\varphi = \gamma_{ij}\,\beta^{\mathrm{P}}_{ij} , \\
\varepsilon_{ipl}\mu H_l \nabla_i \nabla_t u_p\ \kappa\nabla_i \nabla_i\theta - \nabla_i \nabla_i\varphi = 0.
\end{gathered} \tag{3.3}$$

The plastic distortions are connected with the dislocation density tensor α_{ij} and the dislocation flux tensor V_{ijk} by the relations

$$\alpha_{ij} = \varepsilon_{ikl}\,\beta^{\mathrm{P}}_{lj,k} , \quad \tilde{\beta}^{\mathrm{P}}_{ij} = -\,\varepsilon_{imn}\, V_{mnj} . \tag{3.4}$$

Introducing the definition of Green's functions $G_{ij}, G_i, F_i, F, S_i, S$ one can formally write the solution of (3.3) in the form

$$\begin{aligned}
u_m(\mathbf{x}, t) = \iint \Big\{ & G_{nm}(\mathbf{x} - \mathbf{x}', t - t')\, c_{njkp}\, \beta^{\mathrm{P}}_{kp,j'}(\mathbf{x}', t') + \\
& + G_m(\mathbf{x} - \mathbf{x}',\ t - t')\, \gamma_{ij}\, \beta^{\mathrm{P}}_{ij}(\mathbf{x}', t') \Big\}\, \mathrm{d}\mathbf{x}'\, \mathrm{d}t' ,
\end{aligned} \tag{3.5}$$

$$\theta(\mathbf{x}, t) = \iint \Big\{ S_n(\mathbf{x} - \mathbf{x}',\ t - t')\, c_{njkp}\, \beta^{\mathrm{P}}_{kp,j'}(\mathbf{x}', t') +$$

$$+ S(\mathbf{x} - \mathbf{x}', t - t')\, \gamma_{ij}\, \beta^{\mathrm{P}}_{ij}(\mathbf{x}', t') \Big\}\, d\mathbf{x}'\, dt' ,$$

$$\varphi(\mathbf{x}, t) = \int\!\!\int \Big\{ F_n(\mathbf{x} - \mathbf{x}', t - t')\, c_{njkp} \beta^{\mathrm{P}}_{kp,j'}(\mathbf{x}, t') +$$

$$+ F(\mathbf{x} - \mathbf{x}', t - t')\, \gamma_{ij}\, \beta^{\mathrm{P}}_{ij}(\mathbf{x}', t') \Big\}\, d\mathbf{x}'\, dt' .$$

After several transformations we arrive at

$$\theta_{,n} = \int\!\!\int \big[\rho \dot{G}_i - \sigma\mu^2 \varepsilon_{ijq}\, \varepsilon_{ipl} H_l H_q G_p) \varepsilon_{nab} V_{abi} + \tag{3.6}$$

$$\sigma\mu^2 (\kappa S + F) \varepsilon_{ijq} H_q \beta_{ni,j} -$$

$$(c_{jpkq}\, S_{j,p} + S\gamma_{kq}) \big] \varepsilon_{skn}\, \alpha_{sq}\, dV'\, dt' .$$

This formula shows that the temperature-gradient caused by a distribution of moving dislocations are depending explicitly on the tensor of plastic deformation. When we neglect the magnetic coupling, the field can be stated entirely in terms of the dislocation-density and dislocation-flux tensors as in [2]. This is due to the fact that in the magneto-thermo-elastic case the Lorentz force is related to the total distortion. The formulae for the strain and electric field will be analogous, so the following presentation will be confined to the temperature field only.

After several transforamtions, integrations by parts and bearing in mind the definitions of Green's functions, one arrives at the general expression for the temperature field produced by a time-dependent continuous distribution of dislocations. Finally, substituting for Green's funcitons their Fourier transforms with respect to time and space, one obtains

$$\theta_{,n}(\mathbf{x}, t) = - \frac{i}{(2\pi)^4} \int\!\!\int\!\!\int\!\!\int \Big[\varepsilon_{skn} \Big\{ \rho\tau \frac{D_{4q}}{D} + e_{qjk} e_{jpl}\, \sigma\mu^2 H_l H_k \frac{D_{4p}}{D} \Big\} \times \tag{3.7}$$

$$V_{skq}(\mathbf{x}', t) - \varepsilon_{ijq}\, \sigma\mu \Big(\kappa \frac{\mathrm{D}_{44}}{D} + \frac{\mathrm{D}_{55}}{D} \Big) H_q\, \xi_j\, \beta_{ni}(\mathbf{x}', t') -$$

$$\varepsilon_{skn} \Big(\xi_p\, c_{jpkq} \frac{D_{j4}}{D} + \gamma_{kq} \frac{\mathrm{D}_{44}}{D} \Big) \alpha_{sq}(\mathbf{x}', t') \Big] \times$$

$$e^{i\{\xi\cdot(\mathbf{x}-\mathbf{x}') - \tau(t-t')\}} d\xi\, d\tau\, d\mathbf{x}'\, dt' ,$$

where a 5×5 matrix was introduced by

$$
\begin{array}{ccc}
c_{ipjq}\xi_p\xi_q - \rho\tau^2\delta_{ij} + & i\gamma_{ij}\xi_j + & i\varepsilon_{ipq}\sigma\mu H_q\xi_p \\
+\varepsilon_{ijqjpl}\mu^2 H_l H_q \sigma i\tau & + i\varepsilon_{ijq}\mu\sigma\kappa H_q\xi_j & \\
 & & \\
-i\gamma_{ij}\xi_j - & (T_0 i\tau)^{-1}(-\lambda\ \xi_{ip}\xi_{ip} & -\kappa\sigma(i\tau)^{-1}\xi_i\xi_i \\
-i\varepsilon_{ijk}\mu\sigma\kappa H_k\xi_j & +\kappa^2\sigma T_0\xi_i\xi_i + c_\varepsilon i\tau) & \\
 & & \\
\sigma\varepsilon_{pjq}H_q\xi_p\xi_q & \sigma\kappa(i\tau)^{-1}\xi_p\xi_q & \sigma\xi_i\xi_i
\end{array}
$$

D denotes its determinant and $D_{\alpha\beta}$ its $\alpha\beta$–co–factor.

4 Uniform motion of a dislocation

In the frame of this approach, the uniform motion of an infinite straight dislocation is considered. Under the assumption that the dislocation line is oriented along the x_3–axis and moves into the direction of the x_1–axis, introducing the moving reference frame [$\mathbf{y} = (x_1 - vt, x_2, x_3)$, v is the velocity of dislocation], the non–zero components of tensors β^P, α and V should be as follows:

$$\beta_{2q}(\mathbf{y}) = b_q H(-y_1)\delta(y_2),\ \ \alpha_{3i}(\mathbf{y}) = b_i\delta(y_1)\delta(y_2),\ \ V_{13j}(\mathbf{y}) = v\, b_j\ \delta(y_1)\delta(y_2)\ .$$

Substituting this into the general formula (3.6), and carrying out the integrations with respect to t' and $\mathbf{y}'$, one obtains finally:

$$\theta,_n = -\frac{i\,b_q}{(2\pi)^2}\iiiint\Big[\delta_{2n}\Big\{\rho v\xi_1\frac{\bar{D}_{4i}}{\bar{D}} + \varepsilon_{qjk}\varepsilon_{jpl}\,\sigma\mu^2 H_l H_k\frac{\bar{D}_{4p}}{\bar{D}}\Big\} \tag{4.1}$$

$$\varepsilon_{qij}\,\sigma\kappa\Big(\frac{\bar{D}_{44}}{\bar{D}} + \frac{\bar{D}_{55}}{\bar{D}}\Big)H_k\,\frac{\xi_j}{\xi} +$$

$$\varepsilon_{3kn}\Big(\xi_p\, c_{jpkq}\frac{D_{j4}}{\bar{D}} + \gamma_{kq}\frac{D_{44}}{\bar{D}}\Big)\Big]\,e^{i\{\xi_1(x_1 - vt) + \xi_2 x_2)}\,d\xi_1\,d\xi_2\,,$$

where $\bar{D} = D(k_1, k_2, k_3 = 0,\ \tau = k\,v)$.

This formula is suitable for the further numerical calculations. To facilitate the considerations one can discuss several particular cases, assuming selected directions of the primary uniform magnetic field.

References

[1] Maugin, G., *Continuum Mechanics of Electromagnetic Solid,* North-Holland, Amsterdam (1988).

[2] Minagawa, S., Hsieh, R.K.T., Nowacki, J.P. and Shintani, K., *Int. J. Engng. Sci.,* **26**, 11, 1169 (1988).

[3] Minagawa, S. and Oshiro, T., *Int. J. Engng. Sci.*, **25**, 4, 405 (1987).

[4] Minagawa, S., *Phys. Stat. Sol. (B),* **124**, 565 (1984).

[5] Minagawa, S., Shintani, K., *Phil. Mag.*, A, **51**, 277 (1985).

[6] Minagawa, S., Shintani, K., *Phil. Mag.,* A, **56**, 3343 (1987).

[7] Mindlin, R.D., *J. Elasticity,* **2**, 217 (1972).

[8] Nowacki, J.P. and Hsieh, R.K.T., *Int. J. Engng. Sci.*, **24**, 10, 1655 (1986).

[9] Nowacki, J.P., in *Deformable Solids and Structures,* ed. Yamamoto, Y. and Miya, K., North-Holland, Amsterdam (1987).

[10] Nowacki, J.P., *J. Tech. Phys.*, **29**, 1, 99-103 (1988).

[11] Nowacki, J.P., *J. Tech. Phys.*, **26**, 3-4, 315-323 (1985).

[12] Penfield, P. and Haus, H.A., *Electrodynamics of moving media*, MIT Press (1967).

[13] Van de Ven A.A.F., *Interaction of electromagnetic and elastic field in solids,* Report Technishe Hogeshool, Eindhoven, The Netherlands, 1978.

Nowacki J.P.
I.P.P.T.-P.A.N.,
Swietokrzyska 21
00-049 Warsaw
POLAND

M. OSTOJA-STARZEWSKI

Percolation, fractals, and entropy of disorder in damage phenomena

1 Introduction

In this work we give a brief account of our recent researches [3, 6] on damage phenomena. The objective is to develop a model which would account for random microstructural characteristics, scale dependence (of strength, etc.), geometry of damage, and fractal characteristics of damage morphologies. In order to accomplish these goals the analysis can neither be based on continuum damage mechanics nor on phenomenological stochastic models. Rather, it utilizes random fields on graphs to grasp the cooperative nature of damage evolution in the microstructure.

2 Percolation of damage

As a basis of the model we consider a planar graph $G(V, E)$, and a graph $G' = G(V', E')$ dual to it. Here V is the set of vertices and E is the set of edges joining all the interacting vertices. As discussed in [2, 3] the graph representation is useful for materials of granular, cellular, or fibrous types primarily. Thus, the vertices of V correspond now to the centres of mass of the grains, since we restrict our attention to granular microstructures. Furthermore, we admit nearest neighbour interactions only. It follows that E' represents the grain boundaries, while V' the points of their juncture. In Figure 1(a) we give an example of G being a Delaunay network, and in Figure 1(b) we show the corresponding Voronoi tessellation.

A graph-type setting affords a very clear definition of geometrical and/or physical randomness in the microstructure. This is very clearly demonstrated by the case of a discrete system $G(V, E)$ where E is the set of linearly elastic two-force members, and V is the set of frictionless joints. In recent studies [4, 5] we investigated the continuum approximations

$$\bar{\boldsymbol{\sigma}} = \boldsymbol{C}(\mathbf{x}, \omega, \delta)\boldsymbol{\varepsilon} \tag{2.1}$$

of such systems whereby the geometry is generated from the planar Voronoi tessellations. In (2.1) $\bar{\boldsymbol{\sigma}}$ is the (externally applied) Cauchy stress tensor, $\boldsymbol{\varepsilon}$ is the infinitesimal strain tensor, and $\boldsymbol{C}$ is an effective stiffness tensor of an approximating representative volume element ΔV (i.e., scale of the continuum approximation), centred at $\mathbf{x}$ and characterized by the parameter $\sigma = L/d$, in which $L = (\Delta V)^{\frac{1}{2}}$, and d is the

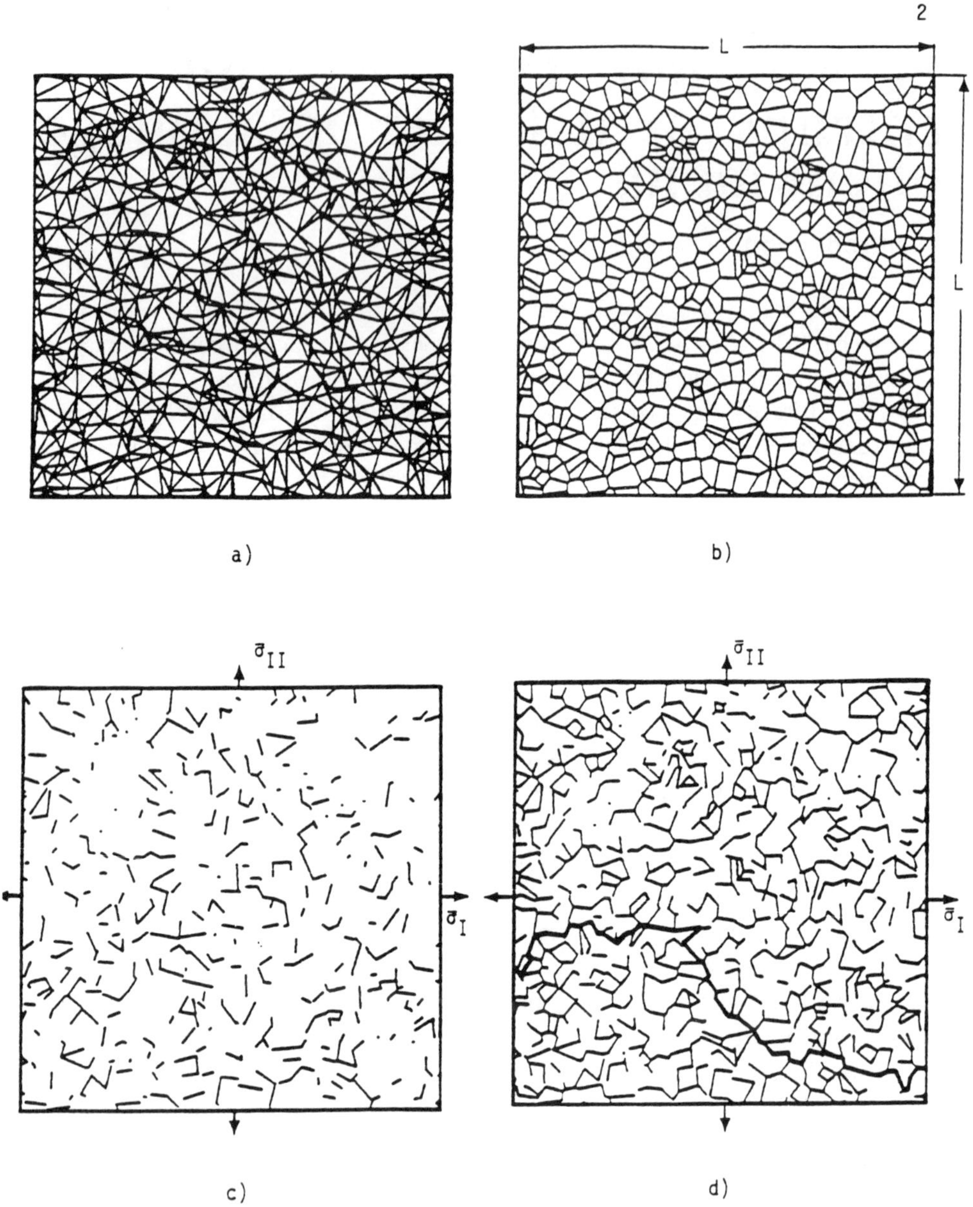

Figure 1. Graph representation of the microstructure; (a) the Delaunay network G; (b) the Voronoi tessellation G'; (c) distribution $z = Z(\omega)$ of damaged edges (Z_1 edges not shown) of E - partial damage; (d) macro-damage; $\bar{\sigma}_I$ and $\bar{\sigma}_{II}$ are two principal stresses

average grain size. The argument ω in (2.1) signifies the actual realization of the microstructure in ΔV, that is the actual realization of the microstructure's geometry and physical properties of all the edges.

Damage is now considered as a microscale phenomenon taking place in the grain boundaries only, that is on the graph G'. Since failure of any grain boundary $e \in E'$ is a random phenomenon, we introduce a random variable Z describing the state of e

$$Z(\omega, e) = \begin{cases} z_1 & \text{if } e \text{ is elastic, i.e., if } {}^{e}f(\boldsymbol{\sigma}, \omega) < 0 \\ z_2 & \text{if } e \text{ is inelastic (damaged), i.e., if } {}^{e}f(\boldsymbol{\sigma}, \omega) \geq 0 \end{cases}$$

in which ${}^{e}f$ is the failure condition. It follows that the damage states of the random medium $B = \{B(\omega); \omega \in \Omega\}$ are described by a binary random field Z on E'

$$Z : \Sigma \times \Omega \times E' \to \{z_1, z_2\} . \tag{2.3}$$

The damage state space Z is the set $\{z_1, z_2\}^{E'}$, and hence $z = Z(\omega)$ is a single realization of damage on E', or a damage state of a single material specimen $B(\omega)$. Σ is the space of values of $\overline{\boldsymbol{\sigma}}$. From now on we will denote an arbitrarily damaged material specimen by $B(z)$, and use B to represent $\{B(z); z \in Z\}$.

As discussed in [2, 3, 6], a consideration of the mean Cauchy stress

$${}^{\alpha}\sigma_{ij} = \sum_{\beta \in N_\alpha} [{}^{\alpha\beta}F_i \, {}^{\alpha\beta}l_j + {}^{\alpha\beta}F_j \, {}^{\alpha\beta}l_i] \tag{2.4}$$

leads to a conclusion that the body B is characterized by a statistical family of failure surfaces $\{{}^{\alpha}S; \alpha \in V\}$, rather than a single surface of determinsitic strength models. It is of interest to find the effective failure surface of any body $B(\omega)$ of B and to describe the scatter in distribution of these surfaces. The effective failure surface is not simply an ensemble average of all the surfaces but is found from the condition of percolation of damage.

Any spatial realization of damage z on E' can be characterized in terms of sets $C_1, C_2, \ldots,$ of connected damaged edges, called clusters. Of the primary interest to us is the cluster C_B spanning the entire body of B. Now we say that

$$\text{if } C_B \begin{array}{c} \text{does not} \\ \\ \text{does} \end{array} \text{ exist, then } B(z) \text{ is in a state of } \begin{array}{c} \text{partial-damage} \\ \\ \text{macro-damage} \end{array} \tag{2.5}$$

and introduce the percolation probability $P(C_B)$

$$P_{\bar{\sigma}}(C_B) = \begin{cases} 0 & \text{then } B \text{ is in a state of partial-damage} \\ x > 0 & \text{then } B \text{ is in a state of macro-damage.} \end{cases} \tag{2.6}$$

The situations of partial-damage and macro-damage are shown in Figures 1(c) and 1(d) respectively. In (2.3) and (2.6) $\bar{\sigma}$ ($\equiv \overline{\boldsymbol{\sigma}}$) was taken as a driving factor in damage evolution; clearly, other choices are possible.

The probability $P(C_B)$ depends on the probability of an edge $e \in E'$ becoming inelastic. Noting that $\boldsymbol{\sigma}$ is a function of $\overline{\boldsymbol{\sigma}}$, we write this probability as

$$P_{\bar{\sigma}}\{z(e) \mid z(\bar{e})\} \equiv P\{Z(\omega, e) = z(e) \mid Z(\omega, \bar{e}) = z(\bar{e}), \boldsymbol{\sigma}, \mathcal{E}f, e_n, {}^{\bar{e}}\{n\}\} . \tag{2.7}$$

There are, in principle, two generic cases of damage percolation phenomena - correlated and uncorrelated - contingent on whether the occurrence of damage events at the microscale

$$z_1(e) \to z_2(e) \tag{2.8}$$

does or does not depend on the states of edges in $\bar{e}$. These cases were analysed and discussed in [3] and [6]; it is one of the principal results of these studies that the percolation approach leads to an explanation of the typically observed size effects in strength of materials.

3 Fractals and multifractals

It follows by the results of percolation theory (see, e.g. [7]) that the cluster of percolating damaged edges should be a fractal for δ sufficiently large. In order to make the ideas more precise we introduce a refinement of (2.2), as follows

$$Z(\omega,e) = \begin{cases} z_1 & \text{if } e \text{ is elastic,} \\ z_2 & \text{if } e \text{ is plastic,} \\ z_3 & \text{if } e \text{ is broken .} \end{cases} \tag{3.1}$$

Clearly, the mechanical states of B are now described by a ternary random field Z on E'. $Z = Z_3$ will signify the formation of microcracks in the grain boundaries, and their clusters will represent cracks of all scales forming in the body domain.

Let C_B denote now a cluster of linked-up cracks, which is, in fact, a profile of a fracture surface. In order to characterize C_B we introduce a surface roughness R and relate it to the scale l of measurement by

$$R = l^{D_s}, \tag{3.2}$$

where D_s is the fracture surface dimension. On the other hand, noting that the fracture energy E is dissipated exactly on the set of all damaged edges, we can introduce the damage process dimension (c is a material constant)

$$E = cl^D . \tag{3.3}$$

The two fractal dimensions D_s and D do not, in general, coincide but actually take values from ranges of spectra because various types of microdamage events occur simultaneously to result in total fracture. Experimental results in support of this statement were obtained in the form of a multifractal pattern of data by Williford [8]. We now present a formulation which shows that our micromechanical model provides a very convenient basis for the study of multifractal patterns. First, we observe from (3.1) that three types of percolations are involved in damage processes

$$\Pi_1 : z_1 \rightarrow z_2 \text{ ductility,} \tag{3.4}$$

$$\Pi_2 : z_2 \rightarrow z_3 \text{ ductile fracture,}$$

$$\Pi_3 : z_1 \rightarrow z_3 \text{ brittle fracture.}$$

Each one of these percolations results in a different spatially random geometric object with different fractal characteristics. We see that brittle fracture occurs thorugh the Π_3 percolation and corresponds to a point A of the multifractal diagram[1] proposed in Figure 2. No plastic edges are present in the body and hence, if we assume that all broken edges belong to the single crack without any branches we find $D_s = D$. That is, the energy has been dissipated on the fracture surface only. One possible realization of this surface is shown by a set of very thick edges for a single specimen $B(z)$ in Figure 1(d). It follows immediately from this figure that, if there are any branches or separate microcracks present in the body, the inequality $D_s < D$ will hold.

The case of ductile fracture is more complicated. There are two fundamental cases to consider:

(i) no brittle fracture events at the microscale occur;

---

[1] The term multifractal is used here in a sense different from the classical formulation; it connotes presence of two fractals, one of dimension D_s and another of dimension D.

(ii) all three percolations occur.

Considering case (i) first, we see that Π_2 occurs on the set generated by Π_1. Energy is dissipated on the set of all z_2's and z_3's and this gives the dimension D of (3.3). Since D_s will correspond to the set of all z_3's along the main crack, the inequality $D_s < D$ will hold.

In case (ii), the main crack is a result of the Π_2 and Π_3 percolations. Again, D_s will correspond to the set of all z_3's along the main crack, and $D_s < D$. However, since the Π_3 percolation represents no fracture toughness, D in this case is smaller than in case (i) above. Thus, point C of the multifractal diagram corresponds to case (i), and point B lying anywhere between A and C, corresponds to case (ii).

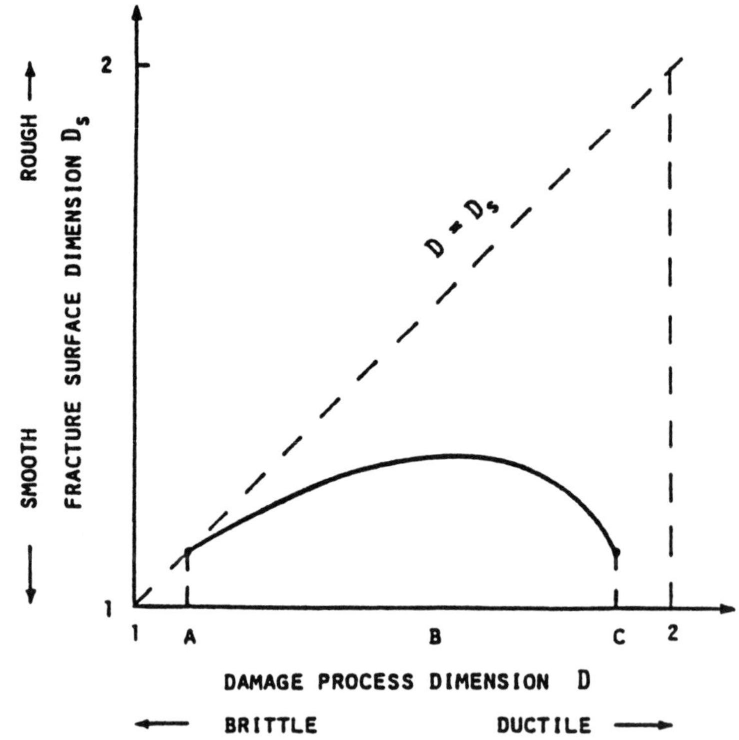

Figure 2. The multifractal diagram for damage and fracture.

4 Entropy of disorder in thermomechanics of damage

Three possible percolations defined in (3.4) are governed by the conditional probability (2.7). In case the length scale of stress concentration is of the order of grain size this probability will simplify to

$$P_{\bar{\sigma}}\{z(e) \mid z(\bar{e})\} = P_{\bar{\sigma}}\{z(e) \mid z(N_e)\}, \quad \forall\, e \in E', \tag{4.1}$$

where N_e is the set of nearest neighbours of edge e. This defines the so-called Markov property of field Z. In view of the equivalence of Markov random fields and Gibbs ensembles [1] it follows that Z is described by a Gibbs probability measure

$$P_{\bar{\sigma}}(z) = Y^{-1} \exp[-W(z)], \tag{4.2}$$

where $W(z)$ is the energy, in the informational sense, of the state of z, and

$$Y = \sum_{z \in Z} \exp[-W(z)] \tag{4.3}$$

is the normalizing constant called the partition function. The uncertainty associated with any damage state z is defined, and found from (4.2), as

$$S_i(z) = \ln[P_{\bar{\sigma}}(z)]^{-1} = W(z) + \ln Y. \tag{4.4}$$

It follows that the informational entropy, or entropy of disorder, of field Z on E' is

$$S_i = - \sum_{z \in Z} S_i(z) P_{\bar{\sigma}}(z) = \langle W \rangle + \ln Y. \tag{4.5}$$

If we call $F = -\ln Y$ the informational free energy, we will find from (4.4)

$$F = \langle W \rangle - S_i \tag{4.6}$$

and conclude that the Gibbs measure minimizes free energy F, and hence maximizes uncertainty as measured by entropy, among all probability measures having the same expected value for the energy as $P_{\bar{\sigma}}$.

The total physical internal energy in the body is

$$U(z) = U^{E}(z) + U^{I}(z), \tag{4.7}$$

where $U^{E}(z)$ is the total elastic strain energy, identical with the Helmholtz free energy $\Psi(z)$, while $U^{I}(z)$ is the total inelastic energy dissipated in the formation of z_2- and z_3-type edges. It has been established in [6] that

$$U^{I}(z) = \upsilon W(z), \tag{4.8}$$

where υ is an average elastic strain energy per edge e at the elastic-inelastic transition. Since for any specimen $B(z)$ we have

$$U(z) = \Psi(z) + T\,S(z) \tag{4.9}$$

it follows that the physical energy $S(z)$ can be related to the uncertainty of state z, according to (4.4), as

$$S(z) = \frac{\upsilon}{T}\,[S_i(z) - \ln Y\,] \tag{4.10}$$

in which T is the absolute temperature.

The direct link made here between the entropy and the state of damage of the material specimen leads to a possibility of derivation [6] of effective constitutive laws in the continuum approximation using the formalism of free energy funciton and the rate of entropy production (dissiaption function) extended to random media.

References

[1] Kindermann, R. and Snell, J.L., Markov random fields and their applications, *Contemporary Mathematics*, Vol. 1, Am. Math. Soc., Providence, R.I., 1980.

[2] Ostoja-Starzewski, M., Graph approach to the constitutive modelling of heterogeneous solids, *Mech. Res. Comm.*, **14** (4), pp. 255-262, 1987.

[3] Ostoja-Starzewski, M., Mechanics of damage in a random granular microstructure: Percolation of inelastic phases, *Lett. Appl. Engrg. Sci.*, **27** (3), pp. 315-326, 1989.

[4] Ostoja-Starzewski, M. and Wang, C., Effective moduli of planar random elastic Delaunay networks, in *Continuum Mechanics and its Applications*, S.K. Malik and G.A.C. Graham, Eds., Hemisphere Publishers, New York, 1989.

[5] Ostoja-Starzewski, M. and Wang, C., Linear elasticity of planar Delaunay networks: Random field characterization of effective moduli, *Acta Mech.*, **80**, pp. 61-80, 1989.

[6] Ostoja-Starzewski, M., Damage in a random microstructure: Size effects, fractals, and entropy maximization, *Appl. Mech. Rev.*, **42** (11, Part 2), pp. 202-212, 1989.

[7] Stauffer, D., *Introduction to Percolation Theory*, Taylor and Francis, London and Philadelphia, 1985.

[8] Williford, R.E., Multifractal fracture, *Scripta Metall.*, **22** (11), pp. 1749-1754, 1988.

Ostoja-Starzewski M.
School of Aeronautics and Astronautics
Purdue University
West Lafayette, Indiana 47907
U.S.A.

V.Z. PARTON

Fracture mechanics of composite materials on account of the microstructure

1 Introduction

The problem of determination of local stress distributions in the neighbourhood of a macrocrack in highly non-homogeneous composites is practically unsolvable in the framework of traditional analytical or numerical methods. This is the reason why a new approach to this problem is suggested. This approach is based on the application of the asymptotic homogenization method [1] and takes into account the additional boudnary layer-type solutions, which describe the local effects occurring in the neighbourhood of a crack [2]. As an example of the application of the suggested method, we consider the problem of a normal rupture crack in a two-component laminated composite. A local distribution of stresses around the crack is obtained and the analysis of the dependence of stress intensity coefficients on the material parameters is performed.

2 Auxiliary problems

Let us consider a plane elasticity problem for a straight-line crack in a composite periodic medium in the case where the crack's length $(2a)$ is much larger than the dimension of a periodic cell ε. We assume that the elastic medium has a two-dimensional inhomogeneity in the plane (x_1, x_2) and the macrocrack lies at the cell boundary. We further assume that the loads at the edges of the crack are normal and self-balanced.

To analyse the stress-strain state around the crack we obtained the asymptotic solution of the elasticity equations for the periodically inhomogeneous half-plane $x_2 > 0$ in the case of mixed boundary conditions at $x_2 = 0$.

Defining the boundary conditions at $x_2 = 0$ amounts to specifying the stresses $\sigma_{i2}(x_1, \pm 0)$, $i = 1,2,3$, in the interval $|x_1| < a$ and mixed conditions on the displacements and stresses for $|x_1| > a$ (Figure 1).

Let us obtain the solutions of two auxiliary problems for this domain to satisfy the mixed boundary conditions at $x_2 = 0$ in the framework of the asymptotic homogenization method.

We consider the equations

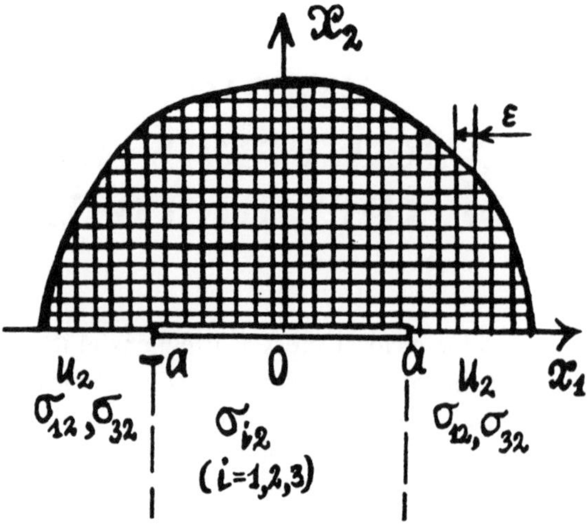

Figure 1.

$$\sigma^{(\varepsilon)}_{i\,\alpha,\,\alpha}(x_1, x_2) = 0,$$

$$\sigma^{(\varepsilon)}_{i\,\alpha} = C_{i\alpha k\beta}(y_1, y_2) U^{(\varepsilon)}_{k,j}(x_1, x_2)\,, \tag{1}$$

where $\sigma^{(\varepsilon)}_{i\,\alpha}$ are the stresses; $U^{(\varepsilon)}_{k}$ are the displacements. The elastic coefficients $c_{i\alpha k\beta}(y_1, y_2)$ are taken to be one-periodic functions of $y_1 = x_1/\varepsilon$, $y_2 = x_2/\varepsilon$; the Latin subscripts range from 1 to 3; the Greek subscripts from 1 to 2; the comma means differentiation with respect to x_1, x_2, and the summation convention for repeated indices is applied.

For the first of the auxiliary problems we assume that

$$\sigma^{(\varepsilon)}_{i\,2}(x_1, 0) = p_i(x_1), \quad i = 1,3, \tag{2}$$

at the boundary $x_2 = 0$. For the second auxiliary problem instead of (2) the mixed boundary conditions are assumed:

$$\sigma^{(\varepsilon)}_{i\,2}(x_1, 0) = p_i(x_1),\; i = 1,2,3,\;\; U^{(\varepsilon)}_{2}(x_1, 0) = V_2(x_1). \tag{3}$$

3 Asymptotic homogenization

For (1), (2) the asymptotic representation of the solution is bound to be:

$$U_n^{(\varepsilon)} = U_n^{(0)}(x_1, x_2) + \varepsilon[N_n^{k\beta}(y_1, y_2) + N_n^{(1)k\beta}(y_1, y_2)]\, U_{k,\beta}^{(0)} + \dots ,$$

$$\sigma_{ij}^{(\varepsilon)} \approx \sigma_{ij}^{(0)} = \left[C_{i\alpha k\beta} + C_{i\alpha n\gamma} \, (N_{n|\gamma}^{k\beta} + N_{n|\gamma}^{(1)k\beta} \right] U_{k,\beta}^{(0)} . \tag{4}$$

Here the vertical line in the subscript means differentiation with respect to y_1, y_2 and $U_n^{(0)}(x_1, x_2)$ is a solution of the homogenized problem

$$<C_{i\alpha k\beta}> U_{k,\alpha\beta}^{(0)} = 0, \; <C_{i\alpha k\beta}> = \int_0^1 \int_0^1 (C_{i\alpha k\beta} + C_{i\alpha n\gamma} N_{n|\gamma}^{k\beta})\mathrm{d}y_1\,\mathrm{d}y_2 . \tag{5}$$

with boundary conditions

$$C_{i2k\beta}^{*} U_{k,\beta}^{(0)} = p_i(x_1) \text{ at } x_2 = 0, \; i = 1,2,3, \tag{6}$$

$$C_{i2k\beta}^{*} = \int_0^1 C_{i2k\beta}(y_1, 0)\mathrm{d}y_1, \; C_{i\alpha k\beta} = c_{i\alpha k\beta} + c_{i\alpha n\gamma} N_{n|\gamma}^{k\beta} . \tag{7}$$

The functions $N_n^{k\beta}(y_1, y_2)$ are solutions of the local problem for a cell, which are one-periodic in y_1, y_2,

$$(c_{i\alpha n\gamma} N_{n|\gamma}^{k\beta})|_{\alpha} = - c_{i\alpha k\beta|\alpha} , < N_n^{k\beta} > = 0. \tag{8}$$

And the functions $N_n^{(1)k\beta}(y_1, y_2)$ are one-periodic only in y_1 and are determined from the boundary-layer problem

$$(c_{i\alpha n\gamma} N_{n|\gamma}^{(1)k\beta})|_{\alpha} = 0, \tag{9}$$

$$c_{i2n\gamma} N_{n|\gamma}^{(1)k\beta} = C_{i2k\beta}^{*} - C_{i2k\beta}(y_1, 0) \text{ at } y = 0, \; i = 1,2,3,$$

$$N_n^{(1)k\beta}(y_1, y_2) \to 0 \text{ as } y_2 \to +\infty.$$

For the second auxiliary problem (1), (3) the following asymptotic representation of the solution is obtained:

$$U_n^{(\varepsilon)} = U_n^{(0)}(x_1, x_2) + \varepsilon\left[N_n^{k\beta}(y_1, y_2) + N_n^{(2)k\beta}(y_1, y_2)\right] U_{k,\beta}^{(0)} + \ldots$$

$$\sigma_{i\alpha}^{(\varepsilon)} \approx \sigma_{i\alpha}^{(0)} = \left[C_{i\alpha k\beta} + C_{i\alpha n\gamma}(N_{n|\gamma}^{k\beta} + N_{n|\gamma}^{(2)k\beta})\right] U_{k,\beta}^{(0)}. \tag{10}$$

Here $U_n^{(0)}(x_1, x_2)$ is the solution of the homogenized problem (5) in the case of the following boundary conditions at $x_2 = 0$:

$$C_{i2k\beta}^{*} U_{k,\beta}^{(0)} = p_i(x_1),\ i = 1, 3;\ U_2^{(0)}(x_1, 0) = V_2(x_1). \tag{11}$$

The functions $N_n^{(2)k\beta}(y_1, y_2)$ are one-periodic only in y_1; they are determined from the following boundary-layer problem:

$$(c_{i\alpha n\gamma} N_{n|\gamma}^{(2)k\beta})|_\alpha = 0,$$

$$c_{i2n\gamma} N_{n|\gamma}^{(2)k\beta} = C_{i2k\beta}^{*} - C_{i2k\beta}(y_1, 0) \text{ at } y_2 = 0,\ i = 1,3, \tag{12}$$

$$N_2^{(2)k\beta}(y_1, 0) = -N_2^{k\beta}(y_1, 0) + \text{const},$$

$$N_n^{(2)k\beta} \to 0 \text{ as } y_2 \to +\infty.$$

It has been proved that the problems (8), (9), (12) have unique solutions.

To determine the microstructure of stresses in the neighbourhood of the vertex of the crack we consider the auxiliary domain IV, containing the end of the crack (Figure 2).

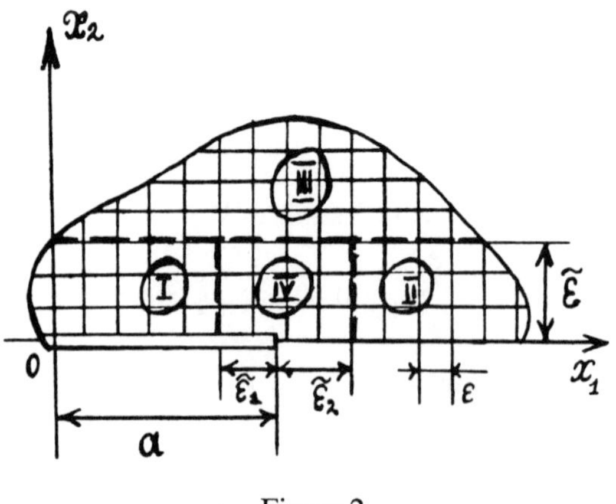

Figure 2

Now we wish to match the solution in domain IV with the above two types of boundary-layer solutions at the vertical boundaries, namely, with the solution of problem (4) in the domain I and the solution of problem (10) in domain II. The functions $N_n^{(1)k\beta}$ and $N_n^{(2)k\beta}$ are the boundary-layer solutions and can, therefore, be neglected in domain III (Figure 2). For the domain IV, which contains only a few periodic cells, the solution can be found numerically. In the case of a normal-rupture crack, the plane-elasticity problem in domain IV has the following form:

$$\sigma_{\alpha\beta,\beta} = 0, \quad \sigma_{\alpha\beta} = c_{\alpha\beta\lambda\mu}(y_1, y_2)u_{\lambda,\mu},$$

$$\sigma_{12} = 0 \text{ at } x_2 = 0, \; a - \tilde{\varepsilon}_1 < x_1 < a + \tilde{\varepsilon}_2,$$

$$\sigma_{22} = p_2(x_1) \text{ at } x_2 = 0, \; a - \tilde{\varepsilon}_1 < x_1 < a,$$

$$u_2 = 0 \text{ at } x_2 = 0, \; a < x_1 < a + \tilde{\varepsilon}_2, \tag{13}$$

$$\sigma_{\mu 1} = \left[c_{\mu 1\lambda\beta} + c_{\mu 1 n\gamma} \left(N^{\lambda\beta}_{n|\gamma} + N^{(1)\lambda\beta}_{n|\gamma} \right) \right] U^{(0)}_{\lambda,\beta} \text{ at } \begin{cases} x_1 = a - \tilde{\varepsilon}_1, \\ 0 < x_2 < \tilde{\varepsilon}, \end{cases}$$

$$\sigma_{\mu 1} = \left[c_{\mu 1\lambda\beta} + c_{\mu 1 n\gamma} \left(N^{\lambda\beta}_{n|\gamma} + N^{(2)\lambda\beta}_{n|\gamma} \right) \right] U^{(0)}_{\lambda,\beta} \text{ at } \begin{cases} x_1 = a + \tilde{\varepsilon}_2, \\ 0 < x_2 < \tilde{\varepsilon}, \end{cases}$$

$$\sigma_{\mu 2} = \left[c_{\mu 2\lambda\beta} + c_{\mu 2 n\gamma} \right) N^{\lambda\beta}_{n|\gamma} \left. \right) \right] U^{(0)}_{\lambda,\beta} \text{ at } \begin{cases} x_2 = \tilde{\varepsilon}, \\ a - \tilde{\varepsilon}_1 < x_1 < a + \tilde{\varepsilon}_2, \end{cases}$$

$$\mu = 1,2.$$

The functions $u_\lambda^{(0)}(x_1, x_2)$ are solutions of the following homogenized problem:

$$< C_{\lambda\alpha\mu\beta} > U^{(0)}_{\mu,\alpha\beta} = 0 \text{ in } x_2 > 0,$$

$$C^*_{12\alpha\beta} U^{(0)}_{\alpha,\beta} = 0 \text{ at } x_2 = 0, \; |x_1| < \infty, \tag{14}$$

$$C^*_{22\alpha\beta} U^{(0)}_{\alpha,\beta} = p_2(x_1) \text{ at } x_2 = 0, \; |x_1 = | < a,$$

$$U_2^{(0)} = 0 \text{ at } x_2 = 0, \; |x_1| > a.$$

The coefficients in (14) are determined by formulae (5), (7). It has proved possible, by considering the local domain IV with known boundary conditions, to compute with sufficient accuracy the stress fields around the crack with an arbitrarily located vertex in the cell of the composite.

4. Application

Let us apply the above approach to the analysis of the stress local distribution and the stress intensity coefficients when the macrocrack is perpendicular to the layers of the two-component laminated composite (Figure 3).

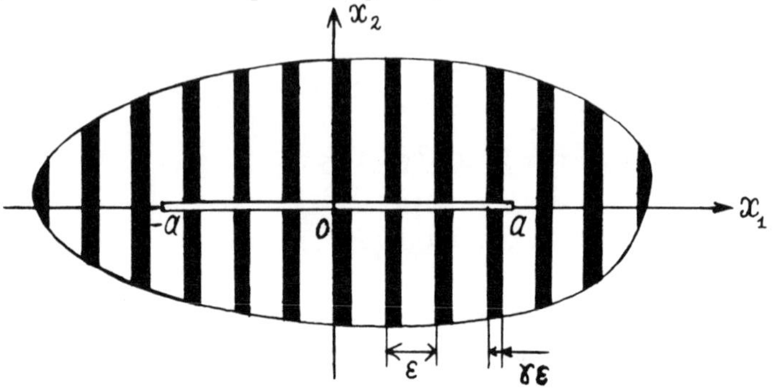

Figure 3.

The cell in this case consists of two layers with material parameters E_1, ν_1 and E_2, ν_2. The solutions of the local problem (8) and homogenized problem (14) in this case can be determined analytically. And the solutions of boundary-layer problems (9), (12) can be determined numerically, by means of the finite-element method.

It was assumed that $\gamma = 0.667$, which means that the thickness ratio of the first and second layers is 2:1. The vertex of the crack was in the middle of the wider layer. It was supposed that $\tilde{\varepsilon}_1 = 2.333\varepsilon$ and $\tilde{\varepsilon}_2 = 2.667\varepsilon$. We looked at two cases with various material parameters. In the first case $E_1 = 20$ GPa, $\nu_1 = 0.3$; $E_2 = 1.5$ GPa, $\nu_2 = 0.446$. In the second case we interchanged these materials.

Thus, in the first case the vertex of the crack lies in the more rigid material, and in the second case in the less rigid one.

In both cases we calculated the stress intensity coefficients. In the first case

$$K_T / \left(p\sqrt{\pi a}\right) = \lim_{x_1 \to a+0} \left[\left(2\pi(x_1 - a)\right)^{1/2} \sigma_{22}(x_1, 0)\right] = 3.21$$

and in the second:

$$K_T / \left(p\sqrt{\pi a}\right) = 1.15.$$

The local distributions of the stresses σ_{11}/p and σ_{22}/p in the neighbourhood of the vertex of the crack for the first case are shown in Figures 4 and 5.

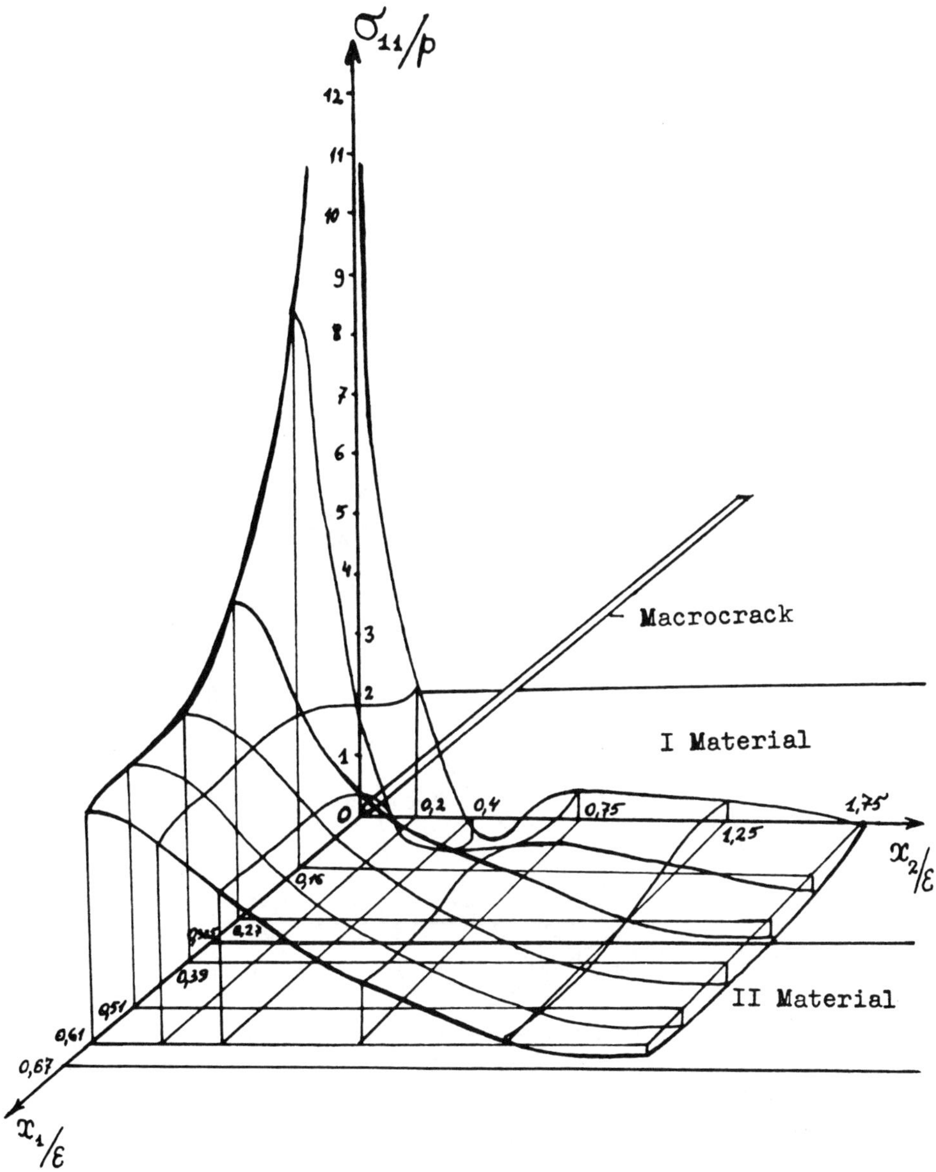

Figure 4. Local distribution of the stress σ_{11}/p in the neighbourhood of the vertex of the crack (first case).

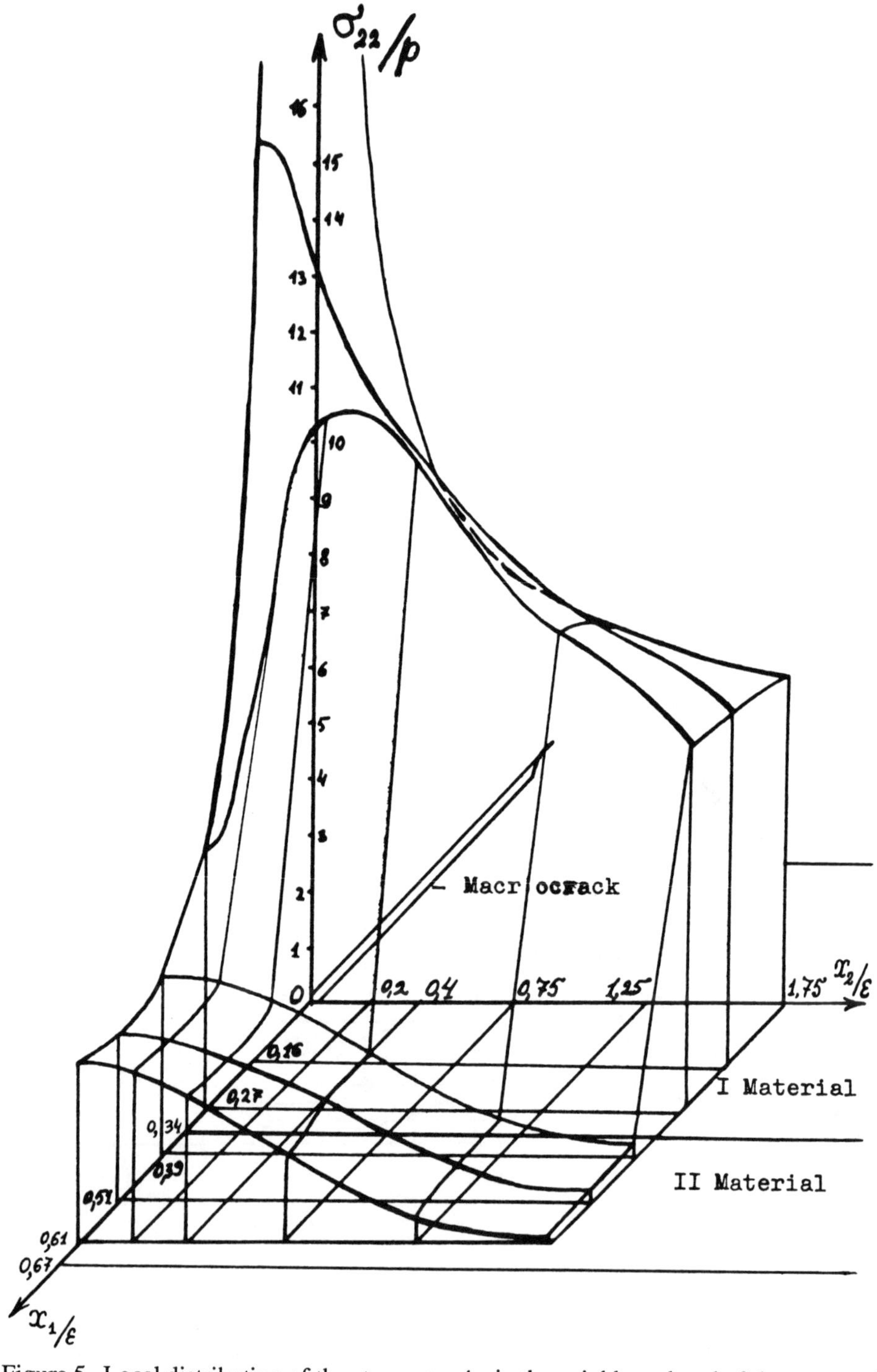

Figure 5. Local distribution of the stress σ_{22}/p in the neighbourthood of the vertex of the crack (first case).

References

[1] Kalamkarov, A.L., Kudryavtsev, B.A. and Parton, V.Z. Asymptotic homogenization method in mechanics of composite materials with regular structure. In: *Itogi Nauki i Techniki VINITI. Ser. Mechanics of Deformable Solids*, Moscow, V.19, 1987, pp. 78-147.

[2] Parton, V.Z., Kalamkarov, A.L. and Kudryavtsev B.A. On the research of microstresses in the neighbourhood of a crack in composite materials. In: *Failure Analysis - Theory and Practice.* Proc. 7th European Conf. on Fracture, Budapest, 1988, V.1, pp. 427-432.

Parton V.Z.
Moscow Institute of Chemical Machines,
Department of Mathematics
Karl Marx Street 21/4
Moscow B-99 107884
U.S.S.R.

J. J. TELEGA

Piezoelectricity and homogenization. Application to biomechanics

Introduction

Various methods of homogenization theory have penetrated into many fields of continuum and structural mechanics. In the present concise contribution we shall consider the problem of finding effective (macroscopic) properties of microscopically heterogeneous piezoelectric bodies. The case of non-uniform homogenization is studied. In this way one can take into account the non-uniformity of say, a, long bone, cf. [1].

To derive elastic, piezoelectric and dielectric macroscopic moduli the variational method of Γ-convergence is used. Two particular interrelated cases of such convergence will be discussed: epi-convergence and epi/hypo-convergence. Performing non-uniform, sometimes called quasi-periodic [11], homogenization we obtain the macroscopic moduli as functions of the macroscopic variables only.

Bones exhibit piezoelectric behaviour and have regular, though non-uniform, internal structure [6, 8]. It seems that homogenization methods can be applied when studying macroscopic behaviour of such biological materials. Some suggestions are given here, yet this interesting and complicated problem requires further studies.

1. Formulation of the homogenization problem and results

Let $\Omega \subset R^3$ denote a bounded, sufficiently regular domain such that $\overline{\Omega}$ stands for the piezoelectric solid considered. $\Gamma = \partial\Omega$ is the boundary of Ω. If $\mathbf{u} = (u_i)$ is a displacement field then $e_{ij}(\mathbf{u}) = u_{(i,j)}$ is the strain tensor; $i, j = 1, 2, 3$. By $\mathbf{D} = (D_i)$, $\mathbf{E} = (E_i)$, $\mathbf{P} = (P_i)$, $\boldsymbol{\sigma} = (\sigma_{ij})$ we denote the electric displacement vector, the electric field, the polarization vector and the stress tensor, respectively. The internal energy is $U = U(x, y, \mathbf{e}, \mathbf{D})$, $x \in \Omega$, $y \in Y$. Here Y is a so-called basic cell [2, 9, 11]. In the linear case we have

$$U(x, y, \mathbf{e}, \mathbf{D}) = \frac{1}{2} a_{ijkl}(x, y) e_{ij} e_{kl} - h_{ijk}(x, y) D_i e_{jk} + \frac{1}{2} \kappa_{ij}(x, y) D_i D_j . \quad (1.1)$$

For each $x \in \Omega$ the functions $a_{ijkl}(x, .)$, $h_{ijk}(x, .)$ and $\kappa_{ij}(x, .)$ are Y-periodic. Usual symmetry relations hold. We assume that, cf. [11]

$$a_{ijkl} \in W^{1,\infty}(\Omega, L^{\infty}(Y)),\ h_{ijk} \in W^{1,\infty}(\Omega, L^{\infty}(Y)),\ \kappa_{ij} \in W^{1,\infty}(\Omega, L^{\infty}(Y)) . \quad (1.2)$$

We also make the following assumptions, cf. [5]

$$\exists\lambda_0 > 0, \lambda_0 (|\mathbf{e}|^2 + |\mathbf{D}|^2 \le U(x, y, \mathbf{e}, \mathbf{D}) \le \rho(x)(a(y) + |\mathbf{e}|^2 + |\mathbf{D}|^2), \quad (1.3)$$

$$|U(x, y, \mathbf{e}, \mathbf{D}) - U(x', y, \mathbf{e}, \mathbf{D})| \le \omega(|x - x'|)(a(y) + U(x, y, \mathbf{e}, \mathbf{D}), \quad (1.4)$$

for each $x, x' \in \Omega$, $y \in Y$, $\mathbf{e} \in \mathbb{M}_s^3$, $\mathbf{D} \in R^3$; moreover $a \in L^1_{\mathrm{loc}}(R^3)$ is Y-periodic, the function ρ is continuous and non-negative, $\omega : R^+ \to R^+$ is an increasing function, continuous at zero and such that $\omega(0) = 0$. $\mathbb{M}_s^3$ stands for the space of symmetric 3×3 matrices.

Typical boundary conditions are imposed

$$\mathbf{u} = 0, \text{ on } \Gamma_0; \ \sigma_{ij}\, n_j = \Sigma_i, \text{ on } \Gamma_1, \quad (1.5)$$

$$\varphi = \varphi_0, \text{ on } \Gamma_2; \ D_i\, n_i = 0, \text{ on } \Gamma_3, \quad (1.6)$$

where $\Gamma = \bar{\Gamma}_0 \cup \bar{\Gamma}_1, \Gamma_0 \cap \Gamma_1 = \emptyset$; $\Gamma = \bar{\Gamma}_2 \cup \bar{\Gamma}_3$, $\Gamma_2 \cap \Gamma_3 = \emptyset$, and $\mathbf{n}$ is the outward unit normal vector to Γ.φ is the electric potential, $E_i = -\varphi_{,i}$, see [10].

For a fixed $\varepsilon > 0$ we set

$$F^{\varepsilon}(\mathbf{u}, \mathbf{D}) = \int_{\Omega} [\, U(x, x/\varepsilon, \mathbf{e}(\mathbf{u}), \mathbf{D}) - b_i u_i]\, \mathrm{d}x - \int_{\Gamma_1} \Sigma_i u_i \, \mathrm{d}\Gamma + \int_{\Gamma_2} \varphi_0\, D_i\, n_i \, \mathrm{d}\Gamma, \quad (1.7)$$

where $\mathbf{u} \in V, \mathbf{D} \in H^0_{\mathrm{div}}$ and

$$V = \{\mathbf{v} = (v_i) \mid v_i \in H^1(\Omega), \ \mathbf{v} = 0 \text{ on } \Gamma_0\}, \quad (1.8)$$

$$H^0_{\mathrm{div}} = \{\mathbf{D} \in H_{\mathrm{div}} \mid D_i\, n_i = 0 \text{ on } \Gamma_3\}. \quad (1.9)$$

The space H_{div} is defined by, see [12]

$$H_{\mathrm{div}} = \{\mathbf{D} = (D_i) \mid D_i \in L^2(\Omega), D_{i,i} \in L^2(\Omega)\}. \quad (1.10)$$

In our case homogenization means a passage to zero with ε. For the sequence of functionals $\{F^{\varepsilon}\}_{\varepsilon > 0}$ defined by (1.7) the method of epi-convergence is very convenient. For details the reader should refer to Ref. [2]. Now the functional

$$G(\mathbf{u},\mathbf{D}) = -\int_\Omega b_i u_i \mathrm{d}x - \int_{\Gamma_1} \Sigma_i u_i \,\mathrm{d}\Gamma + \int_{\Gamma_2} \varphi_0 D_i n_i \,\mathrm{d}\Gamma, \tag{1.11}$$

is the perturbation functional. Let us denote by τ the following topology:

$$\tau = [w - H^1(\Omega, R^3)] \times [s - L^2(\Omega, R^3)].$$

Then τ–epi–limit of the sequence of functionals (1.7) is given by

$$F(\mathbf{u}, \mathbf{D}) = \tau - \lim_{\substack{\varepsilon \to 0}} {}_e F^{\varepsilon}(\mathbf{u}, \mathbf{D}) = \int_\Omega U^h(x, \mathbf{e}(\mathbf{u}), \mathbf{D})\mathrm{d}x + G(\mathbf{u}, \mathbf{D}). \tag{1.12}$$

The macroscopic potential U^h is expressed as follows

$$U^h(x, \mathbf{e}^{\mathrm{h}}, \mathbf{D}^{\mathrm{h}}) = \inf_{\mathbf{v} \in \mathbf{H}_{\mathrm{per}}} \inf_{\mathbf{d} \in \Delta_{\mathrm{per}}} \frac{1}{|Y|}\Big\{ \int_Y [\frac{1}{2} a_{ijkl}(x,y)(e_{yij}(\mathbf{v}) + e^{\mathrm{h}}_{ij})(e_{ykl}(\mathbf{v}) + e^{\mathrm{h}}_{kl}) -$$

$$-h_{kij}(x,y)(d_k(y) + D^{\mathrm{h}}_k)(e_{yij}(\mathbf{v}) + e^{\mathrm{h}}_{ij}) + \frac{1}{2}\kappa_{ij}(x,y)(d_i(y) + D^{\mathrm{h}}_i)(d_j(y) + D^{\mathrm{h}}_j)]\,\mathrm{d}y\Big\}, \tag{1.13}$$

where $\mathbf{e}^{\mathrm{h}} \in \mathbb{M}^3_s$, $\mathbf{D}^{\mathrm{h}} \in R^3$ and $e_{yij}(\mathbf{v}) = \Big(\frac{\partial v_i}{\partial y_j} + \frac{\partial v_j}{\partial y_i}\Big)/2$. The superscript "h" stands for a homogenized quantity and

$$\mathbf{H}_{\mathrm{per}} = \{\mathbf{u} \in H^1(Y, R^3) \mid \mathbf{u} \text{ takes equal values at opposite sides of } Y\}, \tag{1.14}$$

$$\Delta_{\mathrm{per}} = \{\mathbf{d} \in L^2(Y, R^3) \mid \mathrm{div}_y\, \mathbf{d} = 0, \text{ in } Y;\ \int_Y \mathbf{d}(y)\mathrm{d}y = 0;\ d_i N_i \text{ takes opposite values at opposite sides of } Y\}. \tag{1.15}$$

Here $\mathbf{N}$ is the outward unit normal to ∂Y.

The electric enthalpy $H(x, y, \mathbf{e}, \mathbf{E})$ can be calculated as the partial concave conjugate of U, cf. Ref. [10]

$$H(x, y, \mathbf{e}, \mathbf{E}) = \inf\big(-\mathbf{E}\cdot\mathbf{D} + U(x, y, \mathbf{e}, \mathbf{D}) \mid \mathbf{D} \in R^3\big\} = \frac{1}{2} c_{ijkl} e_{ij} e_{kl} - g_{kij} e_{ij} E_k -$$

$$-\frac{1}{2}\varepsilon_{ij} E_i E_j, \tag{1.16}$$

where

$$c_{ijkl} = a_{ijkl} - g_{mij}\, h_{mkl}, \boldsymbol{\varepsilon} = \boldsymbol{\kappa}^{-1}, g_{ijk} = \varepsilon_{il}\, h_{ljk}.$$

We note that the function $H(x, y, .\,, .)$ is a convex-concave one. A proper notion of the variational convergence for a sequence of convex-concave functionals is the epi/hypo-convergence, see [3]. The epi-convergence of bivariate functionals is closely related to the epi/hypo-convergence, see [3, Theorem 3.3]. Denoting by H_{per} the space of periodic scalar functions, we have the following expression for the macroscopic electric enthalpy H^{h}

$$H^{\text{h}}(x, \mathbf{e}^{\text{h}}, \mathbf{E}^{\text{h}}) = \inf_{\mathbf{v}\in \mathbf{H}_{\text{per}}} \sup_{\psi \in H_{\text{per}}} \frac{1}{|Y|}\Big\{\int_Y\Big[\frac{1}{2}c_{ijkl}(x,y)(e_{yij}(\mathbf{v}) + e^{\text{h}}_{ij})(e_{ykl}(\mathbf{v}) + e^{\text{h}}_{kl}) -$$

$$- g_{kij}(x, y)\Big(\frac{\partial\psi}{\partial y_k} + E^{\text{h}}_k\Big)(e_{yij}(\mathbf{v}) + e^{\text{h}}_{ij}) - \frac{1}{2}\varepsilon_{ij}(x, y)\Big(\frac{\partial\psi}{\partial y_i} + E^{\text{h}}_i\Big)\Big]\,\text{d}y\Big\}, \quad (1.17)$$

where $\mathbf{e}^{\text{h}} \in \mathbb{M}^3_s$, $\mathbf{E}^{\text{h}} \in R^3$.

Suppose now that $\mathbf{u}$ and φ solve the infsup problem on the right-hand side of (1.17). We may set

$$u_i(x, y) = \chi_i^{(mn)}(x, y)e^{\text{h}}_{mn} + \Phi_i^{(m)}(x, y)E^{\text{h}}_m, \quad (1.18)$$

$$\varphi(x, y) = \gamma^{(mn)}(x, y)e^{\text{h}}_{mn} + \Lambda^{(m)}(x, y)E^{\text{h}}_m. \quad (1.19)$$

To find the macroscopic moduli first we have to solve two local problems.

Problem $\mathbf{P}_1^{\text{loc}}$

Find $\boldsymbol{\chi}^{(mn)}(x, .) \in \mathbf{H}_{\text{per}}$ and $\gamma^{(mn)}(x, .) \in H_{\text{per}}$ such that

$$\int_Y [c_{ijkl}(x, y)(\delta_{im}\delta_{jn} + \frac{\partial\chi_i^{(mn)}}{\partial y_j}) - g_{ikl}(x, y)\frac{\partial\gamma^{(mn)}}{\partial y_i}]\, e_{ykl}(\mathbf{v})\,\text{d}y = 0, \ \mathbf{v} \in \mathbf{H}_{\text{per}}, \quad (1.20)$$

$$\int_Y [g_{kij}(x, y)(\delta_{im}\delta_{jn} + \frac{\partial\chi_i^{(mn)}}{\partial y_j}) + \varepsilon_{ik}(x, y)\frac{\partial\gamma^{(mn)}}{\partial y_i}]\,\frac{\partial\psi}{\partial y_k}\,\text{d}y = 0, \ \psi \in H_{\text{per}}. \quad (1.21)$$

Problem P_2^{loc}

Find $\mathbf{\Phi}^{(m)}(x,.) \in \mathbf{H}_{per}$ and $\Lambda^{(m)}(x,.) \in H_{per}$ such that

$$\int_Y [c_{klij}(x,y)\frac{\partial \Phi_k^{(m)}}{\partial y_l} - g_{kij}(x,y)(\frac{\partial \Lambda^{(m)}}{\partial y_k} + \delta_{mk})]\, e_{yij}(\mathbf{v})\, dy = 0, \;\; \mathbf{v} \in \mathbf{H}_{per}, \tag{1.22}$$

$$\int_Y [g_{kij}(x,y)\frac{\partial \Phi_i^{(m)}}{\partial y_j} + \varepsilon_{ik}(x,y)(\frac{\partial \Lambda^{(m)}}{\partial y_i} + \delta_{mi})]\frac{\partial \psi}{\partial y_k} dy = 0, \;\; \psi \in H_{per}\,. \tag{1.23}$$

Now we can calculate the macroscopic moduli. Finally we obtain

$$c^h_{rspg}(x) = \frac{\partial^2 H^h}{\partial e^h_{rs}\partial e^h_{pq}} = \frac{1}{|Y|}\Big\{\int_Y [c_{ijkl}(x,y)(\delta_{ip}\,\delta_{jq} + \frac{\partial \chi_i^{(pq)}}{\partial y_j})(\delta_{kr}\,\delta_{ls} + \frac{\partial \chi_k^{(rs)}}{\partial y_l}) +$$

$$+\varepsilon_{ij}(x,y)\frac{\partial \gamma^{(pq)}}{\partial y_i}\frac{\partial \gamma^{(rs)}}{\partial y_j}]\,dy\Big\} = \frac{1}{|Y|}\Big\{\int_Y [c_{ijkl}(x,y)(\delta_{ip}\,\delta_{jq} + \frac{\partial \chi_i^{(pq)}}{\partial y_j})(\delta_{kr}\,\delta_{ls} + \frac{\partial \chi_k^{(rs)}}{\partial y_l}) -$$

$$- g_{kij}(x,y)(\delta_{ip}\,\delta_{jq} + \frac{\partial \chi_i^{(pq)}}{\partial y_j})\frac{\partial \gamma^{(rs)}}{\partial y_k}]\,dy\Big\}, \tag{1.24}$$

$$g^h_{prs}(x) = -\frac{\partial^2 H^h}{\partial e^h_{rs}\partial E^h_p} = \frac{1}{|Y|}\Big\{\int_Y [g_{kij}(x,y)(\delta_{kp} + \frac{\partial \Lambda^{(p)}}{\partial y_k})(\delta_{ir}\,\delta_{js} + \frac{\partial \chi_i^{(rs)}}{\partial y_j}) +$$

$$+\varepsilon_{ij}(x,y)(\delta_{ip} + \frac{\partial \Lambda^{(p)}}{\partial y_i})\frac{\partial \gamma^{(rs)}}{\partial y_j}]\,dy\Big\} = \frac{1}{|Y|}\Big\{\int_Y [g_{kij}(x,y)(\delta_{kp} + \frac{\partial \Lambda^{(p)}}{\partial y_k})(\delta_{ir}\delta_{js} + \frac{\partial \chi_i^{(rs)}}{\partial y_j}) -$$

$$- g_{kij}(x,y)\frac{\partial \Phi_i^{(p)}}{\partial y_j}\frac{\partial \gamma^{(rs)}}{y_k}]\,dy\Big\}, \tag{1.25}$$

$$\varepsilon^{h}_{pq}(x) = -\frac{\partial^2 H^h}{\partial E^h_p \partial E^h_q} = \frac{1}{|Y|}\Big\{\int_Y [\varepsilon_{ij}(x,y)(\delta_{ip} + \frac{\partial \Lambda^{(p)}}{\partial y_i})(\delta_{jq} + \frac{\partial \Lambda^{(q)}}{\partial y_j}) +$$

$$+ c_{ijkl}(x,y)\frac{\partial \Phi^{(p)}_i}{\partial y_j}\frac{\partial \Phi^{(q)}_k}{\partial y_l}]\, dy\Big\} = \frac{1}{|Y|}\Big\{\int_Y [\varepsilon_{ij}(x,y)(\delta_{ip} + \frac{\partial \Lambda^{(p)}}{\partial y_i})(\delta_{jq} + \frac{\partial \Lambda^{(q)}}{\partial y_j}) +$$

$$+ g_{kij}(x,y)(\delta_{kp} + \frac{\partial \Lambda^{(p)}}{\partial y_k})\frac{\partial \Phi^{(q)}_i}{\partial y_j}]\, dy\Big\}. \qquad (1.26)$$

We see that $c^h_{ijkl} = c^h_{klij} = c^h_{jikl}$, $g^h_{ijk} = g^h_{ikj}$, $\varepsilon^h_{ij} = \varepsilon^h_{ji}$.

2 On applications to biomechanics

Since the discovery by I. Yasuda in 1953 that bone is a piezoelectric material, many papers appeared on electromechanical effects in organic materials. An excellent review on those results up to 1978 is given in a paper by Güzelsu and Demiray [8]. In this section general indications as to the possibility of the applications of the formulas for $\mathbf{c}^h$, $\mathbf{g}^h$ and ε^h will be suggested.

The structure of bone as a living organ is very complicated [6, 8]. Only compact parts of bones will be of interest for us in the sequel. The structure and properties of cancellous bones, which are cellular biomaterials, are presented in the book by Gibson and Ashby [7].

For the purpose of studying the macroscopic properties of the compact bone we may asume that it is constituted, mainly, of a crystalline mineral phase (hydroxyapatite), an amorphous mineral phase, a crystalline organic phase (collagen) and an amorphous organic phase and liquids. The crystalline and amorphous mineral phases may be assumed to be lumped together and the amorphous organic phase, which is found in small amounts in the bone, may also be lumped together with collagen. At the microscopic level, compact bone may be considered to consist of hydroxyapatites (minerals), collagen fibres (organic) and liquids (pores). About two-thirds of the weight of bone, or half of its volume, is inorganic material with a composition that corresponds to the formula of hydroxyapatite. It is present as tiny crystals, about 200 Å in length and with an average cross-section of 2500 Å^2. Such a crystal is very strong and stiff. Its Young's modulus along the axis is about 165 GPa (steel - 200 GPa). The hydroxyapatite crystals have the hexagonal (C_{6n}) symmetry and do not exhibit any piezoelectric response; they are centro-symmetrical. Let us write the tensor $\mathbf{g}$ of piezoelectric coefficients in the following matrix form [10]

$$\mathbf{g} = [\,g_{iI}\,],\ i = 1, 2, 3;\ I = 1, 2, \dots, 6, \tag{2.1}$$

where

$$g_{iI} = \begin{cases} g_{ijk}, & \text{if } I \le 3, \\ 2g_{ijk}, & \text{if } I > 3, \end{cases} \tag{2.2}$$

and $I = (j\,k)$; here we set

$$(11) \to 1,\ (22) \to 2,\ (33) \to 3,\ (23) \to 4,\ (31) \to 5,\ (12) \to 6.$$

According to Fukada and Yasuda (see Ref. [123] in [8]) the crystal symmetry of dried collagen fits to the Hexagonal Polar (C_6); then the matrix of piezoelectric coefficients, now dentoed by $\mathbf{g}^c$, has the following form

$$\mathbf{g}^c = [\,g^c_{iI}\,] = \begin{bmatrix} 0 & 0 & 0 & g^c_{14} & g^c_{15} & 0 \\ 0 & 0 & 0 & g^c_{15} & -g^c_{14} & 0 \\ g^c_{31} & g^c_{31} & g^c_{33} & 0 & 0 & 0 \end{bmatrix} \tag{2.3}$$

We see that $g^c_{14} = 2g^c_{123}$, $g^c_{15} = 2g^c_{131}$, $g^c_{31} = g^c_{311}$, $g^c_{33} = g^c_{333}$. It is usually assumed that bone also belongs to the same piezoelectricity symmetry class.

Now a natural question arises: how to find macroscopic coefficients $\mathbf{c}^h$, $\mathbf{g}^h$ and $\boldsymbol{\varepsilon}^h$? Towards this end the formulas (1.24)-(1.26) may be used. It is not necessary to assume the basic cell Y in the form of a rectangular prism, cf. [9]. However, in the case considered such a cell must contain a sufficient number of osteons. A single osteon gives no information about the macroscopic properties of bone. It is worth noting that osteon is also a composite material [5, 8].

Treating the compact bone as a composite material; it is usually assumed that both collagen and hydroxyapatite are locally isotropic in their elastic properties. Accordingly, we write $\bar{Y} = \bar{Y}_1 \cap \bar{Y}_2$, where $|\,Y_\alpha\,| = c_\alpha\,|\,Y\,|$. The constant c_α may even be set equal to $1/2$. Here the subscript 1(2) refers to collagen (hydroxyapatite), while the bar denotes the closure of a set. Hence we have

$$c_{ijkl}(x, y) = \begin{cases} \overset{(1)}{c}_{ijkl}(x, y), & \text{if } y \in Y_1, \\ \overset{(2)}{c}_{ijkl}(x, y), & \text{if } y \in Y_2, \end{cases} \tag{2.4}$$

where

$$c^{(\alpha)}_{ijkl}(x,y) = \lambda^{(\alpha)}(x,y)\delta_{ij}\,\delta_{kl} + \mu^{(\alpha)}(x,y)(\delta_{ik}\,\delta_{jl} + \delta_{il}\,\delta_{jk})\,. \tag{2.5}$$

Here $\lambda^{(\alpha)}$ and $\mu^{(\alpha)}$ stand for the Lamé coefficients. Similarly, for the dielectric coefficients we write

$$\varepsilon_{ij}(x,y) = \begin{cases} \varepsilon^{(1)}_{ij}(x,y), & \text{if } y \in Y_1, \\ \varepsilon^{(2)}_{ij}(x,y), & \text{if } y \in Y_2. \end{cases} \tag{2.6}$$

As a first approximation we may take

$$\varepsilon^{(\alpha)}_{ij}(x,y) = \varepsilon^{(\alpha)}(x,y)\delta_{ij}\,. \tag{2.7}$$

Locally the piezoelectric moduli are given by

$$g_{kij}(x,y) = \begin{cases} \overset{c}{g}_{kij}(x,y), & \text{if } y \in Y_1, \\ 0, & \text{if } y \in Y_2. \end{cases} \tag{2.8}$$

Solving first the local problems $P^{l\alpha}_{\alpha}$, provided that $\mathbf{c}(x,y)$, $\mathbf{g}(x,y)$ and $\boldsymbol{\varepsilon}(x,y)$ are given by (2.4), (2.6) and (2.8), respectively, one can calculate the homogenized or macroscopic coefficients $\mathbf{c}^{h}(x)$, $\mathbf{g}_{h}(x)$ and $\boldsymbol{\varepsilon}^{h}(x)$. Obviously, local problems can, in general, be solved only numerically except for some particular cases. The homogenized coefficents will be anisotropic. The resulting macroscopic anisotropy is strongly influenced by the assumed partition of the basic cell Y into Y_1 and Y_2. Thus it may happen that the non-shear terms of the matrix $\mathbf{g}^{h}(x)$ do not disappear. Such a result confirms experimental findings by Anderson and Eriksson (Ref. [118] in [8]). These authors report values of the non-shear terms of the piezoelectric matrix for dry and wet beef bone, thus completing the earlier results obtained by Fukada and his coworkers [8].

3 Concluding remarks

By using the method of Γ-convergence, macroscopic moduli have been derived for a non-uniformly periodic piezoelectric material. Alternatively, a mathematically formal method of two-scale asymptotic expansions can be applied [4].

To find the macroscopic moduli one has first to solve two local problems. In specific cases solely they can be resolved analytically. For instance, if Y-periodicity occurs in one direction only then the basic cell is an interval (the case of a layered piezoelectric body).

The method of Γ-convergence can likewise be used with piezoelectric structures like plates and shells exhibiting periodic microstructure. It is also applicable to piezoelectric bodies with holes (pores). Somewhat more complicated is the case of a microfissured solid, yet one can derive its overall properties by using the same method.

Possible applications to the derivation of overall eleastic, piezoelectric and dielectric moduli of compact bones have only been sketched. The subject is very interesting and deserves deeper studies by biomechanicians. A next step would be to take into account liquids (like blood) filling pores, viscosity and all existing couplings, yet still within the framework of a linear theory. For such a more complicated, but more realistic, model of bone, homogenization could be performed by the method of asymptotic expansions. Application of homogenization methods to finding effective properties of cancellous bones remains an open problem.

References

[1] Ashman, R.B., Cowin, S.C., Van, Buskirk W.C. and Rice, J.C., A continuous wave technique for the mesurement of the elastic properties of cortical bone, *J. Biomechanics*, **17**, 349-361, 1984.

[2] Attouch, H., *Variational Convergence for Functions and Operators*, Pitman, Boston, 1984.

[3] Attouch, H., Azé, D. and Wets, R.J.B., Convergence of convex-concave functions: continuity properties of the Legendre-Fenchel transform with applications to convex programming and mechanics, Publications AVAMAC, Université de Perpignan, Mathématiques, No. 85-08.

[4] Bielski, W. and Bytner, S., in preparation.

[5] Braides, A., Omogeneizzazione di integrali non coercivi, *Ricerche di Mat.*, **32**, 347-368, 1983.

[6] Fung, Y.C., *Biomechanics: Mechanical Properties of Living Tissues*, Springer-Verlag, New York, 1981.

[7] Gibson, L.J. and Ashby, M.F., *Cellular Solids: Structure and Properties*, Pergamon Press, Oxford 1988.

[8] Güzelsu, N., and Demiray, H., Electromechanical proeprties and related models of bone tissues, *Int. J. Eng. Sci.*, **17**, 813-851, 1979.

[9] Léné, F., Contribution à l'étude des matériaux composites et de leur endommagement, Thèse d'Etat, Université Paris VI, 1984.

[10] Maugin, G.A., *Continuum Mechanics of Electromagnetic Solids*, North-Holland, Amsterdam, 1988.

[11] Suquet, P., Plasticité et homogénéisation, Thèse d'Etat, Université Paris VI, 1982.

[12] Temam, R., *Navier-Stokes Equations: Theory and Numerical Analysis*, North-Holland, Amsterdam, 1979.

Telega J.J.
I.P.P.T.-P.A.N.,
Swietokrzyska 21
PL-00-049 Warsaw
POLAND

PART 3 : MICROSTRUCTURE, THERMODYNAMICS AND GEOMETRY

K.-H. ANTHONY

Defect dynamics and Lagrangian thermodynamics of irreversible processes

1 Introduction

Defect dynamics, i.e., migration, creation and annihilation of defects and reactions between defects of different types are dissipative, irreversible processes. By means of *Lagrange-Formalism* (LF) I aim to establish a dynamical continuum theory of thermomechanical processes in ordered materials inlcuding the dynamics of topological defects.

LF of complex-valued fields applies to thermodynamics of irreversible processes (TIP). For heat transport, diffusion and chemical reactions I verified this statement [1-6]. A short outline will be given in Section 2. The dynamics of topological defects in ordered materials can be regarded as chemical reaction dynamics in a generalized sense [5]. Section 3 is concerned with that topic. Within LF, dissipation is associated with an irreversible energy transfer from kinematical and material degrees of freedom to thermal degrees of freedom. This transfer is formally due to Noether's theorem which is a most important structure element of LF. In Section 2 this statement is discussed for chemical reactions and in Section 3 for defect dynamics. For brevity the discussion in this paper is on the grounds of the field equations rather than Noether's theorem.

Defect dynamics is a prominent example for the concept of internal variables in TIP. Using LF I take explicit account of these variables by means of complex-valued field variables each of which is associated with one particular type of defect. This set of *defect fields* is supplemented by a complex-valued *field of thermal excitation*. Taking over the formal concept of chemical reaction I will show in Section 3 how the theory works in the case of point-like defects. So far the theory is developed for processes in a rigid material matrix, i.e., deformational degrees of freedom are not yet taken into account.

Once a Lagrange-density-function l is established the complete dynamics of the system is defined, i.e., within LF the dynamics is comprised into the smallest entity possible, namely into one scalar function. By means of the well-known procedures of LF all field equations, balance equations and constitutive relations can be derived from the Lagrangian. For these, I refer to former papers [1-7]. In this paper I intend to focus the reader's attention especially to a reaction potential being part of the Lagrangian and taking structural account of the dynamics of chemical reactions or of defects on the atomic level. Generally, it is an essential feature of my Lagrangian approach to take into account the microscopic dynamics as far as possible.

2 Lagrange-Formalism for chemical reaction-dynamics

Let me consider a chemical reaction between N constituents S_k :

$$\nu_1 S_1 + \nu_2 S_2 + \ldots + \nu_m S_m \rightleftarrows \nu_{m+1} S_{m+1} + \ldots + \nu_N S_N . \tag{1}$$

Subsequently the stoichiometric coefficients ν_i are taken positive or negative according to their position at the right-hand or left-hand side of eq. (1). This chemical reaction will take place in the fixed volume V of a rigid material matrix which does not take part in the reaction, but which will take part in the thermal process. To achieve the complete reaction dynamics I allow for heat transport and for diffusion of all constituents S_k.

I introduce complex-valued fundamental field variables: the *thermal excitation field* $\psi(x, t)$, the *matter fields* $\psi_k(x, t)$, $k = 1, \ldots, N$, associated with constituent S_k and $\psi_0(x, t)$ associated with the material matrix. As secondary quantities the *absolute temperature* $T(x, t)$ and the *particle densities* $n_k(x, t)$ of the constituents and of the matrix are defined by

$$T(x, t) = \psi(x, t)\ \psi(x, t)^* \geq 0,$$

$$n_k(x, t) = \psi_k(x, t)\ \psi_k(x, t)^* \geq 0, \quad k = 0, \ldots, N. \tag{2}$$

Each of the complex-valued fields ψ, ψ_k includes two functional degrees of freedom:

$$\psi(x, t) = \sqrt{T(x, t)}\ \exp[i\,\varphi(x, t)],$$

$$\psi_k(x, t) = \sqrt{n_k(x, t)}\ \exp[i\,\Phi_k(x, t)], \quad k = 0, \ldots, N. \tag{3}$$

By means of the phase functions $\varphi(x, t)$ and $\Phi_k(x, t)$ the set of fundamental degrees of freedom is doubled as compared with traditional TIP [8] which is based on the variables T and n_k only. The phase functions get a physical interpretation by considering deviations of the processes from local equilibrium [3]. (LF allows for a quite general dynamics without restricting the processes to local equilibrium.)

Real processes are distinguished as solutions of *Hamilton's variational Principle* [9]:

$$J = \int_{t_1}^{t_2} \int_V l(\boldsymbol{\Psi}, \partial\boldsymbol{\Psi})\,\mathrm{d}^3x\,\mathrm{d}t = \text{extremum} \tag{4}$$

by free and independent variations of all fields in the set Ψ with fixed values at the beginning t_1 and the end t_2 of the process: $\delta\Psi(x, t_{1,2}) = 0$.

The Lagrange-density function l being the kernel of Hamilton's action integral J comprises all information concerning the processes of the system. The symbol Ψ marks the complete set of independent field variables

$$\Psi = \{\psi(x,t),\ \psi_k(x,t),\ \psi^*(x,t),\ \psi_k^*(x,t),\ k = 0, \dots, N\} \tag{5}$$

or equally well

$$\Psi = \{\, T(x,t),\ n_k(x,t),\ \varphi(x,t),\ \Phi_k(x,t),\ k = 0, \dots, N\}. \tag{6}$$

∂ marks the set of all derivatives with respect to time t and the space variables $x = (x^1, x^2, x^3)$.

Using the set (6) the *chemical reaction dynamics* is completely defined by the Lagrangian

$$\begin{aligned} l = -\frac{1}{\omega}\{ & u(T,\mathbf{n})\,\partial_t[\varphi - \varphi_0(t,T)] + \sum_{i=0}^{N} n_i\,\partial_t[\mu_i(T,\mathbf{n})\alpha_i] \\ & + \mathbf{J}(T,\mathbf{n},\nabla T,\nabla\mathbf{n})\cdot\nabla[\varphi - \varphi_0(\dots)] \\ & + \sum_{i=1}^{N} \mathbf{J}_i(T,\mathbf{n},\nabla T,\nabla\mathbf{n})\cdot\nabla[\mu_i(\dots)\alpha_i] \\ & + R(T,\mathbf{n},\boldsymbol{\alpha}) + \partial_t\, G(T,\mathbf{n})]\,. \end{aligned} \tag{7}$$

The following abbreviations are used:

$$\mathbf{n} = \{n_k,\ k = 1, \dots, N\}, \quad \boldsymbol{\alpha} = \{\alpha_k, k = 1, \dots, N\}\,,$$

$$\alpha_i = [\Phi_i - \varphi] - [\Phi_{i\,0}(t,T,\mathbf{n}) - \varphi_0(t,T)],\ i = 0, \dots, N\,,$$

$$\varphi_0(t,T) = -\,\omega t + \frac{T_0}{2T},\ T_0 : \text{reference temperature}, \tag{8}$$

$$\Phi_{i\,0}(t,T,\mathbf{n}) = -\,\omega t + \gamma\frac{n_i}{2T},\ \gamma : \text{dimensional constant}, \tag{9}$$

$$G(T.\,\mathbf{n}) = \frac{T_0}{2T}\int_{T_0/T}^{1} \frac{p(\xi T, \xi\,\mathbf{n})}{\xi^2}\mathrm{d}\xi\,. \tag{10}$$

The yet unspecified functions entering into (7) and (10) get their interpretation during the course of LF (Noether's theorem, see [9, 1–7]). They are free for fitting to experimental data or to results of microscopic theories. The meaning of the various functions is as follows:

$u(t, \mathbf{n})$: density of the internal energy,

$J(T, \mathbf{n}, \nabla T, \nabla \mathbf{n})$: flux density of the internal energy,

$\mu_i(T, \mathbf{n})$: chemical potential of constituent S_i ($i = 1, \dots, N$) and of the material matrix ($i = 0$),

$\mathbf{J}_i(T, \mathbf{n}, \nabla T, \nabla \mathbf{n})$: diffusion flux density of constituent S_i,

$p(T, \mathbf{n})$: hydrostatic pressure defined by

$$u = T\,s - p + \sum_{i=0}^{N} \mu_i\, n_i\,. \tag{11}$$

This formula is well known from thermostatics. In the present context it is used in the sense of local equilibrium. The *entropy density* s entering into (11) is defined in LF by means of a straightforward procedure [4]. In the case of local equilibrium it coincides with the known entropy density of thermostatics. (The term $\mu_0\, n_0$ entering into (11) is associated with the material matrix and its contribution to the thermal process.) Of course in the present context, where I assume a rigid matrix, the pressure p has to be interpreted as a rigid body reaction of the matrix.

Frequency ω entering into (7)–(9): This factor is necessary for dimensional reasons. At this stage of the theory it is not essential insofar as it drops out from all subsequent equations.

$R(T, \mathbf{n}, \boldsymbol{\alpha})$: *Reaction potential* with the property

$$R(T, \mathbf{n}, \boldsymbol{\alpha}) = 0 \ \text{ for all } \ \alpha_i = 0,\ i = 1, \dots, N. \tag{12}$$

This potential controls the chemical reaction (1). It describes the energy transfer between the material and the thermal degrees of freedom: The heat release of the chemical reaciton is defined from this potential. R does not depend on the matrix matter field ψ_0.

Hamilton's Principle taken for (7) results in the *Euler-Lagrange field* equations associated with the variations of φ, Φ_i, T, n_i:

$$\delta\varphi: \qquad \partial_t u_Q + \nabla\cdot \mathbf{J}_Q = \xi_Q , \tag{13}$$

$$\delta\Phi_i , i = 1, \dots , N: \qquad \partial_t (\mu_i n_i) + \nabla\cdot(\mu_i \mathbf{J}_i) = \xi_i , \tag{14}$$

$$\delta T: \qquad . \ . \ . \ . \ . \ . \ . \ . \ . \ . \ . \ . \ . \ . \tag{15}$$

$$\delta n_i , i = 1,\dots,N:: \qquad . \ . \ . \ . \ . \ . \ . \ . \ . \ . \ . \ . \ . \ . \tag{16}$$

Eqs. (15) and (16) can easily be found from eq. (7). They are lengthy partial differential equations containing terms

$$(\varphi - \varphi_0), \quad [(\Phi_k - \varphi) - (\Phi_{k0} - \varphi_0)] ,$$

and its first and second derivatives in such a way as to be fulfilled identically by the particular solution

$$\varphi = \varphi_0(t, T) = - \omega t + \frac{T_0}{2T(x,t)}, \tag{17}$$

$$\Phi_i = \Phi_{i0}(t, T, \mathbf{n}) = - \omega t + \gamma \frac{n_i(x,t)}{2T(x,t)}. \tag{18}$$

This particular solution of eqs. (15), (16) is physically associated with local equilibrium processes. (The Lagrangian (7) is modelled in such a way as to rediscover the traditional TIP [8] in all details concerning heat conduction, diffusion and chemical reactions. Of course this Lagrangian is but a truncated version of a more general Lagrangian which describes the processes even outside of local equilibrium [3].)

The densities, flux densities and production rate densities entering into the field equations (13), (14) are defined by

$$u_Q(T, \mathbf{n}) = u(T, \mathbf{n}) - \sum_{i=0}^{N} \mu_i (T, \mathbf{n})\, n_i , \tag{19}$$

$$\mathbf{J}_Q(T, \mathbf{n}, \nabla T, \nabla \mathbf{n}) = \mathbf{J}(\dots) - \sum_{i=1}^{N} \mu_i (\dots)\, \mathbf{J}_i (\dots) , \tag{20}$$

$$\xi_i (T, \mathbf{n}, \varphi, \Phi_k) = \frac{\partial R(\dots,\boldsymbol{\alpha})}{\partial \alpha_i} + n_i \partial_t \mu_i(\dots) + \mathbf{J}_i (\dots)\cdot\nabla\mu_i(\dots) , \tag{21}$$

$$\xi_Q = -\sum_{i=0}^{N} \xi_i \,. \tag{22}$$

Thus eqs. (13), (14) can obviously be interpreted as *partial energy balances* associated with the thermal degree of freedom ψ (eq. (13)) and with the material degrees of freedom ψ_i of constituents S_i (eq. (14)). u_Q is the *thermal part of the internal energy* density, (density of the "heat content" of the system) whereas $\mu_i\, n_i$ is the *partial chemical energy* associated with constituent S_i or with the material matrix. $\mathbf{J}_Q$ is the *heat flux* density and $\mu_i\, \mathbf{J}_i$ the *partial chemical energy flux* density associated with the diffusion flux $\mathbf{J}_i$. The *partial energy production rate* densities ξ_Q and ξ_i describe the *energy transfer* between the thermal and material degrees of freedom. By definition (22) the energy transfer is balanced as it should be for an isolated system. Formally it is due to the assumption (12). According to the first term in definition (21) ξ_Q can be regarded as an internal heat supply due to the chemical reaction. However, the second and third terms in (22) give rise to an additional energy transfer due to changes of the local state, only. These terms still work if no chemical reaction takes place ($R = 0$).

Carrying out the differentiations in eq. (14) I get the *mass balance equations* of the diffusing and chemically reacting constituents S_i:

$$\partial_t\, n_i + \nabla\cdot\mathbf{J}_i = \sigma_i \,, \tag{23}$$

The densities of the *particle production rates* are given by

$$\sigma_i = \frac{1}{\mu_i(T,\mathbf{n})}\,\frac{\partial R(T,\mathbf{n},\boldsymbol{\alpha})}{\partial \alpha_i}. \tag{24}$$

Here it becomes apparent that the potential R is physically associated with matter transformation during a chemical reaction. Taking account of the *Principle of Multiple Proportions*

$$\sigma_1 : \sigma_2 : \ldots : \sigma_N = \nu_1 : \nu_2 : \ldots : \nu_N \tag{25}$$

and of the *gross mass balance* of a chemical reaction

$$\sum_{i=1}^{N} M_i\, \nu_i = 0 \tag{26}$$

and solving eq. (25) by

$$\sigma_i = \nu_i \, r(T, \mathbf{n}, \boldsymbol{\alpha}) \tag{27}$$

one gets the mass transfer balance of the reaction

$$\sum_{i=1}^{N} M_i \, \sigma_i = 0 \, . \tag{28}$$

Together with (23) this results in a homogeneous balance equation for the total mass. From (24), (27) one has the constitutive relation for the *reaction velocity* r as derived from the reaction potential R:

$$r = \frac{1}{\nu_i \mu_i (T, \mathbf{n})} \frac{\partial R (T, \mathbf{n}, \boldsymbol{\alpha})}{\partial \alpha_i} . \tag{29}$$

Of course r must be independent of the index i. (M_i is the molecular mass of constituent S_i.) Looking at (29) in the reverse order it defines for given functions r and μ_i a set of differential equations for the reaction potential,

$$\frac{\partial R (T, \mathbf{n}, \boldsymbol{\alpha})}{\partial \alpha_i} = \nu_i \mu_i (T, \mathbf{n}) \, r(T, \mathbf{n}, \boldsymbol{\alpha}), \; i = 1, \dots, N \, , \tag{30}$$

which lead to the integrability conditions

$$\frac{\partial}{\partial \alpha_j} (\nu_i \mu_i r) = \frac{\partial}{\partial \alpha_i} (\nu_j \mu_j r), \; i, j = 1, \dots, N \, . \tag{31}$$

For given chemical potentials μ_i eq. (31) establishes a constraint for the reaction velocity r which with regard to (30) is an integrating factor for the expressions $\nu_i \mu_i$.

Due to (25), (26) the reaction potential R fits to the atomic structure of matter.

3. Defect Dynamics

The dynamics of topological, point-like defects in ordered materials is analogous to the chemical reaction dynamics. Thus the main structures of LF can be taken over from the preceding chapter with a few modifications due to particular features of the defect dynamics on the microlevel. Let me demonstrate the main ideas by a few examples.

In crystalline metals there exist single *lattice vacancies* V in thermal equilibrium. By thermal activation these obstacles migrate thorugh the crystal giving rise to vacancy diffusion. In special cases the single vacancies tend to form *double vacancies DV*. The double vacancies themselves tend to decay again into single vacancies. Thus one has to regard the reaction equation

$$2\,V \rightleftarrows 1\,DV \tag{32}$$

which obviously is an analogue of the chemical reaction eq. (1). Of course this reaction is accompanied with an energy release which in solid state physics is called *formation energy* of the DV if (32) is read from left to right. This means that one has again an energy transfer between thermal and "material" degrees of freedom. As in the case of chemical reactions the complete dynamics includes heat conduction, diffusion of the defects V and DV and the reaction (32). I introduce the field $\psi(x, t)$ of *thermal excitation* and two *defect fields* $\psi_V(x, t)$ and $\psi_{DV}(x, t)$ associated with the single and double vacancies respectively. From these complex-valued fields I define the temperature and the *defect densities* n_V and n_{DV} as before (see eq. (2)). The whole theory of Section 2 is taken over. Especially the argument T in the diffusion fluxes $\mathbf{J}_i$ (see 7, 20) is due to the thermally activated diffusion of the defects while the arguments $\mathbf{n}$ may be cancelled for sufficiently small defect densities. By eqs. (24), (27), (29) the reaction potential R defines the defect creation (or annihilation) rates σ_V and σ_{DV} due to the reaction (32). The argument T in the constitutive relation (29) of the reaction velocity r is related with thermally activated DV-formation. The quantity ξ_Q entering into the thermal energy balance (13) and defined by (22), (21) will be the *formation energy of the double vacancies*. Of course the Principle of Multiple Proportions (25) is still preserved for the reaction (32). However, the gross mass balance eq. (26) will be replaced by a *gross nuclei balance*

$$\sum_{i=1}^{N} N_i\, \nu_i = 0 \tag{33}$$

where N_i denotes the number of lattice nuclei incorporated into the lattice defects. In the case of the single and double vacancies these numbers will be $N_V = -1$ and $N_{DV} = -2$, i.e., V and DV are characterized by 1 and 2 missing lattice nuclei.

Asking for the basic creation process of vacancies one has to look, e.g., for the thermally activated, *pair-wise creation of vacancies V and interstital atoms J*. Of course the inverse process is physically relevant for vacancy annihilation if interstitials are available. Again the associated reaction equation

$$O \rightleftarrows 1\ V + 1\ J \tag{34}$$

together with diffusion of the defects V and J and heat conduction is described with the formalism of Section 2. "O" in (34) marks the unperturbed crystal, so to speak the "vacuum state" of the system. In LF the interstitials J will be described by means of a complex-valued *defect field* $\psi_J(x, t)$ and the associated nucleus-number in (33) is chosen as $N_J = +1$ (1 excess nucleus). The quantity ξ_Q (eqs. (22), (21)) is related with the *formation energy of the vacancy-interstitial pairs.* In reality the mobility of interstitials is much greater than that of the vacancies giving rise to a final excess vacancy density. This effect is modelled by appropriate constitutive relations for the respective diffusion fluxes $\mathbf{J}_V$ and $\mathbf{J}_J$. It concerns the respective diffusion coefficients and its temperature dependence. Of course both reactions (32) and (34) can be regarded simultaneously by generalizing the Lagrangian (7). Due to each reaction an associated reaction potential comes into the play.

In material physics *thermally activated formation and decay of one single sort of defects D is known,* too:

$$O \rightleftarrows D. \tag{35}$$

As an example I refer to kinks on macromolecules of high polymers. In this case the Principle of Multiple Proportions (25) and the gross balance (26), (33) become irrelevant. The remainder of LF, however, is still preserved. The quantity ξ_Q (eqs. (22), (21)) is associated with the *formation energy of the defects.* For illustration let me discuss a simple ansatz: In the actual case eqs. (23) reduce to the single balance equation of the defects D :

$$\partial_t n_D + \nabla \cdot \mathbf{J}_\mathrm{D} = \sigma_D . \tag{36}$$

The ansatz

$$\sigma_D = \sigma_D(T, n_D) = a(T) - n_D\, b(T) \tag{37}$$

$(a(T), b(T) > 0)$ for the total production rate of defects D takes into account a thermally activated formation rate $a(T)$ and a thermally activated decay rate $-n_D b(T)$. (For a small defect density n_D, i.e., if there is no interaction between the defects, the decay probability of a defect will be proportional to the actual number n_D of defects.) Using (30) I get the reaction potential for the reaction (35)

$$R = \mu_D(T, n_D)\,(a(T) - n_D b(T)) \cdot \alpha\,, \tag{38}$$

where

$$\alpha = (\Phi_D - \gamma\frac{n_D}{2T}) - (\varphi - \frac{T_0}{2T}), \tag{39}$$

φ and Φ_D are the phase variables associated with the thermal excitation field ψ and the defect field ψ_D respectively. Following eqs. (22), (21) the potential (38) results in

$$\xi_Q = -\xi_D = -\mu_D \sigma_D - n_D \partial_t \mu_D - n_0 \partial_t \mu_0 - \mathbf{J}_D \cdot \nabla \mu_D . \tag{40}$$

Obviously the first term is the formation energy: $-\mu_D \sigma_D < 0$ for defect creation, $\sigma_D > 0$.

During defect formation this energy is transferred from the thermal field ψ to the defect field ψ_D. In the case of defect decay the energy transfer is reversed. Let me further assume a simple constitutive relation for the thermal partial energy density u_Q defined by (19):

$$u_Q = c\,(T - T_0)\,, \tag{41}$$

where c is the constant specific heat. Then using (41) and (37) the fundamental field equations (13) and (36) read as follows ((36) is equivalent to (14)):

$$c\,\partial_t T + \nabla\cdot \mathbf{J}_Q(T, n_D, \nabla T, \nabla n_D)$$

$$= -\mu_D(T, n_D)\cdot(a(T) - n_D\, b(T)) - n_D\, \partial_t \mu_D(\ldots) - n_0\, \partial_t \mu_0(\ldots)$$

$$- \mathbf{J}_D(\ldots)\cdot \nabla \mu_D(\ldots)\,, \tag{42}$$

$$\partial_t n_D + \nabla\cdot \mathbf{J}_D(T, n_D, \nabla T, \nabla n_D) = a(T) - n_D b(T)\,. \tag{43}$$

Without specifying the constitutive relations of the quantities $\mathbf{J}_Q, \mathbf{J}_D, \mu_D, \mu_0$, a and b in more detail eqs. (42), (43) can be solved for the isolated system by a homogeneous process ($\nabla T = 0$, $\nabla n_D = 0$): Inserting (43) into (42) results in

$$c\,(T - T_0) + \mu_D(T, n_D)\cdot n_D + \mu_0(T, n_D)\cdot n_0 = 0 \tag{44}$$

which correlates with total energy conservation of the homogeneous process. Solving (44) with respect to T leads to

$$T = T(n_D)\,. \tag{45}$$

Inserting (45) into (43) results finally in the non-linear reaction equation

$$\partial_t n_D = a(T(n_D)) - n_D b(T(n_D)). \tag{46}$$

Via (45) its solution $n_D(t)$ leads to the solution $T(t)$. In the most simple case, when the creation and decay rates $a(T)$ and $b(T)$ are assumed to be constants, eq. (46) is solved by the simple relaxation process

$$n_D = \frac{a}{b}\,(1 - A\exp(-bt))\,. \tag{47}$$

This simple example demonstrates how LF for defect dynamics works in principle. For realistic processes one has to specify the constitutive functions entering into the Lagrangian (7). Concerning the reaction potential R, I suppose that annealing experiments of quenched materials provide LF with the necessary data.

From the considerations given above it becomes obvious that the formalism allows for synergetic effects, too, especially of the open system. Oscillations in space and time will be due to non-linearities of the constitutive relations and furthermore to inertial effects. The latter ones will be associated with supplementary, second grade terms in a generalized version of the Lagrangian (7) [3]. Especially, deviations from local equilibrium processes get quite interesting in this respect. (Besides its fundamental significance Hamilton's Principle provides a basis for numerically solving those problems.) A further generalization has to take account of deformational degrees of freedom in order to liberate the material matrix from rigidity.

ACKNOWLEDGEMENTS

The investigations reported in this paper are part of a cooperation project between the Theoretical Physics Department at the University of Paderborn and the Institute of Fundamental Technological Research (Warsaw) of the Polish Academy of Sciences. Financial support by the Deutsche Forschungsgemeinschaft and the Polish Academy is highly appreciated.

References

[1] Anthony, K.-H.: A new approach describing irreversible processes. In "Continuum Models of Discrete Systems 4", eds.: O. Brulin, R.K.T. Hsieh, North-Holland, Amsterdam, 1981, p. 481.

[2] Anthony, K.-H.: A new approach to thermodynamics of irreversible processes by means of Lagrange-formalism. In "Disequilibrium and Selforganisation", ed.: C.W. Kilmister. Reidel, Dordrecht (Holland), 1986, p. 75.

[3] Anthony, K.-H. and H. Knoppe: Phenomenlogical thermodynamics of irreversible processes and Lagrange-formalism. Hyperbolic equations for heat transport. In "Kinetic Theory and Extended Thermodynamics", eds.: I. Müller, T. Ruggeri. Pitagora Editrice Bologna, Bologna, 1987, p. 15.

[4] Anthony, K.-H.: Entropy and dynamical stability. A method due to Lagrange-formalism as applied to thermodynamics of irreversible processes. In "Trends in Applications of mathematics to mechanics", eds.: J.F. Besseling, W. Eckhaus.

Springer-Verlag, Berlin, 1988, p. 297.

[5] Anthony, K.-H.: Unification of continuum mechanics and thermodynamics by means of Lagrange-formalism. Present status of the theory and presumable applications. Archives of Mechanics (Warsaw), in press.

[6] Anthony, K.-H.: Phenomenological thermodynamics of irreversible processes within Lagrange-formalism. Submitted to Acta Physica Hungarica. To appear in Vol. 67.

[7] Anthony, K.-H.: Continuum description of liquid crystals. In "Continuum Models of Discrete Systems 3", eds.: E. Kröner, K.-H. Anthony, University of Waterloo Press, Waterloo (Canada), 1980, p. 761.

[8] De Groot, S.R. and P. Masur: "Non-Equilibrium Thermodynamics". North-Holland, Amsterdam, 1969.

[9] Schmutzer, E.: Symmetrien und Erhaltungssätze der Physik", Akademie-Verlag, Berlin, 1972.

Anthony K.-H
Universität Paderborn
FB 6, Theoretische Physik
Warburger Strasse 100
D 4790 Paderborn
F.R.G.

H. GOUIN

Variational method for fluid mixtures of grade *n*

1 Introduction

The knowledge of equations of processes for fluid mixtures in one or several phases is scientifically and industrially very important [3, 13]. Through the use of a continuum theory of mixtures, equations of balance and motion were first derived by Truesdell [19]. The mixture is then considered as a distribution of different continuous media co-existing in the same physical space, at time t. Each constituent of the mixture is identified with its reference space. The thermal and mechanical equations of balance are introduced for each constituent. They keep the usual form of a fluid description but they must include terms of interaction between the different constituents. The study of the average motion is deduced from the sum of momenta, energy and entropy growing for each constituent. Several assumptions are, as yet, doubtful. How must we choose an entropy? What is the form of the second law of thermodynamic? Many authors argue about some of the points of view [1, 4, 8, 15, 17, 20]. In several papers, Bedford and Drumheller [2] presented a variational theory for immiscible mixtures. That the constituents remain physically separate has several implications. It is not possible to use such a method for a more general model.

We suggest a new systematic method that leads to the equations of conservative motion and energy for miscible fluid mixtures without chemical reaction: the mixture is represented by several distinct continuous media that occupy the same physical space at time t. The variational principle used is applied to a Lagrangian representation associated with a reference space for each component. For classical fluids, the method corresponds to Hamilton's principle in which one makes a variation of the reference position of the particle. It was proposed in [9]. For mixtures, the principle appears far more convenient than the method of only taking variation of the average motion into consideration: it separately tests each component and leads to a thermodynamic equation of the conservative motion for each of them. The internal energy is assumed to be a function of successive derivatives of densities and entropies. Because of the form of the principle used, one must consider an entropy for every component, but in the state of mixture. The total entropy of the system is the sum of the partial entropies.

Two general cases of conservative motions for mixtures are considered: isentropic motion and isothermal motion. An important case is the one of *thermocapillary mixtures*. It corresponds to an energy taking into account the gradients of densities and entropies. It usefully represents fluid interfaces. In the same way as for thermocapillary fluids, an additional term with the dimension of heat flux may be added to the equation of energy [5].

In the dissipative case, viscosity, diffusion and the law of Fourier for heat conduction yield a generalization of Clausius–Duhem's inequality which represents the second law of thermodynamics.

2. Conservative motions of a fluid mixture

For clarity, we study a mixture of two fluids, but the method can easily be extended to a mixture of several fluids. The motion of a two-fluid continuum can be represented by two surjective differentiable mappings

$$z \to X_1 = M_1(z) \text{ and } z \to X_2 = M_2(z)$$

(subscripts 1 and 2 are associated with each constituent). X_1 and X_2 denote the positions of each constituent in reference spaces D_{01} and D_{02}; $z = (t, x)$ belongs to W, an open set of the space–time occupied by the fluid in D_t between time t_1 and time t_2. Variations of motion of particles are deduced from

$$X_1 = \Psi_1(x, t\,; \beta_1) \text{ and } X_2 = \Psi_2(x, t\,; \beta_2)$$

where β_1 and β_2 are defined in a neighbourhood of zero; they are associated with a two-parameter family of virtual motions of the mixture. The real motion corresponds to $\beta_1 = 0$ and $\beta_2 = 0$. Associated virtual displacements can be written as

$$\delta_1 X_1 = \left.\frac{\partial \Psi_1}{\partial \beta_1}\right|_{\beta_1 = 0} \quad \text{and} \quad \delta_2 X_2 = \left.\frac{\partial \Psi_2}{\partial \beta_2}\right|_{\beta_2 = 0} \tag{1}$$

The principle tests separately each constituent of the mixture and so is very efficient.

In the case of compressible fluids, writing the equations of virtual displacement associated with only a one–parameter family of virtual motion of the fluid is deduced from the variation defined by Serrin by the way of vector–space isomorphism [10, 18]. For miscible mixtures, only eq. (1), representing a virtual displacement of miscible fluid mixtures, can be considered.

The Lagrangian of the mixture is

$$L = \frac{1}{2}\rho_1 V_1^* V_1 + \frac{1}{2}\rho_2 V_2^* V_2 - \varepsilon - \rho_1 \Omega_1 - \rho_2 \Omega_2 ,$$

where V_1 and V_2 denote the velocity vectors of each constituent, ρ_1 and ρ_2 are the densities, Ω_1 and Ω_2 are the extraneous force potentials depending only on z, ε is the volumic internal energy and $*$ denotes the transposition in D_t. Because of the interaction between constituents, ε does not necessarily divide into energies related to each constituent of the mixture, as is the case for *simple mixtures of fluids* [14].

Hamilton's action between time t_1 and time t_2 is

$$a = \int_{t_1}^{t_2} \int_{D_t} L \, \mathrm{d}v \, \mathrm{d}t \, .$$

The mixture is supposed to be not chemically reacting. Conservation of mass requires that

$$\rho_i \det F_i = \rho_{0i}(X_i) \tag{2}$$

for the density of each constituent. Subscript i belongs to {1,2}. At t fixed, the Jacobian mapping associated with each defined mapping M_i is denoted by F_i; ρ_{0i} is the reference specific mass in D_{0i}. In differential form, eq. (2) is equivalent to

$$\frac{\partial \rho_i}{\partial t} + \operatorname{div} \rho_i \, V_i = 0.$$

The volume internal energy ε is given by the thermodynamic behaviour of the mixture and is inserted in the equations of motion. In the same way that each constituent has a specific mass, two specific entropies s_1 and s_2 are supposed to be respectively associated with each constituent. If specific internal energy depends on derivatives of gradients of densities and entropies up to the $n-1$ order, we say that the mixture is of grade n

$$\varepsilon = \varepsilon\,(s_1, \operatorname{grad} s_1, ..., (\operatorname{grad})^{n-1} s_1 \, ; \; s_2, \; \operatorname{grad} \, s_2 ,..., (\operatorname{grad})^{n-1} \; s_2;$$

$$\rho_1, \operatorname{grad} \rho_1 ,..., (\operatorname{grad})^{n-1} \rho_1 \, ; \rho_2, \operatorname{grad} \rho_2 ,..., (\operatorname{grad})^{n-1} \rho_2).$$

Computations follow the same line as in Ref. [9]. The orders of gradients of ρ_1, ρ_2, s_1, s_2 can be chosen differently, but the same order is kept for simplicity.

Two cases of conservative motions are considered:

Isentropic motions

We assume that the specific entropy of each particle is constant

$$\frac{\mathrm{d}}{\mathrm{d}t} s_i = 0 \qquad (\text{or } \frac{\partial s_i}{\partial t} + \frac{\partial s_i}{\partial x} V_i = 0) \, ;$$

that is to say $s_i = s_{i0}(X_i)$. This defines an isentropic motion. Motion equations are deduced from Hamilton's principle. The two variations of Hamiltonian action are: $\delta_i a = a'(\beta_i) \mid_{\beta_i} = 0$. For $i \in \{1,2\}$, we deduce

$$\delta_i a = \int_{t_1}^{t_2} \int_{D_t} \Big\{ \big(\frac{1}{2} V_i^* V_i - \varepsilon_{,\rho_i} - \Omega_i\big)\, \delta_i \rho_i + \rho_i V_i^* \delta_i V_i - \varepsilon_{,s_i} \delta_i s_i$$

$$- [\, \varepsilon_{,s_{i,\gamma}} \delta_i s_{i,\gamma} + \ldots + \varepsilon_{,s_{i,\gamma_1,\ldots,\gamma_{n-1}}} \delta_i s_{i,\gamma_1,\ldots,\gamma_{n-1}} \,]$$

$$- [\, \varepsilon_{,\rho_{i,\gamma}} \delta_i \rho_{i,\gamma} + \ldots + \varepsilon_{,\rho_{i,\gamma_1,\ldots,\gamma_{n-1}}} \delta_i \rho_{i,\gamma_1,\ldots,\gamma_{n-1}} \,] \Big\}\, dv\, dt\,.$$

Subscript γ corresponds to spatial derivatives associated with gradient terms; as usual, summation is made on repeated subscripts γ. Let us note

$$h_i = \varepsilon_{,\rho_i} + \sum_{p=1}^{n-1} (-1)^p \, (\varepsilon_{,\rho_{i,\gamma_1,\ldots,\gamma_p}})_{,\gamma_1,\ldots,\gamma_p}$$

for the specific enthalpy of the constituent i *of the mixture* and

$$\theta_i = \frac{1}{\rho_i} [\varepsilon_{,s_i} + \sum_{p=1}^{n-1} (-1)^p \, (\varepsilon_{,s_{i,\gamma_1,\ldots,\gamma_p}})_{,\gamma_1,\ldots,\gamma_p}$$

for the temperature of the constituent i *of the mixture*.

An easy computation yields the equation of motion of constituent i of the mixture as

$$\Gamma_i = \theta_i \operatorname{grad} s_i - \operatorname{grad}(h_i + \Omega_i)\,. \qquad (3)$$

Each constituent of the mixture has its own entropy: this allows a theory of mixtures in which each constituent can have its own temperature [12].

Only the whole entropy of the mixture is conserved

$$\int_{D_t} (\rho_1 s_1 + \rho_2 s_2) dv = K\,, \quad \text{where } K \text{ is constant.}$$

Let us use the method applied in [6] for compressible fluids. Assume that the variation of the entropy of each component is the sum of a variation associated with the virtual motion and another one, $\delta_{0i}\, s_i$ related to the particle: thus

$$\delta_i s_i = \frac{\partial s_{i0}}{\partial X_i} \delta_i X_i + \delta_{0i} s_i ,$$

$\delta_{0i} s_i$ and $\delta_i X_i$ are independent and δ_{0i} must verify, with respect to component i of the mixture,

$$\delta_{0i} \int_{t_1}^{t_2} \int_{D_t} (\rho_1 s_1 + \rho_2 s_2) \mathrm{d}v = 0. \tag{4}$$

Classical methods of variational calculus provide the variation of Hamiltonian action. Variations $\delta_i X_i$ yield the motion equation in the form (3). Relation (4) implies that there exists a constant Lagrange multiplier θ_0 such that

$$\delta_{0i} a = \int_{t_1}^{t_2} \int_{D_t} \Big\{ - \varepsilon_{,s_i} \delta_{0i} s_i - \varepsilon_{,s_{i,\gamma}} \delta_{0i} s_{i,\gamma} - \dots - \varepsilon_{,s_{i,\gamma_1,\dots,\gamma_{n-1}}} \delta_{0i} s_{i,\gamma_1,\dots,\gamma_{n-1}}$$

$$+ \rho_i \theta_0 \delta_{0i} s_i \Big\} \mathrm{d}v \, \mathrm{d}t .$$

With variations $\delta_{0i} s_i$ vanishing on the boundary D_t, an integration by parts yields

$$\delta_{0i} a = \int_{t_1}^{t_2} \int_{D_t} \rho_i (\theta_0 - \theta_i) \delta_{0i} s_i \, \mathrm{d}v \, \mathrm{d}t.$$

The principle: "for any $\delta_{0i} s_i$ vanishing on the boundary of D_t, $\delta_{0i} a = 0$" leads to the equation of the temperature

$$\theta_i = \theta_0 . \tag{5}$$

All the constituents have the same temperature θ_0, constant in the flow; the flow is *isothermal.*

This variational principle can be compared with the one expressing that, the equilibrium of compressible fluid kept in a fixed adiabatic reservoir without exchange of mass and entropy with its outside, is obtained by writing the uniformity of temperature and pressure. Equation (3) is replaced by

$$\Gamma_i + \operatorname{grad} (g_i + \Omega_i) = 0 ,$$

where $g_i = h_i - \theta_i s_i$ is a generalization of the *free specific enthalpy* (or chemical potential) of constituent i. Each constituent of the mixture is an isothermal equilibrium state verifies

$$G_i = C_i \, , \tag{6}$$

where $G_i = g_i + \Omega_i$ and C_i is a constant.

The interface in a two-phase mixture is generally schematized by a surface without thickness. Far from critical conditions, a better modelling represents an interface as a thin layer with molecular dimension in which density and entropy gradients can be very large. It follows that if extraneous forces are neglected in a two-phase mixture, and the two different phases are in an isothermal equilibrium state, then the *specific free enthalpies* of each bulk are equal. This result generalizes the classical law [16] to mixtures of grade n.

3 Thermocapillary mixtures

A. Conservative motions for mixtures of thermocapillary fluids

In the case $n = 2$, we write the equations of motion by using the stress tensor. The mixture is said to be a *thermocapillary mixture* if the internal energy ε is written as

$$\varepsilon = \varepsilon(s_1, \operatorname{grad} s_1, s_2, \operatorname{grad} s_2, \rho_1, \operatorname{grad} \rho_1, \rho_2, \operatorname{grad} \rho_2).$$

Expressions

$$\rho = \sum_{i=1}^{2} \rho_i \quad \text{and} \quad \rho\,\Gamma = \sum_{i=1}^{2} \rho_i\,\Gamma_i$$

introduce the total density and acceleration of the mixture. Denote by α the specific energy of the mixture: $\rho\,\alpha = \varepsilon$ such as

$$d\alpha = \sum_{i=1}^{2} \frac{P_i}{\rho\rho_i} d\rho_i + \frac{\rho_i}{\rho}\,\Theta_i\, ds_i + \frac{1}{\rho}\Phi^{i\gamma}\, d\rho_{i,\gamma} + \frac{1}{\rho}\Psi^{i\gamma}\, ds_{i,\gamma}$$

with

$$P_i = \rho\,\rho_i\,\alpha,_{\rho_i}\, , \quad \Theta_i = \frac{1}{\rho_i}\varepsilon\,,_{s_i}\, , \quad \Phi^{i\gamma} = \varepsilon,_{\rho_{i,\gamma}}\, , \quad \Psi^{i\gamma} = \varepsilon,_{s_{i,\gamma}}\, .$$

Four new vectors Φ^i and Ψ^i ($i \in \{1,2\}$) are introduced by the theory. This is an extension of the computation given in [5, 7]. For compressible fluids (with only one constituent), P and Θ are respectively the pressure and the Kelvin temperature. As the fluid is isotropic, the internal energy is in the form

$$\varepsilon = \varepsilon\,(\rho_i, s_i, \beta_{ij}, \chi_{ij}, \gamma_{ij})$$

with

$$\beta_{ij} = \operatorname{grad} \rho_i \cdot \operatorname{grad} \rho_j,\ \chi_{ij} = \operatorname{grad} \rho_i \cdot \operatorname{grad} s_j,\ \gamma_{ij} = \operatorname{grad} s_i \cdot \operatorname{grad} s_j \quad (i, j \in \{1,2\}).$$

It follows that

$$\Phi^i = \sum_{j=1}^{2} C^{ij} \operatorname{grad} \rho_j + E^{ij} \operatorname{grad} s_j \quad \text{and} \quad \Psi^i = \sum_{j=1}^{2} D^{ij} \operatorname{grad} s_j + E^{ij} \operatorname{grad} \rho_j$$

with $C^{ij} = (1+\delta^{ij})\,\varepsilon,_{\beta_{ij}}$, $D^{ij} = (1+\delta^{ij})\,\varepsilon,_{\gamma_{ij}}$, $E^{ij} = \varepsilon,_{\chi_{ij}}$ (δ^{ij} is the Kronecker symbol). Assuming that $\Omega_1 = \Omega_2 = \Omega$, (this is the case for gravity forces), we obtain a general formulation of the equation of conservative motions for thermocapillary mixtures

$$\rho\,\Gamma = \operatorname{Div} \sigma - \rho \operatorname{grad} \Omega \tag{7}$$

$$(\text{or } \rho\,\Gamma_\gamma = \sigma^\nu{}_{\gamma,\nu} - \rho\,\Omega,_\gamma).$$

σ is the generalization of the stress tensor given by successive integrations by parts

$$\sigma = \sum_{i=1}^{2} \sigma_i \quad \text{such as} \quad \sigma_i{}^\nu{}_\gamma = -p_i\,\delta^\nu{}_\gamma - \Phi^{i\nu}\rho_{i,\gamma} - \Psi^{i\nu} s_{i,\gamma}$$

with

$$p_i = P_i - \rho_i \operatorname{div} \Phi^i .$$

For compressible mixtures, Φ^i and Ψ^i are zero and $\sigma^\nu{}_\gamma$ is written $-P\,\delta^\nu{}_\gamma$ where $P = \sum_{i=1}^{2} P_i$ defines the pressure of the mixture.

B. Equation of motion and equation of energy for a thermocapillary mixture

If the motion of the mixture is not conservative, we must add in eq. (7) an irreversible part σ_I to the stress tensor. This irreversible part is associated with terms of viscosity and diffusion

$$\rho\,\Gamma = \operatorname{Div}(\sigma + \sigma_I) - \rho \operatorname{grad} \Omega .$$

In the same way as for the stress tensor σ, we shall write

$$\sigma_I = \sigma_{I1} + \sigma_{I2}.$$

Let

$$M_i = \Gamma_i - \theta_i \,\mathrm{grad}\, s_i + \mathrm{grad}\,(h_i + \Omega_i),\;\; B_i = \frac{\mathrm{d}\rho_i}{\mathrm{d}t} + \rho_i \,\mathrm{div}\; V_i,$$

$$S_i = \rho_i\, \theta_i \frac{\mathrm{d}s_i}{\mathrm{d}t},\;\; E_i = \frac{\partial e_i}{\partial t} + \mathrm{div}\,[(e_i - \sigma_i)\;\mathrm{div}\; U_i - \rho_i \frac{\partial \Omega_i}{\partial t},$$

where

$$e_i = \frac{1}{2}\rho_i\, V_i^2 + \varepsilon_i + \rho_i\,\Omega_i \;\text{ with }\; \varepsilon_i = \frac{\rho_i \varepsilon}{\rho} \;\text{ and }\; U_i = \frac{\mathrm{d}\rho_i}{\mathrm{d}t}\,\Phi^i + \frac{\mathrm{d}s_i}{\mathrm{d}t}\Psi^i$$

$$(\text{with } \frac{\mathrm{d}\rho_i}{\mathrm{d}t} = \frac{\partial \rho_i}{\partial t} + \frac{\partial \rho_i}{\partial x} V_i \;\text{ and }\; \frac{\mathrm{d}s_i}{\mathrm{d}t} = \frac{\partial s_i}{\partial t} + \frac{\partial s_i}{\partial x} V_i).$$

Theorem - For a thermocapillary internal energy and for any motion of the mixture, the relation $\sum_{i=1}^{2} E_i - \rho_i M_i^* V_i - (\frac{1}{2} V_i^2 + h_i + \Omega_i)\, B_i - S_i = 0$ is verified identically.

Corollary 1 - Any conservative motion of an isentropic, perfect thermocapillary mixture verifies the equation of balance of energy

$$\sum_{i=1}^{2} \frac{\partial e_i}{\partial t} + \mathrm{div}\,[(e_i - \sigma_i) V_i] - \mathrm{div}\; U_i - \rho_i \frac{\partial \Omega_i}{\partial t} = 0\,.$$

This results from the simultaneity of equations $S_i = 0$, $B_i = 0$ and $M_i = 0$.

Corollary 2 - For any conservative motion of a thermocapillary, perfectly heat conducting mixture, eq. (5) is equivalent to

$$\sum_{i=1}^{2} \frac{\partial f_i}{\partial t} + \mathrm{div}\,[(f_i - \sigma_i) V_i] - \mathrm{div}\; U_i - \rho_i \frac{\partial \Omega_i}{\partial t} = 0\,, \tag{8}$$

where $f_i = \frac{1}{2}\rho_i\, V_i^2 + \varphi_i + \rho_i\,\Omega_i$ with $\varphi_i = \varepsilon_i - \rho_i\, \theta_0\, s_i$ corresponding to the *specific free energy* of constituent i taken at temperature θ_0.

Equation (8) is called the equation of balance of free energy. In the non-conservative case, the heat flux vector will be denoted by q and the heat supply by r.

Corollary 3 - For any motion of a thermocapillary mixture, the equation of energy

$$\sum_{i=1}^{2} \frac{\partial e_i}{\partial t} + \operatorname{div}\,[(e_i - \sigma_i - \sigma_{Ii})V_i\,] - \operatorname{div} U_i - \rho_i \frac{\partial \Omega_i}{\partial t} + \operatorname{div} q - r = 0$$

is equivalent to the equation of entropy

$$\sum_{i=1}^{2} \rho_i\, \theta_i \frac{d s_i}{d t} - \operatorname{tr}\,(\sigma_{Ii} \frac{\partial V_i}{\partial x}) + \operatorname{div} q - r = 0.$$

Assuming that $\sum_{i=1}^{2} \operatorname{tr}\,(\sigma_{Ii} \frac{\partial V_i}{\partial x}) \geq 0$, it follows a generalized form of *Planck's inequality* :

$$\sum_{i=1}^{2} \rho_i\, \theta_i \frac{d s_i}{d t} + \operatorname{div} q - r > 0\,.$$

At any point in the fluid, every constituent is supposed to have the same (not necessarily uniform) temperature. This requires that the time for relaxation of molecules be short in comparison with the time characterizing the flow. Then, $\theta = \theta_i$.

By use of *Fourier's law* in the form $q \cdot \operatorname{grad} \theta \leq 0$, we deduce a generalization of *Clausius-Duhem's* inequality

$$\sum_{i=1}^{2} \rho_i \frac{d s_i}{d t} + \operatorname{div}\left(\frac{q}{\theta}\right) - \frac{r}{\theta} \geq 0\,.$$

This Clausius-Duhem's inequality is a form of the second principle of thermodynamics. However, we must note that we deduce it just from the writing of a function of dissipation positive or zero and from Fourier's law. Note also that some authors separate q and r in q_i and r_i : this does not seem necessary.

4 Conclusion

Far from critical conditions, interfaces between different phases in a mixture of fluids are layers with strong gradients of densities and entropies. Fluid mixtures of grade n (or, simpler, thermocapillary mixtures) are able to represent these interfaces. At a given temperature, many constitutive equations represent P as a function of densities of the constituents. These equations, useful in the petroleum industry, derive from subtle calculations on mixtures of Van der Waals fluids. Taking into account eq. (6), they allow us to study the existence of a liquid bulk and a vapour bulk for two fluids that can be completely mixed. For equilibrium at pressure P_0 and temperature θ_0, equations fit with Gibb's law of phases that names the system "divariant" [11].

References

[1] Atkin, R.J. and Craine, R.E., *Q.J. Mech. Appl. Math.*, **29**, pp. 209-243 (1976).

[2] Bedford, A. and Drumheller, D.S., *Arch. Rat. Mech. Anal.*, **68**, pp. 37-51 (1978), **71**, pp. 345-355 (1979) and **73**, pp. 257-284 (1980).

[3] Bedford, A. and Drumheller, D.S., *Int. J. Engng. Sci.*, **21**, pp. 863-960, (1983).

[4] Bowen, R.M., *Arch. Rat. Mech. Anal.*, **70**, pp. 235-250 (1979).

[5] Casal, P. and Gouin, H., *C.R. Acad. Sci. Paris*, **306**, II, pp. 99-104 (1988).

[6] Casal, P. and Gouin, H., *Annales de Physique, coll. no. 2*, **13**, pp. 3-12 (1988).

[7] Casal, P. and Gouin, H., *J. Theor. Appl. Mech*, **7**, no. 6, pp. 689-718 (1988).

[8] Drew, D.A., *Ann. Rev. Fluid Mech.*, **15**, pp. 261-291 (1983).

[9] Gouin, H., *C.R. Acad. Sci. Paris*, **305**, II, pp. 833-838 (1987).

[10] Gouin, H., *Research Notes in Mathematics*, **46**, Pitman, London, pp. 128-136 (1981).

[11] Gouin, H., *European Journal of Mechanics, Fluids B.* To appear (1990)..

[12] Green, A.E. and Naghdi, P.M., *Int. J. Engng. Sci.*, **3**, pp. 231-241 (1965).

[13] Ishii, M., *Thermo-Fluid Dynamic Theory of Two-Phase Flow*, Eyrolles, Paris (1975).

[14] Müller, I., *Arch. Rat. Mech. Anal.*, **26**, pp. 118-141 (1967).

[15] Nunziato, J.W. and Walsh, E.K., *Arch. Rat. Mech. Anal.*, **73**, pp. 285-311 (1980).

[16] Rocard, Y., *Thermodynamique*, Masson, Paris, (1967).

[17] Sampaio, R. and Williams, W.O., *J. Appl. Math. and Physics*, **28**, pp. 607-613 (1977).

[18] Serrin, J., *Encyclopedia of Physics*, VIII/1, Springer, Berlin, pp. 144-150 (1959).

[19] Truesdell, C., *Rend. Accad. Lincei, Series 8*, **22**, pp. 158-166 (1957).

[20] Williams, W.O., *Arch. Rat. Mech. Anal.*, **51**, pp. 239-260 (1973).

Gouin H.

Departement de Mathématiques et Mécanique

Faculté des Sciences et Techniques

Av. Escadrille Normandie-Niemen

F-13397 Marseille Cédex 13

FRANCE

R. KOTOWSKI, A. TRZĘSOWSKI, K.-H. ANTHONY

On non-classical diffusion theories

1 Introduction

The aim of this work is the discussion of various attempts to generalize the classical diffusion theory. The classical diffusion theory is based on the so-called Stokes relation between the peculiar velocity and the velocity of the chaotic motion. The latter one is totally defined by the distribution of mass taking part in the diffusion process (Fick's law). We generalize Stokes relation in such a way as to take account of the influence of inertial effects. As a consequence those diffusion processes can be described, the kinetic relaxation time of which exceeds a particular observation time. The problem of the Lagrange formulation for the generalized diffusion theory is also discussed.

2 Classical diffusion theory

Let us consider the balance of mass equation

$$\partial_t \rho + \mathrm{div}\,(\rho\,\mathbf{v}) = 0. \tag{1}$$

In this equation $\rho = \rho(\mathbf{x}, t)$ is the mass density. The velocity $\mathbf{v}$ is called in diffusion theory the diffusion peculiar velocity and is usually represented as a sum of two velocities

$$\mathbf{v} = \mathbf{b} + \mathbf{u}, \tag{2}$$

where $\mathbf{u}$ is the velocity of the chaotic motion

$$\mathbf{u} = -D\,\frac{\nabla \rho}{\rho}. \tag{3}$$

Velocity $\mathbf{b}$ is the drift velocity caused by external or internal sources of stresses in the body, like the electromagnetic field, temperature gradient and so on. In the classical theory, $\mathbf{b}$ is given by the Stokes relation, as a result of an acting force $\mathbf{F}$, in the following form

$$\mathbf{b}(\mathbf{x}, t) = \zeta^{-1}\mathbf{F}(\mathbf{x}, t), \tag{4}$$

where ζ^{-1} is called the mobility of the particle taking part in the diffusion process.

We obtain from eqs. (2), (3) and (4) that eq. (1) now has the form

$$\partial_t \rho - D\,\Delta\rho = -\,\zeta^{-1}\,\mathrm{div}(\rho\,\mathbf{F}), \tag{5}$$

which is the known non-homogeneous diffusion equation with the constant diffusion coefficient D.

We see that the fundamental expressions to describe the diffusion process are the mass balance equation (1) and a certain relation $\mathcal{R}$ between the velocities $\mathbf{v}$ and $\mathbf{u}$ given, for example, by eqs. (2) and (4). The admission of any particular form of this relation means the acceptance of a certain model of the diffusion process. Relation $\mathcal{R}$ depends both on the physical system and the forces acting on the diffusing particle. Let us denote by

$$\tau_* = \mathrm{d}t = t - t_0 > 0 \tag{6}$$

the length of a physically infinitesimally small interval of time which defines the scale of the observation time of the diffusion process. In this paper we will discuss such a relation $\mathcal{R} = \mathcal{R}(\tau)$ which depends on the so called kinematical relaxation time τ

$$\tau = \frac{m}{\zeta}, \tag{7}$$

where m is a mass of diffusing particle [9] and that for

$$\tau \ll \tau_* \tag{8}$$

it tends to Stokes relation (4). For

$$\tau \gg \tau_* \tag{9}$$

the use of the Stokes relation is not justified [5] and a different relation has to be discussed.

3 Quantum analogy

In quantum mechanics the procedure of obtaining the Schrödinger equation from the theory of the diffusion Markov processes is discussed [8]. This method, called the stochastic quantization method (SQM), has been used after certain necessary re-interpreations to describe the diffusion of the lattice point defects in solid bodies [9]. A *non-linear* Schrödinger-like equation was obtained as a result. We obtain this equation again following the different approach presented in papers [3] and [11].

The Born representation

$$\frac{\rho(\mathbf{x}, t)}{\rho_0} = \psi(\mathbf{x}, t)\, \psi^*(\mathbf{x}, t), \tag{10}$$

is used to assure the positive definiteness of the mass density $\rho(\mathbf{x}, t)$ [3] (ρ_0 is an arbitrary constant with the dimension of the mass density). Equation (10) defines the amplitude of the complex function $\psi(\mathbf{x}, t)$ only, and one is free to specify the phase. We define the function $\psi(\mathbf{x}, t)$ in the following way:

$$\psi(\mathbf{x}, t) = e^{R+iS}. \tag{11}$$

The quantity $S(\mathbf{x}, t)$ is the potential part of the peculiar velocity $\mathbf{v}(\mathbf{x}, t)$

$$\nabla S(\mathbf{x}, t) = \frac{1}{2D}\mathbf{v}(\mathbf{x}, t) + \mathbf{q}(\mathbf{x}, t) \tag{12}$$

and

$$\nabla R(\mathbf{x}, t) = -\frac{1}{2D}\mathbf{u}(\mathbf{x}, t), \quad R = \frac{1}{2}\ln\left(\frac{\rho}{\rho_0}\right). \tag{13}$$

The velocities $\mathbf{v}$ and $\mathbf{u}$ in terms of ψ are given by the following formulae

$$\mathbf{v} = iD\left(\frac{\nabla\psi^*}{\psi^*} - \frac{\nabla\psi}{\psi}\right) - 2D\mathbf{q}, \quad \mathbf{u} = -D\,\nabla\ln(\psi\psi^*). \tag{14}$$

Vector $\mathbf{q}$ is defined by eq. (12) up to the transformation

$$\mathbf{q} \to \mathbf{q} + \nabla\lambda, \tag{15}$$

which corresponds to the gauge transformation $S \to S + \lambda$. Let us observe, that S and R are connected with the diffusion process through the Born representation (10) and eqs. (12) and (13), but the physical status of the vector $\mathbf{q}$ is not definitely determined. We give two examples. In quantum mechanics the electromagnetic vector potential $\mathbf{A}$ is an analogue of $\mathbf{q}$ [8] and in the description of those diffusion processes which are coupled with the plastic yielding, vector $\mathbf{q}$ describes the increment of the plastic strain [11]. The decomposition of the vector field $\mathbf{v}$ according to eq. (12) is not unique. This fact is desired for reasons of physical interpretaiton of the quantities ∇S and $\mathbf{q}$.

It can easily be shown that eq. (1) together with definition (10) and velocity $\mathbf{v}$ given by eq. $(14)_1$ are equivalent to the following equation

$$\psi\, H^*(\psi) = \psi^*\, H(\psi), \tag{16}$$

where

$$H(\psi) = i\partial_t\psi + D(\nabla - i\mathbf{q})^2\psi + D\mathbf{q}^2\psi. \tag{17}$$

Equation (16) is fulfilled for any ψ, if and only if $H(\psi)$ is of the form

$$H(\psi) = \psi\, U \tag{18}$$

with $U = U(\psi, \psi^*, \mathbf{q}; \mathbf{x}, t)$ being a real function. If we put

$$V_q(\mathbf{x}, t) = h\left(U - D\mathbf{q}^2\right) \tag{19}$$

and

$$D = \frac{h}{2\,m} \tag{20}$$

(see [9]) then we obtain from eqs. (17) and (19) the Schrödinger-like equation

$$i\,h\,\partial_t\psi + \frac{h^2}{2\,m}(\nabla - i\,\mathbf{q})^2\psi - V_q\psi = 0. \tag{21}$$

This is the Schrödinger-like equation only, because in general it is a non-linear equation and h is not the reduced Planck constant (see discussion in [9]).

Up to now the obtained Schrödinger-like equation is the transformed continuity equation (1). Its connection with the diffusion theory is given by the diffusion constant D (21), but it can in fact be treated quite formally. What we need are the constraints to be imposed on the solutions of eq. (1), which give us the model of a diffusion process in the form of a relation $\mathcal{R}(\tau)$ between $\mathbf{v}$ and $\mathbf{u}$ velocities.

It can be shown ([9], [11]) that one can re-create from eq. (21) the generalized second Newton law (Nelson relation)

$$m\,\mathbf{a} = \mathbf{K}, \tag{22}$$

where we have defined the generalized force $\mathbf{K}$ as

$$\mathbf{K} = \mathbf{F} - \nabla V, \quad \mathbf{F} = h\,(\mathbf{v} \times \text{curl}\,\mathbf{q} - \partial_t\mathbf{q} + D\,\nabla\mathbf{q}^2), \quad V = h\,U. \tag{23}$$

The quantity $\mathbf{a}$ is an (anholonomic) acceleration of the form

$$\mathbf{a} = \dot{\mathbf{v}} + \nabla\varphi_D, \quad \varphi_D = -2D^2\frac{\Delta Q}{Q}, \quad Q = e^R, \tag{24}$$

where the point over $\mathbf{v}$ means the material time derivative $\partial_t + (\mathbf{v}\,\nabla)$. The quantity φ_D is called in [11] the diffusion potential (it is the counterpart of Bohm potential discussed in the hydrodynamical formalism in quantum mechanics).

Equation (22) in the limit $\tau \to 0$ with ζ = const. gives $\mathbf{K} = 0$, and consequently Stokes relation, only if the diffusion coefficient D depends on the relaxation time τ [9].

The relation $\mathcal{R}$ we look for in the generalized diffusion theory is given by eq. (22). We are free to consider various potentials U in eq. (23) and to discuss various generalizations of the classical diffusion theory. In particular, if we take the potential U in the form

$$U = \frac{1}{2\tau}\Big(\ln(\psi\psi^*) + i\ln(\psi/\psi^*) + C(t)\Big) + \frac{1}{h}U_0(\mathbf{x}, t) \tag{25}$$

then we obtain the case considered in our previous paper [9] where we had $\mathbf{q} = 0$. The force $\mathbf{K}$ turns out to be

$$\mathbf{K} = \mathbf{F}_1 - \zeta\mathbf{b}, \quad \mathbf{F}_1 = -\nabla U_0. \tag{26}$$

If the interactions of diffusing particles are neglected then the probability density $p(\mathbf{x}, t)$ of finding a particle with mass m in an infinitesimal volume element of a body can be represented as

$$p = \frac{\rho}{M}, \tag{27}$$

where M is the total mass of the diffusing matter taking part in the diffusion process, and then eq. (1) is equivalent to the continuity equation

$$\partial_t p + \operatorname{div}(p\mathbf{v}) = 0, \tag{28}$$

which gives us a microscopic interpretation of the presented model in line of SQM. The essence of SQM relies on the simultaneous consideration of two stochastic processes: forward and backward processes. As a result we discuss two Fokker–Planck equations describing these two processes and the continuity equation (28) appears to be the necessary condition for these two Fokker–Planck equations to be fulfilled simultaneously. In the case of diffusion of a particle it is more profitable to consider it as a "sum" of two physically different processes: as a diffusional arrival to and as a diffusional departure from the considered point of a body [11].

Nelson relation (22) introduced in [9] and extensively discussed in [10] and [11] is interpreted on the base of stochastic considerations as a kind of constraint imposed on the mean velocities of the diffusion Markov process. The forces appearing in this relation do not have the meaning of the physical cause of the motion of the mass particle. From the physical point of view they are restrictions on the too broad class of trajectories admissible by the Markovian diffusion process. If one accepts the above remarks, then **a** has the meaning of the effective acceleration [11].

4 Diffusion and Lagrange formalism

There exist in the literature a number of variational formulations for the differential equations of the type of the diffusion equation. In the Lagrange formalism one needs something more: the equations following from the Noether theorem should make physical sense. The mathematical theory of embedding any linear integrodifferential equation in a variational statement was given by Edelen in [4]. He proposed, in particular, the following Lagrangian to describe a diffusion process (we change his notation here to be consistent with our approach)

$$\ell = \frac{1}{\omega}\Big(\frac{1}{2}(\rho\partial_t\varphi - \varphi\partial_t\rho) - D\nabla\rho\nabla\varphi\Big). \tag{29}$$

Here $\varphi(\mathbf{x}, t)$ is an additional variable which allows for this embedding and ω is a dimensional constant. We obtain the following Euler-Lagrange equations (ELE) from this Lagrangian

$$\partial_t\varphi + D\Delta\varphi = 0, \quad \partial_t\rho - D\Delta\rho = 0. \tag{30}$$

Equation $(30)_2$ is the known homogeneous diffusion equation. If we put $\varphi(\mathbf{x}, t) = \rho(\mathbf{x}, -t)/\rho_0$ then eq. $(30)_1$ is a usual diffusion equation too. The way in which the additional variable φ is introduced into the Lagrangian (29) does not assure the fulfilling of the balance of mass equations along the solutions of the ELE (30) but the stationary case $\rho(\mathbf{x}, t) = \rho(\mathbf{x})$.

The approach, which makes possible the Lagrange formulation of the irreversible processes and in particular assures the reproduction of the balance of mass equation was proposed by Anthony in [1] and developed by him in the series of papers (see the references in the paper of Anthony of the same Proceedings). He has observed that the equation of heat conduction, even in its simplest form with the right-hand side equal to zero, has not a Lagrange formulation. It was shown in [7] that for the Schrödinger-like equation with the potential given by eq. (25) we have in a certain sense a similar situation and that it does not exist as a Lagrangian formulation. In what follows we try to find, at the beginning quite formally, a different formulation of the constitutive relation $\mathcal{R}(\tau)$ which allows a Lagrangian description of non-classical diffusion theory and which reduces to Stokes relation if $\tau \to 0$.

The elastic force $\mathbf{F}_{el}$ acting on a diffusing particle and being responsible for the "potential energy" caused by an elastic field is represented as

$$\mathbf{F}_{el} = - V_0 \nabla \sigma_{el}, \quad \sigma_{el} = 3K\,\theta, \tag{31}$$

where θ is the elastic dilatation and V_0 is the element of volume transported by the diffusing particle [6]. The plastic force $\mathbf{F}_{pl}$ responsible for inertial effects can be introduced, e.g., by the assumption of its dependence on an internal variable $\dot{\phi}$ ($[\,\dot{\phi}\,] = s^{-1}$) in a similar way as above

$$\mathbf{F}_{pl} = \mathbf{F}_{pl}(\dot{\phi}) = - V(\rho)\, \nabla \sigma_{pl}. \tag{32}$$

Here

$$V(\rho) = \frac{\rho_0}{\rho}\, V_0, \quad \rho_0 = \frac{m}{V_0}, \quad \sigma_{pl} = \kappa\, \dot{\phi}\,. \tag{33}$$

We interpret $\dot{\phi}$ as the "velocity of plastic dilatation". Let us discuss the influence of this plastic force on the diffusion process. Equation (5) in this case takes the form

$$\partial_t \rho - D\, \nabla \rho = \kappa\, \tau\, \nabla \dot{\phi}\,. \tag{34}$$

The kinematical relaxation time τ is given by eq. (7). Let us consider a candidate to be a Lagrangian which gives us eq. (34) as ELE

$$\ell = \frac{1}{\omega}\left(\rho \partial_t \chi - \frac{\theta}{2} (\Delta \chi)^2 - D \nabla \rho \nabla \chi \right). \tag{35}$$

This Lagrangian gives the following set of ELE

$$\partial_t \rho - D \Delta \rho = \theta \Delta \chi, \quad \partial_t \chi + D \Delta \chi = 0. \tag{36}$$

Let us note that the physical interpretations of the additional variables in variational formulations of diffusion-type equations (variables φ and χ in Lagrangians (29) and (35)) depend on the context. In the Anthony approach, for example, φ is interpreted as a variable describing the deviation of the system from the local equilibrium state. The interpretation of φ for the homogeneous diffusion equaiton is given by the comment following eq. $(30)_1$. In the case of eq. (36) we formulate another interpretation.

If we compare eq. $(36)_1$ with eq. (34) we find that

$$\theta = \kappa \quad \text{and} \quad \chi = \tau\phi \,. \tag{37}$$

If we compare eqs. (4), (5) and $(36)_1$ we find that

$$\mathbf{b} = -\frac{\kappa\,\tau}{\rho}\nabla\phi = -\frac{V(\rho)}{l_0^3}\mathbf{B}, \tag{38}$$

where $\mathbf{B} = l_0^2\,\nabla\phi$, and $l_0 = m/\kappa\tau$ is a characteristic length of a diffusion process. We make use of eqs. (2) and (3) and we obtain that Noether theorem produces the following constitutive equations for

- mass

$$w = \frac{\Lambda}{\omega}\rho\,,\ \mathbf{j}_w = \frac{\Lambda}{\omega}\rho\mathbf{v}\,, \tag{39}$$

- energy

$$\varepsilon = \frac{1}{2}\frac{\rho^2}{\rho_0}(\mathbf{v}^2 - \mathbf{u}^2),\quad \mathbf{j}_\varepsilon = \frac{\rho}{\rho_0}D(\partial_t\rho + \rho\mathbf{v}\nabla)(\mathbf{v} - \mathbf{u}), \tag{40}$$

- momentum

$$\mathbf{p} = \frac{\rho^2}{\rho_0}(\mathbf{v} - \mathbf{u}),$$

$$\sigma_{ij} = \frac{\rho^2}{\rho_0}\Big\{\Big(D\nabla(\mathbf{v} - \mathbf{u}) - \frac{1}{2}(\nabla(\mathbf{v} - \mathbf{u}))^2 - \mathbf{u}(\mathbf{v} - \mathbf{u})\Big)\,\delta_{ij}\ \ v_iv_j - u_iu_j\Big\}. \tag{41}$$

We have made the assumption here that $\rho_0 = A\omega\kappa$, and we have scaled A to $A \equiv 1$, because ω was a quite arbitrary constant.

5 Conclusions

We have shown in the paper two generalizations of the classical diffusion theory, which allows consideration of relaxation effects. In the first one, based on SQM, we have obtained the Schrödinger-like new diffusion equation, which gave us the possibilities of studying further generalizations as compared to paper [9]. In the second approach based on the ideas proposed in [2] we have succeeded in giving a Lagrangian formulation of the diffusion process caused both by the gradient of the diffusing mass particles and the forces of the plastic type. Let us observe that ELE (36) are formally equivalent to ELE in [2], dealing with the transport of heat. It is easily seen if we assume that $\dot{\varphi} = \partial_t \varphi$ and then

$$\partial_t \rho - D\Delta\rho = -\Lambda\partial_{tt}\varphi, \ \Lambda = \frac{\kappa\tau}{D}. \tag{42}$$

This equation can be obtained from the modified Lagrangian where one puts $\theta(\Delta\chi)^2 \to -\Lambda(\partial_{tt}\varphi)^2$ (compare [2]). It appears however that the balance equations following from this modified Lagrangian do not have an interpretation in velocities **u**, **v**, **b** and **B** which appear in our approach. That is why this alternative Lagrangian is not suitable for the description of the diffusion process.

Acknowledgements

This paper is the result of a cooperation programme between the Institute of Fundamental Technological Research of the Polish Academy of Sciences and the Institute of Theoretical Physics, University Paderborn, FRG. Financial support by the Polish Academy of Sciences and the Deutsche Forschungsgemeinschaft is highly appreciated.

References

[1] Anthony, K.-H. A new approach describing irreversible processes, in: *Continuum Models of Discrete Systems*, Vol. 4, O. Brulin, R.K/T. Hsieh, Eds., North-Holland, Amsterdam, 1981, pp. 481-494.

[2] Anthony, K.-H., and Knoppe, H. Phenomenological thermodynamics of irreversible processes and Lagrange-formalism, hyperbolic equaitons for heat transport, in: *Conference Proceedings on Kinetic Theory and Extended Thermodynamics*, T. Ruggeri, I. Müller, Eds., Bologna, 1987.

[3] Collins, R.E. Quantum Theory: A Hilbert space formalism for probability theory, *Foundations of Physics*, **7**, 475-494, 1977.

[4] Edelen, D.G.B. *Nonlocal Variations and Local Invariance of Fields*, Elsevier, New York, 1969.

[5] Klimontovitch, J.L. *Statistical Physics*, Nauka, Moscow, 1982 (in Russian).

[6] Kosevich, A.M. *Foundations of Crystal Lattice Mechanics*, Nauka, Moscow, 1972 (in Russian).

[7] Kotowski, R. On the Lagrange Functional for Dissipative Processes (to be published in Archives of Mechanics, 1989).

[8] Nelson, E. Derivation of the Schrödinger equation from Newtonian mechanics, *Phys. Rev.*, **150**, 1079-1085, 1966.

[9] Trzęsowski, A. and Kotowski, R. Nonlinear diffusion and Nelson-Brown movement, *Int. J. Theor. Phys.*, **24**, 533-556, 1985.

[10] Trzęsowski, A. and Kotowski, R. Nonlinear diffusion and Schrödinger equation, in: *Continuum Models of Discrete Systems*, Vol. 5, A.J.M. Spencer, Ed., A.A. Balkema, 1987, pp. 145-150.

[11] Trzęsowski, A. Locally equilibrium diffusion processes, *Int. J. Theor. Phys.*, **28**, 543-586, 1989.

Kotowski R.
I.P.P.T.-P.A.N.,
Świętokrzyska 21
PL-00-049 Warsaw
POLAND

Trzęsowski A.
I.P.P.T.-P.A.N.,
Świętokrzyska 21
PL-00-049 Warsaw
POLAND

Anthony K.-H.
Fachbreich 6/Naturwissenschaft I
Physik, Universität Gesamthochschule
Warburgerstrasse 100/Postfach 1621
D-4790 Paderborn
F.R.G.

M.A. MELEHY

On the internal pressure and uniqueness of entropy in a new thermodynamic formulation

1 Introduction

Since the time of Ancient Greece, pressure has been perceived to be an observable quantity, commonly measurable outside a system. For example, the pressure, p, just under a liquid surface would be considered to be the surrounding pressure. In the nineteenth century, that view of pressure was incorporated into thermodynamics. For any isothermal multiphase system, for example, the criteria of (thermal) equilibrium [3], have been assumed to involve the constancy of p throughout the system. Classical thermodynamic equations relate every term in the Gibbs equation to pressure; i.e., to p.

Since the early 1960s, this author has attempted to solve some interfacial problems involving conduction electrons. In those studies [8–14], the use of the concept of internal pressure, P, as a thermodynamic quantity was extensively explored. For conduction electrons and holes, the internal pressure is the same as the kinetic pressure. For ideal, classical gases in the vapour state, Maxwell [7] has shown that, P reflects both the particle momentum, and thermal kinetic energy per unit volume. For such gases, the external pressure, $p = P$. For other systems, in which the inter-particle, attractive van der Waals' and/or Coulomb forces are significant, P and p can be profoundly different. Examples include liquids and conduction electrons in metals.

Using the internal/kinetic pressure has led to a connection between Newtonian mechanics and thermodynamics, and, in turn, to a thermodynamic generalization of Maxwell's [7] diffusion force, f_D. In its mechanical form, f_D was used by Einstein [2] to derive the well-known diffusion-mobility relationship. An important question to raise: how would f_D interact in any system containing interfaces, with electric, magnetic, and gravity fields, so that the state of thermal equilibrium would be characterized by the principle of microscopic reversibility? Answering this fundamental question has led to the evolution in the last quarter-century of the thermodynamic formulation of generalized fields (TFGF) [4, 8–13]. In that formulation, the interaction in non-equilibrium of all force fields is governed by a global relation, which expresses the conservation of energy in the entire system.

The TFGF was applied to semiconductor diodes and solar cells. Theory has accurately been corroborated by extensive experimental data reported by some 35 authors [5,9–13, 20] between 1951–87.

For reasons relevant to the principle of microscpic reversibility, at (thermal) equilibrium the global relation renders every identifiable force per particle completely decoupled from all other ones [9–11]. Under such equilibrium conditions, when all processes are purely reversible, the interaction of all force fields in a system is governed by Carnot's theorem [1], which is shown to require bound electric charges to reside at nearly all surfaces and interfaces. This interesting result may lead to new explanations for numerous diversified phenomena such as: surface tension, capillarity, adhesion, catalysis, and surface wetting. Interfacial charges may explain why mechanical and electrical degradation of materials occurs as they age, and why rubbing some materials generate static electricity, and why clouds can get electrified. It is no surprise then to find that smoke and aerosol particles are electrically charged, or that electric fields exist in biological membranes as reported by Nördenstrom [15].

In this paper, we shall first address the question: has the concept of internal/kinetic pressure accidentally led to results that are corroborated by experiment for the special cases described above, or is it a basic thermodynamic quantity that may underlie a new formulation? Towards that goal, the existence of a connection between quantum mechanics and thermodynamics will be pointed out. It will then be shown that the internal/kinetic pressure is consistent with the uniqueness of entropy, which is required by the second law. For simplicity, cases treated here will be those obeying classical statistics, but may have any isotropic quantum mechanical band. Generalization of the treatment can be done [14], regardless of the quantum mechanical band, for conduction electrons in solids and holes in semiconductors, even when quantum statistics applies.

2 The Gibbs and Gibbs-Duhem equations

Consider N particles of a specific constituent in a generally multicomponent system that is in equilibrium. Let the particles be in a volume, V, and have a uniform (partial) pressure, $\mathcal{P}$. For those N particles, let the internal energy be U, entropy be S, and absolute temperature be T. If reversible heat, $T\mathrm{d}S$, is added from an outside source to the N particles, and if V quasi-statically expands by $\mathrm{d}V$, and if $\mathrm{d}N$ particles enter V, then the conservation of energy requires that the change in U to be given by

$$\mathrm{d}U = T\,\mathrm{d}S - \mathcal{P}\,\mathrm{d}V + \mu\,\mathrm{d}N\,. \tag{1}$$

This is Gibbs' equation [3]. The term ($\mathcal{P}\,\mathrm{d}V$) is the mechanical work done by the N-particle system on its surroundings, and ($\mu\,\mathrm{d}N$) is the amount of energy introduced by $\mathrm{d}N$ particles that entered V. The parameter μ is called the chemical potential.

From the condition of uniformity, it can be shown [19] that

$$U = T\,S - \mathcal{P}\,V + \mu N\,, \tag{2}$$

$$s = \frac{1}{T}\left(u + \frac{\mathcal{P}}{n} - \mu\right), \tag{3}$$

$$s = \frac{S}{N}, \; u = \frac{U}{N}, \; n = \frac{N}{V}. \tag{4}$$

Here s is the entropy per particle, u is internal energy per particle, and n is the particle concentration.

Taking the differential of U [eq. (2)], and subtracting it from eq. (1) leads to the well-known Gibbs-Duhem equation [16]:

$$\mathrm{d}\mu = -\, s\, \mathrm{d}T + \frac{1}{n}\mathrm{d}\,\mathcal{P}. \tag{5}$$

3 Thermodynamic/quantum mechanical definitions of $\mathcal{P}$ and s

If n and $\mathcal{P}$, in eq. (5), can be expressed as exclusive functions of μ, and T, then by keeping T or μ constant, we get

$$\mathrm{d}\mu = \left(\frac{1}{n}\right)_T (\mathrm{d}\,\mathcal{P})_T\,, \tag{6}$$

$$\mathcal{P}(\mu, T) - \mathcal{P}(\mu_0, T) = \int_{\mu_0}^{\mu} \left[n(\mu, T)\right]_T \mathrm{d}\mu\,, \tag{7}$$

$$s = \frac{1}{n}\left(\frac{\partial\,\mathcal{P}}{\partial\,T}\right)_{\mu}, \tag{8}$$

where $\mathcal{P}(\mu_0, T)$ is an arbitrary constant to be evaluated.

It is important to recognize that eqs. (7) and (8) provide fundamental definitions for $\mathcal{P}$ and s, based on the first and second law of thermodynamics. The exclusive functional dependency of n and $\mathcal{P}$ on μ and T is achievable by quantum mechanical methods, a matter to be briefly discussed next.

4 Thermodynamic pressure for classical conduction electrons

We shall determine here whether the pressure, $\mathcal{P}$, is P, or p, for classical conduction electrons. As is well known [17], those particles exist in non-degenerate semiconductors. For such electrons, the Fermi-Dirac satistics reduces to the Maxwell-Boltzmann's statistics [17] and, therefore, n may be expressed by

$$n(\mu, T) = \int_0^{\infty} g(\varepsilon)\, \mathrm{e}^{(\mu-\varepsilon)/\gamma kT}\, \mathrm{d}\varepsilon\,. \tag{9}$$

Here k is Boltzmann's constant, and $g(\varepsilon)$ is the volume–energy density of quantum states which may generally be any function of the electron kinetic energy, ε. The constant positive parameter, γ, is unity for conduction electrons. Its introduction here is needed for an idealized model to be discussed in Section 6.

The average kinetic energy, $\langle \varepsilon \rangle$, per particle may be defined as

$$\langle \varepsilon \rangle = \frac{1}{n} \int_0^\infty \varepsilon \, g(\varepsilon) \, e^{(\mu-\varepsilon)/\gamma kT} \, d\varepsilon \, . \tag{10}$$

Equation (9) indicates that regardless of temperature, n will tend to zero, if μ tends to $-\infty$. Since the pressure of any system of particles must vanish, if their concentration vanishes, then $\mathcal{P}(-\infty, T) = 0$, and eq. (7) may be rewritten as

$$\mathcal{P}(\mu, T) = \int_{-\infty}^{\mu} \left[n(\mu, T) \right]_T d\mu \, , \tag{11}$$

$$\mathcal{P}(\mu, T) = \int_{-\infty}^{\mu} \left\{ \int_0^\infty \left[g(\varepsilon) \, e^{(\mu-\varepsilon)/\gamma kT} \right]_{\mu,T} d\varepsilon \right\}_{\varepsilon,T} d\mu$$

$$= \int_0^\infty \left\{ g(\varepsilon) \int_{-\infty}^{\mu} \left[e^{(\mu-\varepsilon)/\gamma kT} \right]_{\varepsilon,T} d\mu \right\}_{\mu,T} d\varepsilon \, . \tag{12}$$

In view of eq. (9), therefore, we have

$$\mathcal{P} = \gamma \, k \, n \, T \, . \tag{13}$$

Clearly, with $\gamma = 1$, $\mathcal{P}$ in eq. (13) represents no other than the kinetic pressure, P, for classical conduction electrons. This kind of pressure is significantly different from the electron external pressure, which is essentially zero.

It should be noticed here that the value of $\mathcal{P}$ is valid regardless of what $g(\varepsilon)$ may be. By contrast, in general, the value of $\langle \varepsilon \rangle$ depends on $g(\varepsilon)$. That dependency disappears only for a standard band. For that particular case,

$$g(\varepsilon) = C\sqrt{\varepsilon} \, , \langle \varepsilon \rangle = \frac{3}{2} \gamma \, k \, T \, , \tag{14}$$

where C is a constant.

5 Internal/kinetic pressure and uniqueness of entropy

From eqs. (8), (9), and (13), it follows that

$$s = \frac{1}{n}\frac{\partial}{\partial T}\left\{ \gamma kT \int_0^\infty \left[g(\varepsilon)\, e^{(\mu-\varepsilon)/\gamma kT} \right]_{\mu,T} d\varepsilon \right\}_\mu . \tag{15}$$

Carrying out the mathematical steps, and in view of eqs. (9), (10), and (13), we get

$$s = \frac{1}{T}\left[\langle \varepsilon \rangle + \frac{P}{n} - \mu \right]. \tag{16}$$

For entropy uniqueness eqs. (3) and (16) must be identical; i.e.,

$$u = \langle \varepsilon \rangle = \frac{1}{n} \int_0^\infty \varepsilon\, g(\varepsilon)\, e^{(\mu-\varepsilon)/\gamma kT} d\varepsilon . \tag{17}$$

The above procedure has provided an important definition for internal energy, and has confirmed that the use of internal/kinetic pressure for conduction electrons is consistent with the uniqueness of entropy and the second law. Such results can be shown [14] to hold, for degenerate conduction electrons in solids and holes in semiconductors. Generalizing these conclusions to other systems should also be possible.

6 An idealized model for some classical systems

An expression for $n[\mu, T]$ for all systems, is quite complicated to achieve. A useful case, however, may be that of ideal, classical liquids and ideal classical insulating solids which we define to be those for which eq. (9) holds. The parameter, γ, is one-third the number of degrees of freedom. The justification for this idealized model is based on a number of reasons: First, because of the equipartition of energy over the various degrees of freedom, for liquids, the quantity ε/γ in the exponent [eq. (9)] is simply the translational energy, ε_t, which is the same as for the case of classical electrons. The question then is whether the exponent should be $(\varepsilon_t - \mu/\gamma kT)$, or $(\varepsilon_t - \mu/kT)$. The condition of uniqueness of entropy described by eqs. (3) and (16) requires that the exponent to be as in eq. (9).

Second, for a solid insulator, such as diamond ($\gamma = 2$), eqs. (9) and (14) lead near 300° K to a specific heat, C_v, at constant volume, of 6 calories/gm-mole/K. This result is in agreement with experiment [6]. For water, however, since $n = 3.34567 \times 10^{22}/\text{cm}^3$, and if $\gamma = 6$, from eq. (14), then $C_v = 0.99358$ calorie/gm/K, which is a quite accurate result.

Third, in some classical liquids or insulating solids, less than 1% of the particles have kinetic energy above about 6 kT per particle, associated with each three degrees of freedom. In such cases, the energy of a particle for most particles sweeps only a narrow range of energy, which helps make the approximation accurate.

Fourth, as crude as eq. (9) may be, for other than classical conduction electrons, for which $\gamma = 1$, that equation is consistent with entropy uniqueness, and is free of the following paradox.

7 Phenomenological pressure and entropy uniqueness: A paradox

Consider a liquid-vapour system in equilibrium. In that state, it has commonly been assumed that T, p, and μ are all constant throughout the system [3, 19]. When such conditions are imposed on eq. (5), the difference between the values of the entropy per particle in the vapour, s_V, and that in the liquid, s_ℓ, will be

$$s_V - s_\ell = \left(\frac{1}{n_V} - \frac{1}{n_\ell}\right) \frac{d\,p_V}{d\,T} > 0 \,. \tag{18}$$

Equation (18) is the well-known Clausius-Clapeyron equation [19]. Since n_ℓ (of the liquid) $> n_V$ (of the vapour) and $dp_V/dT > 0$, then $(s_V - s_\ell) > 0$; p_V being the vapour pressure.

For the liquid-vapour system, imposing on eq. (3) the commonly-assumed conditions of equilibrium, described above, we have

$$s_V - s_\ell = \frac{1}{T}\left[u_V - u_\ell + p_V \left(\frac{1}{n_V} - \frac{1}{n_\ell}\right)\right]. \tag{19}$$

Calculation of u_ℓ should be made using data on specific heat C_V. Since such data are not easy to find, u_ℓ is computed approximately from eq. (14), as outlined above. For water, in view of eqs. (14) and (17), and vapour pressure data [18], we have

$$p_V = k\, n_V T, \;\; u_V = \alpha\; kT, \;\; u_\ell = \frac{3}{2}\gamma kT \,. \tag{20}$$

But $n_\ell >> n_V$ and for water, for example, $\gamma = 6$ and $\alpha \approx 5.5$. Therefore, from eqs. (19) and (20), we get for water $s_V - s_\ell < 0$, which is not consistent with eq. (18).

8 Thermodynamic generalization of Maxwell's diffusion force

It is important to recognize that, although eqs. (3) and (5) are derived assuming uniformity within the small volume V, these two equations will still remain valid even in a non-uniform system, provided that u, $\mathcal{P}$, n, and μ are all evaluated at equilibrium, and at any one point in the system. For, one can divide the system into a large number of adjacent small volumes. Within each volume exact uniformity is attained in the limit as V tends to zero.

As explained for eq. (1), the term ($\mu\ dN$) represents the amount of energy introduced into V by dN particles entering V. Thus, μ is but the energy introduced per particle entering V. For the uniqueness of entropy, μ must exclude electric, magnetic and gravity effects. Now suppose that r particles were transported along a differential length, $\mathrm{d}\boldsymbol{\ell}$, and in so-doing crossed a volume, dV, containing dN particles. Suppose that where the particles entered dV, $\mu = \mu_0$, and where they left it, $\mu = \mu_0 + \mathrm{d}\mu$. The net amount of energy, $\mathrm{d}U_\mathrm{D}$, given to the dN non-transported particles and the rest of the host system will be

$$\mathrm{d}U_\mathrm{D} = r\,\mu_0 - r(\mu_0 + \mathrm{d}\mu) = -\,r\,\mathrm{d}\mu = -\,r\,\nabla\mu\cdot\mathrm{d}\boldsymbol{\ell}\,. \tag{21}$$

We then define the actual/virtual force per particle by $\mathbf{f}_\mathrm{D} = -\,\nabla\mu$. If $\mathrm{d}\mu$, dT, and dP in eq. (5) are differentials over a length, $\mathrm{d}\boldsymbol{\ell}$, in a system, then

$$\mathbf{f}_\mathrm{D} = -\,\nabla\mu = \mathbf{f}_\mathrm{d} + \mathbf{f}_\mathrm{T},\ \mathbf{f}_\mathrm{d} = -\,\frac{1}{n}\nabla\mathrm{P},\ \ \mathbf{f}_\mathrm{T} = \mathrm{s}\,\nabla\mathrm{T}\,. \tag{22}$$

The force, $\mathbf{f}_\mathrm{d}$, per particle has been called [9, 10, 11] a diffusion field. It is mechanical in the Newtonian sense. It was first derived by Maxwell [7] for ideal, classical gases in the vapour state. Maxwell's procedure involves calculating the time-rate of increase of momentum (per unit mass) of the particles.

It has been shown [9-11] that $\mathbf{f}_\mathrm{d}$ can be associated with inelastic collisions, and consequent heat dissipation. Under such conditions, $\mathbf{f}_\mathrm{d}$, is called passive or dissipative, and leads to Fick's law and the Einstein diffusion-mobility relationship [2]. The work done per transported particle under the action of $\mathbf{f}_\mathrm{d}$ is equal to the heat dissipated in the host medium.

It has also been shown [9-11] that $\mathbf{f}_\mathrm{d}$ can be associated with purely elastic collisions, such as at all interfaces and surfaces. Under such important conditions, $\mathbf{f}_\mathrm{d}$ is called active, or motive. This fundamental force is intrinsic; i.e., is exerted between a transported particle, and the non-transported particles and the rest of the host system. Thus, the line integral of $\mathbf{f}_\mathrm{d}$ represents the statistical work done on a particle, which equals the amount of decrease of internal energy in the host system.

The intrinsic force, $\mathbf{f}_\mathrm{T}$, per particle, however, is virtual, and is termed thermomotive field. The line integral between two non-isothermal points represents the average amount of heat given to each transported particle, and, therefore, lost by the host medium.

9 Electrification of surfaces and interfaces: A consequence of Carnot's theorem

Consider a differential length, $\mathrm{d}\boldsymbol{\ell}$, joining two points, A and B, within an interface between two different media. Let the system be at equilibriuim. Imagine that r particles of a specific constituent are transported quasi-statically over $\mathrm{d}\boldsymbol{\ell}$, beginning at A. Let that process proceed at constant T, while the particles are subjected to a resultant intrinsic active force, $\mathbf{f}_{\mathrm{ta}}$, per particle. Thus, the internal energy of the host system will decrease by an amount $(r\,\mathbf{f}_{\mathrm{ta}}\cdot\mathrm{d}\boldsymbol{\ell})$. To keep T constant, an amount of reversible heat, $(r\,\mathbf{f}_{\mathrm{ta}}\cdot\mathrm{d}\boldsymbol{\ell})$, has to be added to the system. Now let the system as a whole be raised in temperature quasi-statically to $(T+\mathrm{d}T)$, while the particles are at point B. At $(T+\mathrm{d}T)$, let the particles be quasi-statically transported back to point A over $\mathrm{d}\boldsymbol{\ell}$, after which the system temperature is brought back to T, while the particles are at point A.

According to Carnot's theorem, the sum of all changes in entropy for the particles must vanish, as the system is brought back to its initial state. Mathematically, it can be shown that

$$\frac{\partial}{\partial T}\left(\frac{\mathbf{f}_{\mathrm{ta}}}{T}\right) = \frac{\partial}{\partial T}\left(\nabla s\right). \tag{23}$$

Integrating eq. (23) with respect to T introduces an arbitrary function of position, which, by invoking the third law, can be shown to vanish. We, therefore, have

$$\mathbf{f}_{\mathrm{ta}} = T\,\nabla s\,. \tag{24}$$

At equilibrium, in non-magnetic materials, only two types of intrinsic forces can exist in an interface: an active (motive) diffusion field, $\mathbf{f}_{\mathrm{da}}$, and an active electric force, $\mathbf{f}_{\mathrm{ea}}$, per particle, which is $q\mathbf{E}$. Here q is the charge per particle and $\mathbf{E}$ is the electric field. Thus,

$$\mathbf{f}_{\mathrm{ta}} = \mathbf{f}_{\mathrm{da}} + \mathbf{f}_{\mathrm{ea}} \tag{25}$$

and from eqs. (16), (22), (24), and (25), we have

$$\mathbf{f}_{\mathrm{ea}} = \nabla\left(<\varepsilon> + \frac{P}{n}\right) = q\,\mathbf{E}\,. \tag{26}$$

For the case of a standard quantum mechanical band [eq. (14)],

$$E = \frac{5}{3}\,\frac{1}{q}\nabla<\varepsilon>. \tag{27}$$

Clearly, within a homogenous medium, $<\varepsilon>$, P, and n are all uniform, and therefore $E = 0$. But at surfaces, membranes, and interfaces, $<\varepsilon>$, and P/n can have large gradients, which will lead to significant electric fields and bound electric charges. Equations (26) and (27) suggest that this fundamental interfacial property exists regardless of media phases, and degree of electrical conductivity.

References

[1] Carnot, Sadi., *Reflexions sur la Puissance Motrice du Feu.* Original publication, Paris, 1824. English Translation by Robert Fox: *Reflexions on the Motive Power of Fire*, Manchester University Press, and Lilian Barber Press, New York, 1986.

[2] Einstein, A., *Ann. Physik,* **17**, 549 (1905).

[3] Gibbs, J.W., On the equilibrium of heterogeneous substances, Transactions of the Connecticut Academy, III. pp. 108-248, Oct. 1875-May, 1876, and pp.343-524, May, 1877-July, 1878; reprinted in *The Scientific Papers of J.W. Gibbs*, Vol. 1, pp. 62-67, Dover Publications, New York.

[4] Heerden, van, P.T., Foundations of the thermodynamic theory of generalized fields (book review). *American Journal of Physics*, **44**, pp. 895-896, 1976.

[5] Jain, F.C., and Melehy, M.A., High-forward-voltage phenomenon in injection GaAs/Ge heterojunctions. *Applied Physics Letters*, **27**, 36-38, 1975.

[6] Kittel, C., *Introduction to Solid State Physics,* pp. 204, John Wiley, New York, 1971.

[7] Maxwell, J.C., Illustrations of the Dynamical Theory of Gases, Philosophical Magazine, January and July, 1860; reprinted in *The Scientific Papers of James Clerk Maxwell;* pp. 377-405, vol. I, edited by W.D. Niven, Dover Publications, New York.

[8] Melehy, M.A., Proposed distribution of electron energy in p-n junctions and the theory of injection. *Nature,* **202**, 864-868, 1964; Nature of electron equilibrium in heterogeneous solids and the proposed electron energy distribution in degenerate p-n junctions. *Nature*, **205**, 456-464, 1965; Thermodynamics of new generalized transport laws for liquids, gases and electrons in matter. *Nature,* **209**, 670-677, 1966.

[9] Melehy, M.A., An introduction to the generalized field theory: I. Forward conduction in p-n junctions and heterojunctions. *International Journal of Electronics*, **24**, 41-68, 1968; Melehy, M.A., An introduction to the generalized field theory: II. Forward conduction in Schottky diodes. *International Journal of Electronics*, **29**, 525-532, 1970.

[10] Melehy, M.A., On the theory of generalized fields in nonequilibrium thermodynamics: I. Some general aspects of the steady state; On the theory of generalized fields in nonequilibrium thermodynamics: II. Unified theory of electrical conduction in p-n junctions, heterojunctions and Schottky diodes. *Proc. 1969 Pittsburgh International Symposium on "A critical review of thermodynamics"*, edited by E.B. Stuart, B. Garl-Or, and A.J. Brainard, Baltimore: Mono, pp. 345-405, 1970.

[11] Melehy, M.A., *Foundations of the thermodynamic theory of generalized fields*, Baltimore: Mono 1973.

[12] Melehy, M.A., On the theory of homojunction and heterojunction solar cells. *Internation J. of Electronics*, **44**, 211-217, 1978.

[13] Melehy, M.A., Roots and ramifications of a unified theory of electrical conduction in p-n junctions, heterojunctions, and solar cells, *International J. Electronics*, **63**, 555-571 (1987).

[14] Melehy, M.A., On the foundations of the thermodynamic formulation of generalized fields, an invited paper, 17th International Conference on Thermodynamics and Statistical Mechanics - Satellite Meeting on Non-Equilibrium Thermodynamics, Iguazu Falls, Argentina, August 7-10, 1989; to be published in the proceedings.

[15] Nördenstrom, B.E.W., *Biologically Closed Electric Circuits* Nordic Medical Publications, Uppsala, 1983.

[16] Reif, F., *Fundamentals of Statistical Mechanics and Thermal Physics*, pp. 315, McGraw-Hill, New York, 1965.

[17] Shockley, W., *Electrons and Holes in Semiconductors,* D. Van Nostrand, 1950.

[18] Weast, R.C., *Handbook of Chemistry and Physics*, pp. D-157, 54th edition, Chemical Rubber Publishing Co., Cleveland, OH, 1973-74.

[19] Wilson, A.H., *Thermodynamics and Statistical Mechanics*, pp. 44, Cambridge University Press, Cambridge, 1966.

[20] (In preparation for publication).

Melehy M.A.
Elect. and Sys. Eng. Dept.
University of Connecticut
260 Glenbrook Road
Storrs CT 06269-3157
U.S.A.

W. MUSCHIK

A variational principle in thermodynamics of discrete systems and Landau equations

1 Introduction

In thermostatics it is well known that, if constraints are given, special potentials of discrete systems will be minimal. The aim of this work is to give a continuum theoretical foundation of such a variational principle which is also valid in non-equilibrium. We further derive from this principle Landau equations for liquid crystals in alignment tensor formulation.

2 Evolution criterion

Starting out with the *global dissipation inequality* of the system $G(t)$

$$\frac{\mathrm{d}}{\mathrm{d}\,t}\int_{G(t)} \rho\, s \,\mathrm{d}V + \oint_{\partial G(t)} \boldsymbol{\Phi} \cdot \mathrm{d}\mathbf{f} - \int_{G(t)} \rho\, \varphi \,\mathrm{d}V \geq 0 \tag{1}$$

(ρ = mass density, s = specific entropy , $\boldsymbol{\Phi}$ = entropy flux density, φ = entropy supply) the global entropy flux through ∂G is

$$-\oint_{\partial G(t)} \boldsymbol{\Phi} \cdot \mathrm{d}\mathbf{f} = \dot{Q}/T^*. \tag{2}$$

Here

$$\dot{Q} = -\oint_{\partial G(t)} \mathbf{q} \cdot \mathrm{d}\mathbf{f} \tag{3}$$

($\mathbf{q}$ = heat flux density) is the global heat exchange between the system and its vicinity being in equilibrium at thermostatic temperature T^* and at pressure p^* which are both independent of time. Because

$$T^* = \text{const, on } \partial G, \tag{4}$$

(1) by (2), (4), and (3) results in

$$\frac{\mathrm{d}}{\mathrm{d}\,t}\int_{G(t)} \rho\, T^*\, s \,\mathrm{d}V - \int_{G(t)} \rho\, T^*\, \varphi \,\mathrm{d}V \geq -\oint_{\partial G(t)} \mathbf{q} \cdot \mathrm{d}\mathbf{f}. \tag{5}$$

Inserting the *global balance of internal energy*

$$\frac{\mathrm{d}}{\mathrm{d}\,t}\int_{G(t)} \rho\varepsilon\,\mathrm{d}V = -\oint_{\partial G(t)} \mathbf{q}\cdot\mathrm{d}\mathbf{f} + \int_{G(t)} (\boldsymbol{T}:\nabla\mathbf{v} + \rho r)\mathrm{d}V \tag{6}$$

(ε = specific internal energy , $\boldsymbol{T}$ = stress tensor, $\mathbf{v}$ = material velocity, r = supply of internal energy) we get

$$\frac{\mathrm{d}}{\mathrm{d}\,t}\int_{G(t)} (\rho\, T^*\, s - \rho\varepsilon)\mathrm{d}V - \int_{G(t)} \rho T^*\,\varphi\,\mathrm{d}V \geq$$

$$\geq -\oint_{\partial G(t)} \mathbf{v}\cdot\boldsymbol{T}\cdot\mathrm{d}\mathbf{f} + \int_{G(t)} (\mathbf{v}\,\nabla : \boldsymbol{T}^{\mathrm{T}} - \rho r)\mathrm{d}V. \tag{7}$$

Because

$$\boldsymbol{T} = -\,p^*\,\boldsymbol{1}, \text{ on } \partial G \tag{8}$$

is induced by the vicinity of the system we have by use of Reynolds' transport theorem

$$\oint_{\partial G(t)} \mathbf{v}\cdot\boldsymbol{T}\cdot\mathrm{d}\mathbf{f} = -\,p^*\,\frac{\mathrm{d}}{\mathrm{d}\,t}\int_{G(t)} \mathrm{d}V. \tag{9}$$

The *local momentum balance*

$$\rho\dot{\mathbf{v}} - \nabla\cdot\boldsymbol{T}^{\mathrm{T}} = \rho\,\mathbf{f} = -\,\rho\,\nabla\gamma \tag{10}$$

($\rho\mathbf{f}$ = conservative force density, γ = potential) results in

$$\mathbf{v}\nabla : \boldsymbol{T}^{\mathrm{T}} = \rho\mathbf{v}\cdot\dot{\mathbf{v}} + \rho\mathbf{v}\cdot\nabla\gamma \tag{11}$$

which yields

$$\int_{G(t)} \mathbf{v}\nabla : \boldsymbol{T}^{\mathrm{T}}\,\mathrm{d}V = \frac{\mathrm{d}}{\mathrm{d}\,t}\int_{G(t)} (\frac{1}{2}\rho\mathbf{v}^2 + \rho\gamma)\mathrm{d}V. \tag{12}$$

By (9) and (12), (7) becomes

$$\frac{\mathrm{d}}{\mathrm{d}\,t}\int_{G(t)} (\rho T^* s - \rho\varepsilon - p^* - \frac{1}{2}\rho\mathbf{v}^2 + \rho\gamma)\mathrm{d}V \geq$$

$$\geq \int_{G(t)} (\rho T^*\,\varphi - \rho r)\mathrm{d}V. \tag{13}$$

Using the sufficiently general constitutive equation

$$\varphi = r/T^* \tag{14}$$

the inequality (13) yields an *evolution criterion*

$$\frac{\mathrm{d}}{\mathrm{d}\,t}\int_{G(t)} [\,\rho\varepsilon - \rho T^* s + \rho\gamma + \frac{1}{2}\rho \mathrm{v}^2 + p^*\,]\,\mathrm{d}V \leq 0. \tag{15}$$

In contrast to the usual procedure this evolution criterion does not need the hypothesis of local equilibrium (no Gibbs fundamental equation was used) and no stability criteria [3]; it is material independent (except for (14)), and it is valid during the whole of the non-equilibrium process and not only near equilibrium. Consequently the integral in (15) takes a minimum at all times compared with its earlier values. In equilibrium it is in its absolute minimum.

3 Landau equations

We now exploit the evolution criterion (15) for *liquid crytals in equilibrium.* The result will be, the Euler-Lagrange equations of the evolution criterion are the Landau equations determining the fields of mass density, temperature, and alignment tensor.

The *local balance equations* are

mass: $$\dot{\rho} + \rho\nabla\cdot\mathbf{v} = 0 \tag{16}$$

momentum: $$\rho\dot{\mathbf{v}} - \nabla\cdot\boldsymbol{T}^{\mathrm{T}} - \rho\mathbf{f} = \mathbf{0}, \tag{17}$$

internal energy: $$\rho\dot{\varepsilon} + \nabla\cdot\mathbf{q} - \boldsymbol{T}:\nabla\mathbf{v} - \rho r = 0, \tag{18}$$

spin: $$\rho\,\dot{\mathbf{S}} - \nabla\cdot\boldsymbol{M}^{\mathrm{T}} - \mathbf{g} + \mathbf{t}^* = \mathbf{0}, \tag{19}$$

($\mathbf{S}$ = spin density, $\boldsymbol{M}$ = momentum stress tensor, $\mathbf{g}$ = momentum force density, $\mathbf{t}^*$ = stress vector belonging to the antisymmetric part of $\boldsymbol{T}$)

alignment tensor [2]: $$\dot{\boldsymbol{a}} - 2\,\Omega\times\boldsymbol{a} - \nabla\cdot(\mathcal{M}_5:\boldsymbol{a}) - \mathcal{J}_4:\boldsymbol{a} - \boldsymbol{A} = \boldsymbol{0} \tag{20}$$

($\boldsymbol{a}$ = alignment tensor, Ω = angular velocity of rigid body motion of the distribution function of the microscopic director with respect to the material frame, $\mathcal{M}_5$, $\mathcal{J}_4$, $\boldsymbol{A}$ = functions and their gradients of the moments of orientational and rotational diffusion).

In contrast to director theories we can prove the

Proposition [1]: The balance equations (16) to (19) applied to liquid crystals are explicitly unattached by $\boldsymbol{a}$. The spin balance is independent of the alignment tensor balance.

The unknown fields in (16) to (20) are ρ, ε , $\mathbf{v}$ and $\boldsymbol{a}$. The quantities $\boldsymbol{T}$, $\mathbf{q}$, r, S, $\boldsymbol{M}$, Ω, $\mathcal{M}_5$, $\mathcal{J}_4$, and $\boldsymbol{A}$ are determined by constitutive equations. $\mathbf{f}$ and $\mathbf{g}$ are given external forces and moments, from which the constitutive maps in the alignment tensor balance also depend.

The material independent formulated *equilibrium conditions* which we need as constraints for exploiting (15) are:

(i) The velocity field represents a rigid body motion,

$$\mathbf{v} = \mathbf{v}^0 + \omega \times \mathbf{x}. \tag{21}$$

(ii) The volume of the system is constant.

(iii)(Objective) material time derivatives, the heat flux density, and the supply of internal energy are zero.

(iv)The gradient of the temperature field is zero.

(v) Orientational and rotational diffusion and Ω are zero.

By these equilibrium conditions each term in (16) and (20) vanishes, whereas (17), (18), and (19) result in

$$\nabla \cdot \boldsymbol{T}^{\mathrm{T}} - \rho \nabla \gamma = \mathbf{0}, \tag{22}$$

$$\nabla \cdot \boldsymbol{M}^{\mathrm{T}} + \mathbf{g} - \mathbf{t}^* = \mathbf{0}, \tag{23}$$

$$\mathbf{t}^* \cdot \omega = \mathbf{0}. \tag{24}$$

Because of independence of changing the frame

$$\mathbf{t}^* = \mathbf{0} \tag{25}$$

is valid, and we get the following variational problem:

$$\int \{\rho\varepsilon - \rho T^* s + \rho\gamma + \lambda \cdot (\nabla \cdot \boldsymbol{T}^{\mathrm{T}} - \rho\nabla\gamma) + \mu \cdot (\nabla \cdot \boldsymbol{M}^{\mathrm{T}} + \mathbf{g})\} \mathrm{d}V \rightarrow \min. \tag{26}$$

Here the Lagrange multipliers are functions of position: $\lambda(x)$, $\mu(x)$. The thermodynamical state space of a fluid in equilibrium according to (iv) is

$$(\rho, T, \boldsymbol{a}, \nabla\rho, \nabla\boldsymbol{a}). \tag{27}$$

The potential and the momentum force density are functions of position and of the alignment tensor

$$\gamma = \gamma\,(\mathbf{x}, \boldsymbol{a}, \nabla\boldsymbol{a}), \quad \mathbf{g} = \mathbf{g}(\mathbf{x}, \boldsymbol{a}, \nabla\boldsymbol{a}). \tag{28}$$

If we denote the integrand in (26) by G, the Euler–Lagrange equations belonging to (26) of the unknown fields ρ, T, and $\boldsymbol{a}$ are after inserting the equilibrium conditions (22) and (23)

$$G_T = 0 = \rho\frac{\partial\varepsilon}{\partial T} - \frac{\partial s}{\partial T} = \rho\frac{\partial\varepsilon}{\partial T}\,(1 - T^*\frac{\partial s}{\partial\varepsilon}) \tag{29}$$

$$G_\rho - \nabla\cdot G_{\nabla\rho} + \nabla\nabla : G_{\nabla\nabla\rho} = \mathbf{0}$$

$$= \varepsilon + \rho\frac{\partial\varepsilon}{\partial\rho} - T^*\,s - \rho\,T^*\,\frac{\partial s}{\partial\rho} + \gamma$$

$$- \nabla\cdot\{\rho\frac{\partial\varepsilon}{\partial\nabla\rho} - \rho\,T^*\,\frac{\partial s}{\partial\nabla\rho}\}, \tag{30}$$

$$G_{\boldsymbol{a}} - \nabla\cdot G_{\nabla\boldsymbol{a}} + \nabla\nabla : G_{\nabla\nabla\boldsymbol{a}} = \boldsymbol{0}$$

$$= \rho\frac{\partial\varepsilon}{\partial\boldsymbol{a}} - \rho\,T^*\,\frac{\partial s}{\partial\boldsymbol{a}} + \rho\frac{\partial\gamma}{\partial\boldsymbol{a}}$$

$$- \nabla\cdot\{\rho\frac{\partial\varepsilon}{\partial\nabla\boldsymbol{a}} - \rho\,T^*\,\frac{\partial s}{\partial\nabla\boldsymbol{a}} + \rho\frac{\partial\gamma}{\partial\nabla\boldsymbol{a}}\}. \tag{31}$$

Introducing the density of the free energy by

$$\psi := \varepsilon - T^*\,s + \gamma \tag{32}$$

we get Landau equations

$$\psi_\varepsilon = \mathbf{0}, \quad (\rho\psi)_\rho - \nabla \cdot (\rho\psi)_{\nabla\rho} = \mathbf{0}, \tag{33}$$

$$(\rho\psi)_{\boldsymbol{a}} - \nabla \cdot (\rho\psi)_{\nabla \boldsymbol{a}} = \boldsymbol{0} \tag{34}$$

From (29) results for the temperature according to (4) and (iv)

$$\frac{\partial s}{\partial \varepsilon} =: \frac{1}{T} = \frac{1}{T^*} \tag{35}$$

The solution of (34) yields the alignment tensor field

$$\boldsymbol{a} = \boldsymbol{a}\,(\mathbf{x}, \rho, T^*, \nabla\rho) \tag{36}$$

and after inserting (36) into (33) we get the mass density field

$$\rho = \rho(\mathbf{x}, T^*). \tag{37}$$

According to (36) the Landau equations reduce the dimension of the equilibrium state space (27) to $(\rho, T^*, \nabla\rho)$ of which the alignment tensor field depends in equilibrium. Consequently $\boldsymbol{a}$ is an internal variable in equilibrium.

Equations (33) and (34) are not equations for determining ψ. Therefore we have to know or to choose ψ. This choice is not totally arbitrary because up to here no use of the local dissipation inequality was made. Second law and material symmetry restrict the constitutive equation of ψ.

4 Summary

Starting out with the global formulation of the dissipation inequality an evolution criterion is also derived valid for non-equilibrium states of the discrete system considered. No additional pressuppositions as to the hypothesis of local equilibrium or stability conditions are necessary. The evolution criterion is applied to liquid crystals in external fields in equilibrium. The Landau equations for the alignment tensor field and for the mass density field are the Euler-Lagrange equations of the variational principle.

References

[1] Blenk, S. and Muschik, W. Orientational Balance Equations for Liquid Crystals: Formal Theory of Mixtures for Molecules with Microscopic Director, to be published.

[2] Ehrentraut, H., Muschik, W. and Hess, S. Alignment Tensor Balance Equations, to be published.

[3] Glansdorff, P. and Prigogine, I. *Thermodynamic Theory of Structure, Stability and Fluctuations*, Wiley, London, 1971, Chapter IX, Section 8.

Muschik W.
Institut für Theoretische Physik
Technische Universität Berlin
Hardenbergstrasse 36
D-1000 Berlin 12
F.R.G.

PART 4 : NON-LINEAR EXCITATIONS AND COHERENT STRUCTURES

A.A. AL ASSA'AD AND J.S. DARROZÈS

Non-linear waves in liquid-gas mixtures: An asymptotic theory

1 Introduction

Ideal liquid–gas mixtures represent a subclass of fluid mixtures characterized by the absence of dissipative effects such as viscosity, heat conduction, diffusion and phase changes (creation of vapour) [1]. Based on Eringen's micromorphic theory of continua [2, 3], new local balance laws, constitutive relations, field equations and an equation of state for bubbly liquids with single velocity and temperature fields have recently been developed.

Since the discovery of the K.dV solitons, many other soliton systems have been found. Efforts have been made to derive such systems from real physical situations. In this way, various perturbation methods are developed to reduce complicated non-linear physical systems to simply solvable systems [4, 5, 6]. In this paper we consider one-dimensional acoustical waves in an ideal liquid–gas mixture initially at rest. We investigate how slowly varying parts of the wave train such as the amplitude are modulated by non-linear interactions. A reductive perturbation method [4, 5, 6] is applied to the analysis of weak non-linear waves propagating through the mixture. It is shown that the non-linear Schrödinger equation can be derived. To start with, we present a *résumé* of constitutive laws of the mixture, which are compatible with the material objectivity and the entropy principle.

2 Constitutive laws and field equations

The basic system of equations of balance of a heat conducting mixture of a liquid and gas bubbles with single velocity and temperature are, in the absence of external fields [1]

(i) *Conservation of mass*

$$\frac{\partial \rho}{\partial t} + (\rho\, u_k)_{,k} = 0\,; \tag{2.1}$$

(ii) *Conservation of microinertia*

$$\frac{\partial I}{\partial t} + I_{,k}\, u_k - 2\,\nu\, I = 0\,; \tag{2.2}$$

(iii) *Conservation of linear momentum*

$$\rho\left(\frac{\partial u_k}{\partial t} + u_l\, u_{k,l}\right) - t_{kl,k} + \rho f_l = 0\,; \tag{2.3}$$

(iv) *Balance of moment of momentum*

$$t_{ml} - s_{ml} - \rho I\left(\frac{\partial \nu}{\partial t} + \nu_{,k}\, u_k + \nu^2\right)\delta_{lm} = 0\,; \tag{2.4}$$

(v) *Balance of energy*

$$\rho\frac{\mathrm{d}e}{\mathrm{d}t} = t_{kl}\,D_{kl} + (s_{kl} - t_{kl})\nu_{(kl)} - q_{k,k} + \rho h \tag{2.5}$$

and, in absence of phase change, the balance of the quality $x = \alpha(\rho_g/\rho)$ of the gas

$$\frac{\partial x}{\partial t} + x_{,k}\, u_k = 0\,, \tag{2.6}$$

where

$\rho \equiv$ mass density,
$I \equiv$ microinertia tensor,
$f_k \equiv$ body force,
$t_{kl} \equiv$ stress tensor,
$e \equiv$ internal energy density,
$s_{kl} \equiv$ microstress average tensor.
$q_k \equiv$ density of heat flux,
$h \equiv$ heat source,
$u_k \equiv$ velocity vector,
$\nu_{kl} = \nu\,\delta_{kl} \equiv$ gyration,
tensor,

These equations involve the constitutive quantities t_{kl}, s_{kl}, q_k, and e related to the fields ρ, I, x, u_k, ν ,..., etc., in a manner determined by the mixture behaviour. The mass density ρ, the gyration ν and the velocity field u_k are such as [1]

$$\rho = \alpha\rho_g + (1 - \alpha)\rho_\ell\,, \tag{2.7}$$

$$u_{k,k}\,\mathrm{d}v = \frac{\mathrm{d}}{\mathrm{d}t}(\mathrm{d}v) = 3\nu\,\mathrm{d}v = \frac{1}{1-\alpha}\left(\frac{\mathrm{d}\alpha}{\mathrm{d}t}\right)\mathrm{d}v\,, \tag{2.8}$$

where α is the volumetric fraction of the gas, ρ_g and ρ_ℓ are, respectively, the mass densities of the gas and the liquid. Insertion of (2.7) and (2.8) into eqs. (2.1) and (2.2) leads to

$$\frac{d\rho}{\rho} = -\frac{d\alpha}{1-\alpha}, \quad \frac{dI}{I} = \frac{2}{3}\frac{d\alpha}{1-\alpha}. \tag{2.9$_{1,2}$}$$

The equations of balance (2.1)-(2.6), supplemented by constitutive equations, become field equations which relate t_{kl}, s_{kl}, e and q_k to the thermodynamic fields ρ, I, x, u_k and ν, where θ is the absolute temperature. In particular, the mixture is called an *ideal mixture*, if the constitutive equations are in the form

$$\mathcal{S} = \mathcal{S}^*(\rho, \theta, I, x), \tag{2.10}$$

where $\mathcal{S}$ denotes any one of the constitutive quantities, and if it does not exhibit any dissipative effect. The most general form of the constitutive relations (2.10) which are compatible with the material objectivity and the entropy inequality is [1]

$$t_{kl} = t^*_{kl} = -p(\rho, \theta, I, x)\delta_{kl}, \tag{2.11$_1$}$$

$$e = e_I^*(\rho, \theta, I, x) \text{ and } \psi^* = \psi^*_I(\rho, \theta, I, x), \tag{2.11$_2$}$$

$$\eta^* = -\frac{\partial \Psi^*}{\partial \theta}, \tag{2.11$_3$}$$

$$s_{kl} = s^*_{kl} = -\pi(\rho, \theta, I, x)\delta_{kl}, \tag{2.11$_4$}$$

where ψ is the free energy density, η the entropy, and the pressure p and the extra-pressure π are given by

$$p = \rho^2\left(\frac{\partial \psi_I^*}{\partial \rho}\right)_{\theta, I, x}, \quad \pi = -2\rho I\left(\frac{\partial \psi_I^*}{\partial I}\right)_{\rho, \theta, x} \tag{2.12$_{1,2}$}$$

and, they are related to the pressure p_g of the gas through the equation of state [1]

$$p + \frac{\pi}{3} = \frac{r\theta}{\alpha}\{\rho - (1-\alpha)\rho_\ell\} \tag{2.13}$$

when the gas behaves as an ideal gas. On the other hand,

$$de = c_{\alpha, x}(\theta)d\theta \tag{2.14}$$

$$\rho c_{\alpha,x}\frac{\mathrm{d}\theta}{\mathrm{d}t} = -\left(p+\frac{\pi}{3}\right)D_{kk} = -p_g D_{kk}\,.$$

By (2.7) and (2.9) we have (the subscript 0 refers to an undisturbed state)

$$\rho = \rho_0\,\frac{1-\alpha}{1-\alpha_0},\quad \rho_g = \frac{\alpha_0}{\alpha}\,\frac{1-\alpha}{1-\alpha_0}\rho_{g_0}\,; \tag{2.15$_{1,2}$}$$

then by (2.8)

$$\frac{\theta}{\theta_0} = \left(\frac{\alpha}{\alpha_0}\,\frac{1-\alpha_0}{1-\alpha}\right)^{\gamma} \quad \text{with } \gamma = \frac{\alpha_0 r\,\rho_{g_0}}{\rho_0 c_{\alpha,x}}, \tag{2.16$_1$}$$

$$p_g = p_{g_0}\left(\frac{1-\alpha_0}{\alpha_0}\,\frac{\alpha}{1-\alpha}\right)^{\gamma-1} \quad \text{where } p_{g_0}\, r\theta_0\,. \tag{2.16$_2$}$$

These equations are, formally, similar to those obtained in gas dynamics. Field equations of *non-linear one-dimensional motion of an ideal mixture* are

$$\beta_t + (u\beta)_x = 0 \tag{2.17$_1$}$$

$$\frac{\rho_0}{\beta_0}\beta(u_t + uu_x) + p_x = 0\,, \tag{2.17$_2$}$$

$$p_g - p = \frac{\rho_0 I_0}{9\beta_0^{5/2}}\beta^{5/2}\,(u_{xt} + uu_{xx} + \frac{1}{3}u_x^2)\,, \tag{2.17$_3$}$$

with p_g given by (2.16)$_2$, $\beta = 1-\alpha$, and where $u(x, t)$ is the x-component of the velocity field $\mathbf{u}$, and indices denote a partial derivative.

3 A non-linear Schrödinger equation: Wave modulation

It was shown [1] that, by means of a coordinates stretching,

$$X = \varepsilon^n(x - c_0 t),\ T = \varepsilon^{n+1}t \tag{3.1}$$

of field equations (2.17), the system of weak non-linear waves propagating through an ideal mixture can be reduced to the well-known Korteweg-de Vries equation

$$u^{(1)}{}_T + p\, u^{(1)} u^{(1)}{}_X + q u^{(1)}{}_{XXX} = 0\,, \tag{3.2}$$

where

$$p = \left[1 - \frac{1}{2}(\gamma - 2\beta_0)(1 + \beta_0)\frac{\rho_0 c_0^2 \beta_0}{1 - \beta_0}\right]$$

$$q = \frac{\rho_0 I_0 c_0^3}{18}(1 - \beta_0^2)$$

and the acoustic speed given by

$$c_0 = \frac{p_0}{\rho_0 \alpha_0}\left(1 + \frac{\alpha_0 p_0}{\rho_0 c\,\theta_0}\right)^{1/2}, \tag{3.3}$$

where $c = c_{\alpha,x}$ is supposed to be uniform. However, this formulation is restricted to the propagation of waves of small wave number and low frequency. If the amplitude of the wave is small but finite, non-linear terms give rise to a modulation of the amplitude as well as higher harmonics. When the amplitude varies slowly over the period of the oscillation, a stretching transformation allows us to separate the system into a rapidly varying part associated with the oscillation and a slowly varying one such as the amplitude. Then, a formal solution is given in an asymptotic expansion, and we derive, in the lowest order, an equation to determine the modulation of the amplitude which, in our case, becomes the non-linear Schrödinger equation.

For plane disturbances in the form $\exp[i(kx - \omega t)]$, eq. (2.17) leads to the dispersion relation

$$\omega^2 = c_0{}^2 k^2 [1 + (I_0/9)k^2]^{-1}\,, \tag{3.4}$$

where k and ω are, respectively, the wave number and the frequency. However, superposition of these non-linear plane waves is not a solution; hence, for waves to transmit signals we assume, for β, u, p and p_g, the form $\varphi(x, t)\exp[i(kx - \omega t)]$, modulated by a slowly varying function φ. In addition, we require that the solutions include steady pulsive waves growing spatially at infinity. This condition may be obtained as follows: putting $\omega = Uk$ to solve (3.4) for k yields that k becomes complex if $|U|$ exceeds the critical value $U_0 = \pm c_0[1 + (I_0/9)k^2]^{1/2}$. This condition implies that the phase velocity is equal to the group velocity, i.e., $\partial\omega/\partial k = \omega/k$ $(= \pm U_0)$, and we have the critical frequency $\omega = I_0{}^{1/2}/3c_0$ and wave number $|k_0| = U_0{}^{-1}\omega_0$, respectively. In the subsequent discussion we consider a wave with small but finite amplitude propagating with a velocity equal to the critical velocity. Hence, we may assume that k is positive and that the wave has right-hand polarization.

We first introduce the scale transformation in terms of a slowness parameter ε through the equations

$$\xi = \varepsilon(x - Ut), \quad \tau = \varepsilon^2 t, \tag{3.5}$$

where U differs from U_0 by a small number $\varepsilon^2\lambda$, e.g., $U = U_0 + \varepsilon^2\lambda$, and assume the following expansions

$$\begin{pmatrix} \beta \\ u \\ p \\ p_g \end{pmatrix}(x, t) = \sum_{n=0}^{\infty} \varepsilon^n \sum_{\ell=-\infty}^{\ell=+\infty} \begin{pmatrix} \beta^{(n,\ell)} \\ u^{(n,\ell)} \\ p^{(n,\ell)} \\ p_g^{(n,\ell)} \end{pmatrix} (\xi,\tau) \exp[i(kx - \omega t)], \tag{3.6}$$

where $u^{(0,\ell)} = 0$. The expansion parameter ε is assumed to be much less than unity, and the coefficients $\beta^{(n,\ell)}$, $u^{(n,\ell)}$, $p^{(n,\ell)}$ and $p_g^{(n,\ell)}$ are assumed to vary slowly over a wavelength $\sim k^{-1}$ during one cycle $\sim\omega^{-1}$. For β, u, p and p_g to be real we must have

$$\begin{aligned} \beta^{(1,-1)} &= \beta^{(1,1)*}, \\ u^{(1,-1)} &= u^{(1,1)*}, \\ p^{(1,-1)} &= p^{(1,1)*}, \\ p_g^{(1,-1)} &= p_g^{(1,1)*}. \end{aligned}$$

Since we are interested in the modulation of a quasi-monochromatic wave, the $\beta^{(1,1)}$, $u^{(1,1)}$, $p^{(1,1)}$ and $p_g^{(1,1)}$ are to be set equal to zero except for $\ell = \pm 1$; that is, we can assume

$$\begin{pmatrix} \beta^{(1)} \\ u^{(1)} \\ p^{(1)} \\ p_g^{(1)} \end{pmatrix} = \begin{pmatrix} \beta^{(1,1)} \\ u^{(1,1)} \\ p^{(1,1)} \\ p_g^{(1,1)} \end{pmatrix} \exp[i(kx - \omega t)] + \text{c.c.}, \tag{3.7}$$

where c.c. denotes the complex conjugate. Substituting eqs. (3.5)–(3.6) into (2.17) and equating the coefficients of each power of ε, we obtain the following reduced system of perturbation equations

$O(\varepsilon)$:

$$\mathcal{D}(k,\omega)u^{(1,1)} = 0\,; \tag{3.8$_1$}$$

$O(\varepsilon^2)$:

$$\mathcal{D}(\ell k, \ell\omega) \sum_{\ell=-\infty}^{+\infty} u^{(2,\ell)}\exp[i(kx-\omega t)] + i\left[c_0^2\frac{k^2}{\omega}(U-c_g) + \right.$$

$$\left. + U\omega\left(1+\frac{I_0}{9}k^2\right) + k\left(\frac{I_0}{9}\omega^2 - c_0^2\right)\right] u_\xi^{(1,1)}\exp[i(kx-\omega t)]$$

$$= \left(4G_0\frac{k^3}{\omega} + \frac{14}{27}I_0\,\omega k^3\right) u^{(1,1)2}\exp[2i(kx-\omega t)] + \text{c.c.}\,; \tag{3.8$_2$}$$

$O(\varepsilon^3)$:

$$i\left[U\omega\left(1+\frac{I_0}{9}k^2\right) + k\left(\frac{I_0}{9}\omega^2 - c_0^2\right) + \frac{\omega^2}{k}(U-c_g)\right] u_\xi^{(2,1)} +$$

$$+ \frac{I_0}{9}\frac{\omega^3}{c_0^2k^2}\,u_{\xi\xi}^{(1,1)} - 2i\omega\,u_\tau^{(1,1)} =$$

$$= \omega k\left(\omega k - 1 - c_0^2\frac{k^2}{\omega^2}\right)[u^{(2,0)}u^{(1,1)} + u^{(2,2)}\,u^{(1,-1)}]\,, \tag{3.8$_3$}$$

where

$$\mathcal{D}(\ell k, \ell\omega) = c_0^2\,\ell^2\,k^2 - \ell^2\,\omega^2 - \frac{I_0}{9}\,\ell^4\,k^2\,\omega^2 \tag{3.9}$$

$$G_0 = 4\left(\frac{D_0}{\rho_0}\beta_0^2 - c_0^2\right)$$

$$D_0 = p_{g_0}\left[\frac{\gamma(\gamma-1)}{2}\beta_0^{-2} + \frac{(\gamma-1)(\gamma-2)}{2}(1-\beta_0)^{-2} - (\gamma-1)^2\beta_0^{-1}(1-\beta_0)^{-1}\right]$$

and

$$c_g = \omega^3/c_0^2k^3 \tag{3.10}$$

denotes the group velocity. It is easy to see that the lowest order equation $(3.8)_1$ yields the dispersion relation

$$\mathcal{D}(k, \omega) = c_0{}^2k^2 - \omega^2 - \frac{I_0}{9} k^2 \omega^2 = 0 . \tag{3.11}$$

To satisfy eq. $(3.8)_2$, the coefficients of each mode should be set equal to zero individually. If $\mathcal{D}(\ell k, \ell\omega) \neq 0$ for $\ell > 1$, it follows from $(3.8)_2$ that

$$\mathcal{D}(2k,2\omega)u^{(2,2)} = \left(G_0 \frac{k^3}{\omega} + \frac{14}{27} k^3 \omega \right) (u^{(1,1)})^2 \tag*{$(3.12)_1$}$$

$$u^{(\ell,2)} = 0 \text{ for } \ell > 2 \tag*{$(3.12)_2$}$$

$$\left[c_0^2 \frac{k^2}{\omega} (U - c_g) + \right.$$

$$\left. + U\omega \left(1 + \frac{I_0}{9} k^2 \right) + k \left(\frac{I_0}{9} \omega^2 - c_0^2 \right) \right] u_{\xi}^{(1,1)} = 0 . \tag*{$(3.12)_3$}$$

If

$$U = c_g \tag{3.13}$$

non-trivial solutions are possible for $u^{(1,1)}$. The equation governing the slow variation of $u^{(1,1)}$ can be obtained from order $O(\varepsilon^3)$. The first two terms on the left-hand side of $(3.8)_3$ automatically cancel out due to (3.11) and (3.13) giving the equation

$$iu_{\tau}^{(1,1)} + \frac{1}{6} \left(\frac{dc_g}{dk} \right) u_{\xi\xi}^{(1,1)} +$$

$$+ k \left(\omega k - 1 - c_0^2 \frac{k^2}{\omega^2} \right) [u^{(2,0)} u^{(1,1)} + u^{(2,2)} u^{(1,-1)}] = 0 . \tag{3.14}$$

Since $u^{(2,2)}$ is related to $u^{(1,1)}$ through $(3.12)_1$, eq. (3.14) becomes a closed equation for $u^{(1,1)}$, provided that $u^{(2,0)}$ is related to $u^{(1,1)}$. This is given by the constant terms; that is, by the $\ell = 0$ mode of the $O(\varepsilon^4)$ problem as follows

$$(U^2 - c_0{}^2)u^{(2,0)} = u_{\xi\xi}^{(1,1)} u^{(1,-1)} , \tag{3.15}$$

which yields after integration with respect to ξ

$$u^{(2,0)} = \frac{1}{c_g^2 - c_0^2} u^{(1,1)} u^{(1,1)*} + \delta , \tag{3.16}$$

where δ is an integration constant. Substituting $(3.12)_1$ and (3.16) into (3.14), and replacing $u^{(1,1)}(\xi,\tau)$ by the amplitude $A(\xi, \tau)$, we finally arrive at the non-linear Schrödinger equation

$$i A_\tau + \frac{1}{6}\left(\frac{dc_g}{dk}\right) A_{\xi\xi} +$$

$$+ k\left(\omega k - 1 - c_0^2 \frac{k^2}{\omega^2}\right)\left[\left(\frac{1}{c_g^2 - c_0^2} + \frac{H(k, \omega)}{\mathcal{D}(2k, 2\omega)}\right) A A^{2^*} + \delta A\right] = 0 \tag{3.17}$$

where

$$H(k, \omega) = 4G_0 \frac{k^3}{\omega} + \frac{14}{27} I_0 \, \omega \, k^3 .$$

4 Conclusions

The study considered in this paper models, in a classical way, the problem of non-linear waves in a bubbly liquid considered as a micromorphic continuum. The theory does not present more complex changes in the shape of the bubbles. However, the results presented here are the simplest that would include the characterization of the local effects. The theory extending that given here is now developed to characterize the thermomechanical behaviour of a liquid-gas mixture including both stretching of bubbles and phase change.

This model may now be applied to specific problems. We have used the model in studies of non-linear waves propagating through an ideal mixture and a non-linear Schrödinger equation is derived. The mathematical significance of the appearance of solitons in the mixture is to be explored.

References

[1] A. Al Assa'ad and J.S. Darrozes, Toward a continuum theory of liquid-gas mixtures, *Recent Advances in Engineering Science: the Eringen Symposium*, Berkeley, 1988. *Lect. Notes Engineering* Vol. 39, 21-38, (S.L. Koh and C.G. Speziale, ed.) Springer-Verlag, Berlin 1989.

[2] A.C. Eringen, Simple micro-fluids, *Int. J. Eng. Sci.* **2**, 205-217, 1964.

[3] A.C. Eringen and E.S. Suhubi, Nonlinear theory of simple microelastic solids. Part I., *Int. J. Eng. Sci.* **2**, 189-202, 1964.

[4] A. Jeffrey and T. Kawahara, *Asymptotic Methods in Nonlinear Wave Theory*. Pitman, London, 1982.

[5] A. Jeffrey and T. Kakutani, Weak nonlinear dispersive waves: A discussion centred around the Korteweg-de Vries equation, *SIAM. Rev.* **14**, 582-643, 1972.

[6] C.H. Su and C.S. Gardner, Korteweg-de Vries equation and generalization III: Derivation of the Korteweg-de Vries equation and Burgers equation, *J. Math. Phys.* **10**, 536-539, 1969.

Al Assa'ad A.A.
Groupe Phénomènes d'Interface
Ecole Nationale Supérieure de
Techniques Avancées,
Centre de l'Yvette
Chemin de la Hunière
F-91120 Palaiseau
FRANCE

Darrozès J.-S.
Groupe Phénomènes d'Interface
Ecole Nationale Supérieure de
Techniques Avancées,
Centre de l'Yvette
Chemin de la Hunière
F-91120 Palaiseau
FRANCE

A. ASKAR

Dynamics of single alpha and double DNA helices: Continuum and discrete analyses

1 Introduction

The goal in this article is to present a detailed model for biological polymers with helices forming the primary structure. In particular the alpha-helix with its single helix and the DNA with its double helices will be studied. The dynamical modes in these molecules determine their primary functions. For the modal anlayses linear approaches are of value, while for transport phenomena as well as processes requiring energy localization, non-linearity brings the essential ingredient. The aforementioned polymers are shown in Figure 1 [1, 2]. It is seen that in addition to the basic helices, there exist other important bonds which are of hydrogenic type. For the alpha-helix, hydrogen bonds connect the peptides along the helix which are a step apart. For the DNA the two helices are bound through the base pairs which are joined via hydrogen bonds. The intermolecular bonds forming the sugar, the phosphate molecules and the bases as well as the intramolecular bond between these are much stronger than the hydrogen bonds. These considerations form the basis of the model. The various molecular groups will be considered as rigid particles; the bonds along the helices will be approximated through quadratic potentials while the weaker hydrogen bonds will be accounted for by non-linear interaction potentials. The major effort here goes to the description of the helix by accounting for the various interactions between the molecular groups taken as point masses. The total deformation energy of the polymer will then be derived by adding the appropriate hydrogen bond energies to the helix energy for the specific alpha-helix or the DNA cases. For this latter the dipole interaction between the subsequent base pairs is also introduced as this is an important contribution. In this work, the molecular groups of sugar, phosphate and base pairs are taken as point masses.

2 The deformation of the helix: discrete analysis

There exist three main types of interaction between the sugars and the phosphates forming the helical backbone. These are (i) the stretch interaction due to the change of the equilibrium distance between two consecutive molecular groups; (ii) the bending interaction which involves the change in the angle between three consecutive points and (iii) the torsional interaction which involves the changes in the dihedral angle between four consecutive points along the helical backbone. The dihedral angle is defined as the

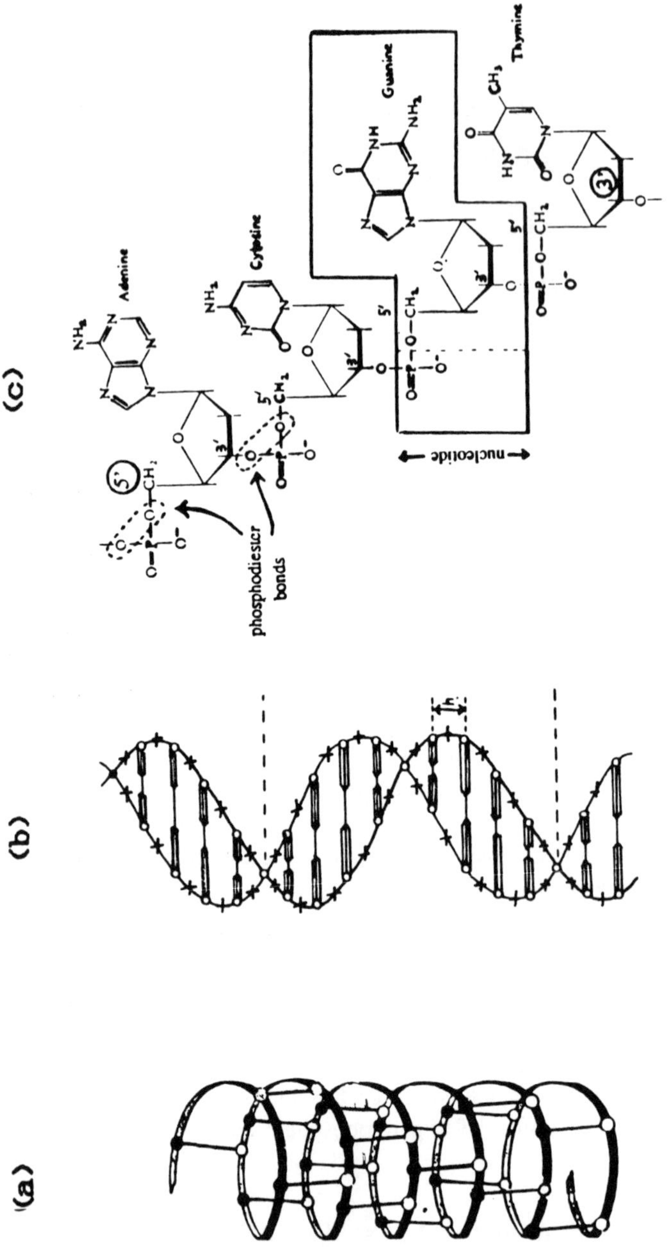

Figure 1. Illustration of (a) the Alpha–helix and (b), (c) the DNA double helices.

(a) o peptide O atoms; •, peptide N atom; —— Hydrogen bond;

(b) o Sugar; + phosphate; Base; —— Hydrogen bond;

(c) Details of a fragment of DNA.

angle between the two planes formed respectively by the first, second, third and second, third, fourth particles in a succession. The dihedral angle can equivalently be defined as the angle between the normals of these two planes. In this work, the latter definition is found more suitable for the kinematics description. The bending and dihedral of angles are presented in Figure 2. The molecular groups are much more tightly bound against

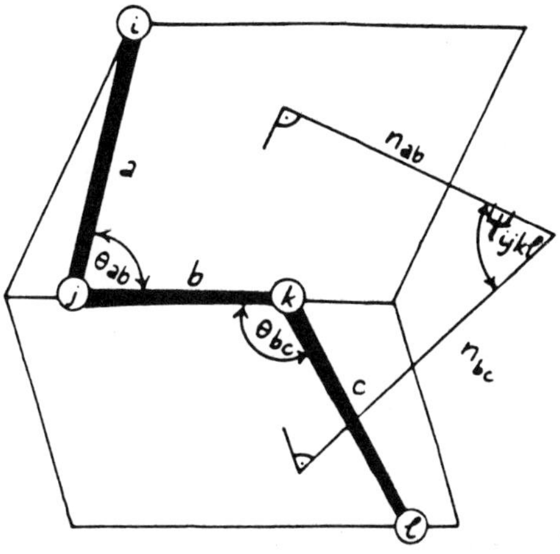

Figure 2. Bending angle θ and dihedral angle ψ.

the stretching of the intermolecular distances than against the two angular changes. It is therefore a fair assumption to take the intramolecular distances as unchanged during the deformations of the helix. In the present work, for ease in the notation we denote the actual positions of the molecules as $\mathbf{x}_k$ without recourse to an additional index to distinguish the sugars from the phosphates. This distinction is easily achieved for the odd and even indices k denoting, respectively, the sugars and the phosphates. The equilibrium positions of the same particles are likewise denoted by the capital letters $\mathbf{X}_k$. The intermolecular distances, the bending angles and the dihedral angles at equilibrium and after the deformation are likewise denoted, respectively, as ΔS, Θ, Ψ and Δs, θ, ψ. In terms of the particle positions, the kinematic variables are defined by taking $\Delta S = \Delta s$ along with S to be the same throughout the chain as

$$\Delta s = | \mathbf{x}_k - \mathbf{x}_{k-1} | ,$$

$$\cos\theta_k = (\mathbf{x}_k - \mathbf{x}_{k-1}) \cdot (\mathbf{x}_{k-1} - \mathbf{x}_k)/\Delta S^2 , \tag{1}$$

$$\cos\psi_k = [(\mathbf{x}_k - \mathbf{x}_{k-1}) \cdot (\mathbf{x}_{k+1} - \mathbf{x}_k)] [(\mathbf{x}_{k+2} - \mathbf{x}_{k+1}) \cdot (\mathbf{x}_{k+1} - \mathbf{x}_k)]/\Delta S^4$$

$$- (\mathbf{x}_k - \mathbf{x}_{k-1}) \cdot (\mathbf{x}_{k+2} - \mathbf{x}_{k+1})/\Delta S^2.$$

Similar expressions at equilibrium are given for the capitalized quantities in terms of the $\mathbf{x}$'s. In evaluating the dihedral angles, first the normals of the two consecutive planes are evaluated thorugh cross-products and subsequently the cosine of the angles between these two normal vectors is evaluated by their inner product. In the calculations the simple equivalence of the product of four vectors as $(\mathbf{A} \times \mathbf{B}) \cdot (\mathbf{C} \times \mathbf{D}) = (\mathbf{A} \cdot \mathbf{C})\,(\mathbf{B} \cdot \mathbf{D}) - (\mathbf{A} \cdot \mathbf{D})\,(\mathbf{B} \cdot \mathbf{C})$ is used. In our calculations $\mathbf{B} = \mathbf{C}$ and the vectors are of unit lengths.

The helix energy is evaluated in terms of the above quantities by the addition of the bending and torsional contributions as [3]

$$V_{\text{hel}} = \frac{1}{2} \sum_{k} B(\cos \theta_k - \cos \Theta_k)^2 + T(\cos \psi_k - \cos \Psi_k)^2. \tag{2}$$

Above, certainly more general expressions than quadratic, in particular for the torsional interactions may be necessary for large configurational changes. However for a broad class of problems, the basic non-linear contributions come from the hydrogen bond stretching and the linear dynamics of the helix about its equilibrium configuration would be adequate.

3 The deformation of the helix: continuum analysis

For the discrete analysis, the particle positions were the dynamical variables. The displacements of the particles from their equilibrium positions prove to be more suitable for the continuum analysis to be undertaken here. The displacement of a particle is obtained from x and X as

$$\mathbf{x}_k = X_k + \mathbf{u}_k. \tag{3}$$

In evaluating the various quantities, the structure of the helix is introduced through the Fresnet formulae. The unit tangent, normal and binormal vectors are used as the natural basis. Hence the displacement vector has the components u, v, w along these three local basis vectors. The continuum representation of the deformation energy in (3) is obtained in the usual long wave approximation [4] of the discrete variables through the equations as

$$\mathbf{u}_{k\pm 1} \simeq \mathbf{u} \pm \mathbf{u}' \,\Delta S + \frac{1}{2} \mathbf{u}'' \,\Delta S^2 + \dots , \tag{4}$$

where the primes indicate the derivatives with respect to S, the arc length parameter and ΔS is the finite distance between two mass points on the helix. Particular care should be paid to taking the expansions about the centre of the two, three and four particle groups. With the suitably centred expansions, the kinematic variables in (2) become

The kinetic energies associated with the two helices are obtained as in (9) using the $\mathbf{x}$'s as well as the $\mathbf{y}$'s simultaneously.

For the DNA, in addition, the rotations of the base pairs in subsequent stacks along the helices involve an energy commensurate with those entering into the helix deformations and the hydrogen bond stretchings between these. This particular energy is expressed as [5, 6]

$$V_{\text{Base}} = \sum_k \frac{1}{2} R_1(\varphi_\ell - \varphi_{\ell-1})^2 + R_2(1 - \cos\varphi_\ell). \tag{12}$$

Above φ_ℓ denotes the rotation of the base pairs in the l th step of the helix with respect to its equilibrium configuration about the helix axis. A suitable approximation is to suppress the independent rotations of the base pairs from the helices and evaluate φ_ℓ from the positions of the helix points they join as

$$\varphi_\ell = | [(\mathbf{x}_\ell - \mathbf{y}_\ell) - (X_\ell - Y_\ell)] \times \mathbf{k} \, / \, | \mathbf{X}_\ell - \mathbf{Y}_\ell | . \tag{13}$$

Above k denotes the unit vector along the helix axis. With the rotation of the base pairs taken into account, it also becomes necessary to account for their librational energy. For I denoting the rotatory inertia of the base pairs about the helix axis, we have

$$L \cdot E \, \frac{1}{2} \sum_\ell I_\ell \, \dot{\varphi}_\ell^2 . \tag{14}$$

In this work the rotations as propeller-twist of the base pairs are neglected. The appropriate Lagrangian and the accompanying dynamical equations are obtained straightforwardly following the above descriptions. Similarly, the continuum representation follows in a straightforward manner from the above equations in terms of two displacements in connection with the two position vectors x and y.

5 Conclusion

As described above, a single position variable or equivalently a single displacement vector defines the dynamics of the alpha-helix. This model may be complemented by introducing the rotations of the molecular groups of peptides. For the DNA, the above derivations may be taken with various levels of sophistication. First, the rotations of the base pairs may be taken as independent dynamical variables or be related to the displacements of the surrounding sugar groups they connect. This approximation is given in eq. (13). Similarly, the dynamics of the two helices may be introduced through the symmetric and the antisymmetric variables $\mathbf{x} + \mathbf{y}$ and $\mathbf{x} - \mathbf{y}$ [5]. This particular decomposition permits the independent study of phenomena involving or excluding the hydrogen bond stretching [5]. A further level of sophistication can be achieved by considering the rotations of the sugar and phosphate groups along with the propeller-twist motion of the base pairs in the DNA. Finally, with the rich dynamics involved, simpler uni-dimensional and uni-modal studies may be of value by separating the axial extensional, rotational and transversal flexural modes as well as concentrating on the non-linear dynamics of the hydrogen bonds.

References

[1] W. Saenger, *Principles of Nucleic Acid Structure*, Springer-Verlag, New York (1984).

[2] R.E. Dickerson, The DNA helix and how it is read, *Sci. Amer.* **249**, 94 (1983).

[3] T. Schlick, Modeling and Minimization Techniques for Predicing Three-Dimensional Structures of Large Biological Molecules, PhD Thesis (1987), Courant Institute of Mathematical Sciences, New York University, NY 10012.

[4] A. Askar, *Lattice Dynamical Foundations of Continuum Theories of Solids: Elasticity, Piezoelectricity, Viscoelasticity, Plasticity,* World Scientific, Singapore (1988).

[5] M. Peyrard and A.R. Bishop, Dynamics of Nonlinear Excitations in DNA, 6th Interdisciplinary Workshop on Nonlinear Structures in Physics, Mechanics and Biological Systems, Montpellier, France (1989).

[6] S. Yomosa, Soliton excitations in DNA double helices, *Phys. Rev.* **A27**, (1983).

Askar A.
Department of Mathematics
Bogazici Universitesi
Bebek
Istanbul
TURKEY

O.M. BRAUN, Yu.S. KIVSHAR AND I.I. ZELENSKAYA

Frenkel-Kontorova model with long-range interparticle interactions

1 Introduction

The Frenkel-Kontorova (FK) model [4] is successfully used to explain the dynamics of a number of physical objects, such as dislocations in solids, domain walls in magnetics, charge density waves in one-dimensional conductors, adsorbed layers on surfaces, and so on (see, e.g., Ref. [5]). The discrete Hamiltonian of the FK model

$$H = \sum_n \left\{ \frac{1}{2}\dot{u}_n^2 + 1 - \cos u_n + \frac{1}{2} \sum_{k \geq 1} \left[V(u_{n+k} - u_n) + \right.\right.$$

$$\left.\left. + V(u_n - u_{n-k}) \right] \right\} \tag{1}$$

describes a chain of unit-mass atoms in a periodic potential with interparticle interactions $V(u)$. The periodic potential relief has the period $a_s = 2\pi$ and the amplitude $\varepsilon_s = 2$. In the well-known FK model [4] it is assumed that the interaction potential $V(u)$ is a local function (nearest neighbours interact only) and it is also harmonic, $V(u) = \frac{1}{2} (\ell/\pi)^2 (u - a)^2$. Here a is the equilibrium distance between atoms, and ℓ characterizes the intensity of the interparticle interaction. For the case $a = 2\pi\ q$ ($q = 1,2, \ldots$) the ground state of the system corresponds to the periodic structure of the atoms with a period a which is commensurate with the a_s. In the continuum approximation (the latter valid for $V(a) \gg 1$) simple transformations of the discrete Hamiltonian (1) yield the well-known Hamiltonian of the sine-Gordon (SG) model

$$H_{SG} = \int (dx/a) \left[\frac{1}{2} u_t^2 + 1 - \cos u + \frac{d^2}{2} u_x^2 \right], \tag{2}$$

where the indices t and x stand for partial derivatives with respect to time and space coordinate, $x = an$. The parameter d determines the characteristic space scale, $d = 2lq$. The SG equation which follows from (2) has a number of non-linear solutions; the simplest is the kink (antikink)

$$u(x) = 4 \tan^{-1} \exp[- \sigma(x - X)/d]. \tag{3}$$

The kink ($\sigma = +1$) (or antikink, $\sigma = -1$) corresponds to the minimally possible contraction (extension) of the commensurate, structure of the atoms. The function $-u_x/a$ describes the density of excess (deficient) atoms in comparison with the commensurate structure, so that a kink is related to a fractional quantity of atoms, σ/q.

A kink is a particle-like excitation with some coordinate X, an effective mass m_k and a topological charge. It is important that the kinks are deformable quasi-particles; they interact with an energy W_{int} which may be obtained by asymptotics of the kink shape, for example, for SG kinks we have the result, $W_{\text{int}}(x_0) \sim \sigma_1 \sigma_2 \exp(-x_0/d)$, x_0 being the distance between the kinks.

In this paper we consider the extended version of the FK model which describes a chain of atoms in a periodic potential with long-range interparticle interactions. The long-range character of the interaction leads to new effects even in the simplest case when only next nearest neighbour interactions are included [5]. A physically important example of the long-range interaction is the dipole-dipole repulsion of atoms chemisorbed on a metal surface [2]. That interaction may be described approximately by the power potential in eq. (1)

$$V(u) = V_0\,(2\pi/|u|)^n, \quad n \geq 1. \tag{4}$$

2 Analytical results

In the continuous approximation the Hamiltonian (1) leads to the extended SG equation which is an integro-differential one. Indeed, if we substitute $x \to y = x + u(x)$, $\mathrm{d}y = (1 + u_x)\mathrm{d}x \approx (1 + u_y)\mathrm{d}x$, the interaction energy takes the form

$$H_{\text{int}} = \frac{1}{2}\sum_{n \neq k} V(u_n - u_k) \to \frac{1}{2}\iint \frac{\mathrm{d}x\,\mathrm{d}x'}{a^2} u_x(x)\,V(x - x')\,u_{x'}(x')\,, \tag{5}$$

where we neglect the unessential constants and change our notation. The expression (5) has the simple physical sense: It describes the interaction of excess atoms with the density u_x/a. In the case of a local potential, we have to insert $V(x) = ad^2\,\delta(x)$ and then eq. (5) transforms to that for the SG model [see eq. (2), the last term].

For the non-local power potential (4) the integral in eq. (5) diverges. Therefore, we must cut off the integration region excluding the point $x = x'$ on the distance $\tilde{a}$ (we assume below $\tilde{a} = a$). After straightforward transformations we may simplify the interaction Hamiltonian (5), and obtain a non-local Hamiltonian of the SG type. The motion equation corresponding to the Hamiltonian may be written in the following form:

$$u_{tt} + \sin u - u_{xx} = A \frac{\partial}{\partial x} \int_{\alpha}^{\infty} \frac{dx'}{(x')^n} [u_x(x+x') + u_x(x-x')\,]. \tag{6}$$

In eq. (6) we express the parameter V_0 from (4) as a function of the parameter d and introduce the new parameter α :

$$\alpha \equiv a/d, \quad A = \alpha^{n-1} / n(n+1)\,. \tag{7}$$

The resulting eq. (6) differs from the SG equation by the last term which describes non-local properties of kinks. Equation (6) has the static kink solution with the boundary conditions, $u(x = +\infty) - u(x = -\infty) = -2\pi\sigma$. However, the asymptotics of the kink "tails" is power and substantially differs from those of the SG kink (3). Indeed, in the region of the kink "tail" ($|x| >> 1$) eq. (6) can be linearized near the value $u(|x| = \infty)$. The simple analysis of that linearized equation shows that the kink has the asymptotic form

$$u(x) \approx u(|x| = \infty) + 2\pi n A \,/\, |x|^{n+1}\,, \quad |x| >> 1. \tag{8}$$

The result (8) stipulates the power law of interaction between kinks in the system with long-range interparticle interactions

$$W_{\text{int}}(x) = W_{\text{int}}^{(0)}(x) + W_{\text{int}}^{(1)}(x)\,, \tag{9}$$

where

$$W_{\text{int}}^{(0)}(x) = \sigma_1\, \sigma_2\, V_0\, |a/x|^n\, q^{-(2n+1)} \tag{10}$$

is the energy of interaction of excess atoms which gives the main contribution into (9). The addendum $W^{(1)}$ may be calculated as the derivation of kink asymptotics (8)

$$W_{\text{int}}^{(1)}(x) \sim |x|^{-(n+2)}\,, \quad |x >> 1. \tag{11}$$

The FK model with a long-range interaction between kinks is similar to the long-range one-dimensional Ising model. Namely, for rational concentrations of atoms,

$$\theta = p/2\pi q$$

when one simple cell of the chain contains p atoms and q wells of the external periodic potential, the ground state of the system at the zero temperature $T = 0$ is related to a commensurate structure of atoms. When the concentration θ is increased, these structures are changed from one to another through an infinite series of phase transitions (see, e.g., Ref. [1]). For $T \neq 0$ the long-range order in a one-dimensional chain of atoms is impossible. However, it remains the short-range order for distances $x \leq L$, where the correlation length L is determined by the relations

$$L \simeq \theta_{\mathrm{pair}}^{-1}, \quad \theta_{\mathrm{pair}} \sim a^{-1} \exp(- \varepsilon_{\mathrm{pair}}/k_B T).$$

$\varepsilon_{\mathrm{pair}}$ being the energy of the kink-antikink pair creation.

As a result, the function $\varepsilon_{\mathrm{p}}(0)$, i.e., the Peierls energy for the kink, takes at $T = 0$ the form of the Devil's staircase; namely, for each rational concentration (12) (p and q are arbitrary integers) the value ε_{p} undergoes a step $\delta\varepsilon_{\mathrm{p}}$. In the case $T \neq 0$ most of the steps disappear and, as a result, the curve $\varepsilon_{\mathrm{p}}(\theta)$ will be smoothed. The steps will remain only for simple concentrations, $p = 1$ and $q = 2,3, \ldots$, when the correlation length L exceeds the kink width d.

We have calculated the values of $\delta\varepsilon_{\mathrm{p}}$ and $\varepsilon_{\mathrm{pair}}$ analytically in the case of small coupling between atoms. In the latter case $V(2\pi) << 1$ and displacements of atoms from their equilibrium positions in the potential minima may be neglected. Then, for the concentration range $(1 + N)^{-1} < p/q < N^{-1}$, $N \geq 1$ is an integer, including the kink for $\theta = 1/2\pi\,(N + 1)$ and the antikink for $\theta = 1/2\pi N$, from simple geometrical considerations we may obtain the amplitude of the Peierls potential,

$$\varepsilon_{\mathrm{p}} \approx 2 + 2V(2\pi N + \pi) - V(2\pi N) - V(2\pi N + 2\pi).$$

In the same approximation the difference $\delta\varepsilon_{\mathrm{p}}$ of the Peierls energies for an antikink and kink as well as the energy $\varepsilon_{\mathrm{pair}}$ are dependent on the length of the elementary cell only

$$\delta\varepsilon_{\mathrm{p}} \approx [\, 2V(a + \pi) - V(a) - V(a + 2\pi) \,] -$$

$$[\, 2V(a - \pi) - V(a - 2\pi) - V(a) \,],$$

$$\varepsilon_{\mathrm{pair}} \approx V(a - 2\pi) + V(a + 2\pi) - 2V(a).$$

3 Numerical simulations

We have compared our analytical results with the computer calculations. We used a chain of 20 atoms with the dipole–dipole interparticle repulsion ($n = 3$ in eq. (4)). The calculation procedure and the method of molecular–dynamic simulations were standard. All our numerical results are obtained for the parameters $\ell = \pi/2$ and $a = 2\pi$. First of all, in Figure 1 the interaction energy for kinks in the model under consideration is prsented as a function of the parameter $Q = x/a$. The value $\Delta E_\sigma(Q)$ is the difference of the total energy and the energy of the kinks at infinity. This value is proportional to the value $W^{(1)}$ considered above.

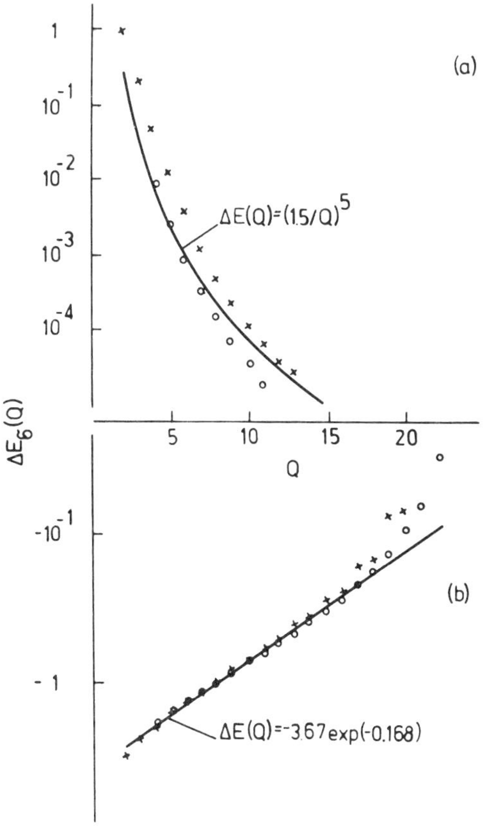

Figure 1. Interaction energy of the kinks in the chain as a function of distance between them, for the dipole ($n = 3$) (a), and the Coulomb ($n = 1$) (b) mechanisms of atomic interactions in the chain. Results for kink indicated by crosses, and for antikink by circles.

Using the molecular–dynamic method for numerical simulations, we also calculated the Peierls energy ε_p and the kink effective mass m versus the atomic concentration θ. The results are shown in Figure 2.

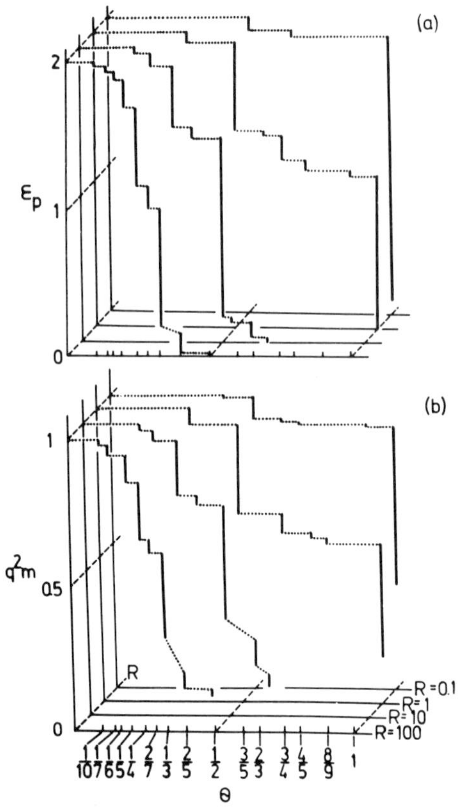

Figure 2. Peierls energy $\varepsilon_p(\theta)$ and the value q^2m as functions of the coverage parameter θ for the dipole interaction of atoms $(n = 3)$ at various values of the parmeter $R = V_0/\varepsilon_s \equiv V(2\pi)/2$.

4 Conclusions

Thus, in the case of a long-range interparticle interaction the analogous long-range interaction arises for kinks. That interaction stipulates a number of new physical effects in systems which can be described by the extended SG model. For example, it is important to note that the described dependence $\varepsilon_p(\theta)$ may be experimentally observed in a number of quasi-one-dimensional systems. In particular, the value ε_p is identical to the activation energy of the surface diffusion of atoms adsorbed on stepped or "furrowed" crystal surfaces [2, 3]. The study of surface diffusion in such structures may detect the step-like dependence of physical parameters as functions of the concentration.

The above results differ qualitatively from those of the local SG model and the discrete FK model [1] and should play an important role in forming the ground state of quasi-one-dimensional systems with long-range interactions, as well as their dynamical and thermodynamical properties.

References

[1] Bolshov, L.A., *Acta Univ. Wratislaviensis,* **37**, 73 (1980).

[2] Bolshov, L.A., Napartovich, A.P., Naumovets, A.G. and Fedorus, A.G., *Usp. Fiz. Nauk.*, **122**, 125 (1977).

[3] Braun, O.M., Kivshar, Yu.S. and Kosevich, A.M., *J. Phys.*, **C21**, 3881 (1988).

[4] Kosevich, A.M., *Foundations of Crystal Lattice Mechanics* (in Russian), Nauka, Moscow (1972).

[5] Marianer, S. and Bishop, A.R., *Phys. Rev.*, **B37**, 9893 (1988).

Braun O.
Institute of Physics
Ukrainian Acad. of Science
Prospect Nauki 46
252028 Kiev
U.S.S.R.

Kivshar Yu.
Institute of Physics
Ukrainian Acad. of Science
Prospect Nauki 46
252028 Kiev
U.S.S.R.

Zelenskaya I
Institute of Physics
Ukrainian Acad. of Science
Prospect Nauki 46
252028 Kiev
U.S.S.R.

A.P. MAYER AND A.A. MARADUDIN

Effects of non-linearity and dispersion on the propagation of surface acoustic waves

1 Introduction

The present investigation is concerned with the propagation of surface acoustic waves in the weakly non-linear regime. It assumes the existence of weakly non-linear continuous periodic waves, which can be decomposed in a sinusoidal fundamental and higher harmonic components, the amplitudes of the latter being small. By decreasing a parameter which is monotonic with the fundamental amplitude, these non-linear waves develop continuously into the linear surface wave solutions. In perturbation calculations of the properties of these waves, the amplitude A of the fundamental component, divided by its wavelength, may then be used as an expansion parameter. In order to ensure the presence of these properties, we require that the non-linearity is balanced by spatial dispersion.

Since the equations of elasticity theory do not contain a length scale, spatial dispersion for surface acoustic waves has to be introduced externally. In the following section, three different mechanisms for the introduction of dispersion are presented. If the new lengths relevant for the dispersion are small compared to the wavelength of the fundamental component of the non-linear wave, its influence on the wave propagation enters mainly via a modification of the boundary conditions for the displacement field at the surface.

Another effect of spatial dispersion consists in localizing a linear wave of purely shear horizontal polarization at the surface. The influence of non-linearity on the propagation of surface waves of shear horizontal polarization is different from the familiar Rayleigh wave case, since the second harmonic is not excited resonantly.

As geometry for quantitative calculations, we choose the (100) direction (x-direction) for the surface wave propagation on a (001) surface of a cubic elastic medium. The surface is located at $z = 0$ and the medium fills the lower half space. A non-linear dispersion relation is derived, from which, using the method of Karpman and Krushkal′ [4], the non-linear Schrödinger equation is obtained for slow variations of the complex amplitude A and the possibility of the formation of envelope solitons and self-channelling is examined. Under certain conditions, the second harmonic of a Love wave or of a surface wave on a periodic grating becomes leaky. This leads to an amplitude dependent attenuation of the weakly non-linear wave.

2 Spatial dispersion

The equations of motion and boundary conditions for the displacement field $\mathbf{u}(\mathbf{x}; t)$ in a semi-infinite elastic medium with a stress-free surface are

$$\rho\, \ddot{u}_\alpha(\mathbf{x}; t) = T_{\alpha\beta\,|\,\beta}\,(\mathbf{x}; t)\,, \tag{2.1}$$

$$T_{\alpha z}(\mathbf{R}, 0; t) = 0\,, \tag{2.2}$$

where $\mathbf{x} = (\mathbf{R}, z)$ and ρ is the density of the elastic medium. The stress tensor $\boldsymbol{T}$ is given in terms of the displacement gradients by

$$T_{\alpha\beta} = C_{\alpha\beta\mu\nu}\, u_{\mu\,|\,\nu} + \frac{1}{2} S_{\alpha\beta\mu\nu\gamma\delta}\, u_{\mu\,|\,\nu}\, u_{\gamma\,|\,\delta} + O(u^3)\,. \tag{2.3}$$

For the connection of the tensor $\boldsymbol{S}$ with the elastic moduli see Ref. [5]. In the following, it is shown how these equations can be extended to include dispersion in a physical way.

2.1 Coating by a thin film

By coating the semi-infinite elastic medium with a film of different material, a new length, the film thickness d is introduced. If the displacement field varies little within the film, i.e., if $d/\lambda \ll 1$ with λ the wavelength of the surface wave, the displacement field in the film may be eliminated to first order in d/λ in favour of an effective boundary condition for the displacement field in the substrate at the surface of the form [6, 9]

$$T_{\alpha z}(z = 0\_) = -\,d\left\{ \rho_F\, \ddot{u}_\alpha(z = 0\_) - D^{(F)}_{\alpha\varphi\mu\theta}\, u_{\mu\,|\,\varphi\theta}(z = 0\_) \right\} + O(d^2)\,, \tag{2.4}$$

where the coefficients $(D^{(F)}...)$ depend on the linear elastic moduli of the film and ρ_F is the density of the film material.

2.2 Discreteness of the underlying lattice

With decreasing wavelength, the surface acoustic wave is influenced by the atomic structure of the elastic material. To low order in a/λ with a the lattice constant, this effect can be included into the equations of motion and boundary conditions of the displacement field. This has been demonstrated for the (001) surface of a simple cubic crystal on the basis of a lattice dynamical model which couples nearest neighbour planes only [1, 6]. The boundary conditions obtained to first order in a/λ are of the form

$$T_{\alpha z}(z = 0) = a\; D^{(L)}_{\alpha\beta\mu\nu}\, u_{\mu\,|\,\beta\nu}(z = 0) + O(a^2)\,, \tag{2.5}$$

where, within the model of Ref. [1], the coefficients $(D^{(L)}...)$ can be expressed in terms of the linear elastic moduli. The lowest order correction to the equations of motion is of second order in a/λ and involves fourth derivatives of the displacement field.

2.3 Surface corrugation

New length scales can also be introduced by corrugation of the surface. For simplicity, we choose the periodic profile

$$\zeta(x) = \zeta_0 \cos(Gx) . \tag{2.6}$$

Upon expanding the stress tensor at the corrugated surface with respect to the corrugation profile, the following effective boundary condition is obtained to first order in ζ_0:

$$0 = T_{\alpha z}(\mathbf{R}, z = 0) + \zeta(x)\, T_{\alpha z \mid z}(\mathbf{R}, z = 0) - \zeta'(x) T_{\alpha x}(\mathbf{R}, z = 0) . \tag{2.7}$$

A common feature of all three ways of introducing dispersion is that in the case of sagittal polarization, the difference $\omega(q) - v_R q$, i.e., the correction to the non-dispersive frequency, is of second order in q unlike the case of bulk waves, and to leading order in q, the dispersion enters the calculation only via the boundary conditions (v_R is the Rayleigh wave phase velocity). In the case of shear horizontal polarization, however, the frequency correction is of third order in q and is also influenced by the additional term in the equation of motion.

3 Non-linear dispersion relation

For the case of a semi-infinite elastic medium covered by a thin film, a dispersion relation for weakly non-linear waves of both sagittal and predominantly shear horizontal polarization of the form

$$\omega = \omega_0(\mathbf{q}) + \omega^{(2)}_{\mathbf{q}} \mid A \mid^2 \tag{3.1}$$

is derived, where A is the amplitude of the fundamental in the linear limit,

$$\mathbf{u}^{(1)}(\mathbf{x}; t) = e^{i(qx-\omega t)} \sum_r c_r\, \mathbf{w}^{(r)}(q \mid z) A + \text{c.c.} \tag{3.2}$$

in the case of sagittal polarization, with $\mathbf{w}^{(r)}(q \mid z)$ containing the z dependence of the Rayleigh wave displacement field [8], $(w_y = 0)$, or

$$u^{(1)}_\alpha(\mathbf{x}; t) = e^{i(qx-\omega t)+\beta z}\ \delta_{\alpha y} A + \text{c.c.} \tag{3.3}$$

in the case of predominantly shear horizontal polarization.

The calculation is performed in two steps. First, the second harmonic $\mathbf{u}^{(2)}$ is calculated in a way analogous to the theoretical treatment of the generation of the second harmonic of Rayleigh waves by Normandin et al. [7] and by Tiersten and Baumhauer [10]. In the geometry chosen, the second harmonic is of sagittal polarization for the fundamentals (3.2) and (3.3). It can be decomposed into a particular solution of the equations of motion of second order in A with (3.2) or (3.3) inserted in the terms quadratic in u, and a homogeneous solution of the form

$$\mathbf{u}^{(2,h)}(\mathbf{x};\, t) = e^{2i(qx-\omega t)} \sum_r E(r)\mathbf{w}^{(r)}(2q \mid z)A^2 + \text{c.c.} \tag{3.4}$$

to satisfy the boundary conditions to second order in A. Unlike the situation in Refs. [7, 10], there is no secular term in x because of the presence of spatial dispersion. In the case of a sagittal fundamental, we have

$$E(r) = O(1/(qd)), \tag{3.5}$$

which demonstrates the effect of the dispersion to inhibit the growth of the second harmonic in the case of purely sagittal polarization.

In the second step, the correction $\omega_2^2 = 2\omega_0\omega^{(2)} \mid A \mid^2$ to the squared frequency is calculated from the equations of motion and boundary conditions of third order in A. The explicit calculation of the displacement field to third order in A can be bypassed by making use of the fact that the part of the displacement field proportional to $\exp(-i\omega t)$ solves the equations of motion and boundary conditions separately and projecting the terms of the equations of motion proportional to $\exp(-i\omega t)$ (marked by a tilde) on the first-order solution:

$$\int d^3x\, \tilde{u}_\alpha^{(1)^*} \left\{ \rho\omega_2^2\, \tilde{u}_\alpha^{(1)} + \rho\omega_0^2\, \tilde{u}_\alpha^{(3)} + C_{\alpha\beta\mu\nu}\, \tilde{u}^{(3)}_{\mu \mid \beta\nu} + \tilde{T}^{(N)}_{\alpha\beta \mid \beta} \right\} = 0\,, \tag{3.6}$$

where

$$T^{(N)}_{\alpha\beta} = S_{\alpha\beta\mu\nu\gamma\delta}\, u^{(1)}_{\mu \mid \nu}\, (u^{(2)}_{\gamma \mid \delta} + u^{(0)}_{\gamma \mid \delta}) + \frac{1}{6} S_{\alpha\beta\mu\nu\gamma\delta\sigma\lambda}\, u^{(1)}_{\mu \mid \nu}\, u^{(1)}_{\gamma \mid \delta}\, u^{(1)}_{\sigma \mid \lambda}\,. \tag{3.7}$$

Here, $\mathbf{u}^{(0)}$ is a static contribution to the solution of second order in A. Integrating by parts and using the boundary conditions, we can eliminate $\mathbf{u}^{(3)}$ and obtain the following expression for the frequency correction, which contains the linear solution, the second harmonic, and the static part:

$$\omega^{(2)} \mid A \mid^2 = N^{-1} \int d^3x\, \underset{\sim}{u}^{(1)^*}_{\alpha \mid \beta}\, \tilde{T}^{(N)}_{\alpha\beta}\,, \tag{3.8}$$

with

$$N = 2\rho\omega_0 \int d^3x \, \tilde{u}_\alpha^{(1)^*} \underset{\sim}{u}_\alpha^{(1)} .$$

In the case of purely sagittal polarization, the term in (3.7) involving the second harmonic should be dominant because of (3.5). Neglecting, therefore, the other two contributions to (3.7), we may, to leading order in qd, decompose

$$\omega^{(2)} = \frac{v_R \, q^2}{d \, M_s} Q , \tag{3.9}$$

where M_s is the sagittal mismatch factor

$$M_s = \frac{\rho_F}{\rho_s} - \frac{c_F}{c_s}, \quad c_j = c_{11}^{(j)} - c_{12}^{(j)^2} / c_{11}^{(j)} , \tag{3.10}$$

and the dimensionless quantity Q depends only on the second and third order elastic moduli of the substrate. In the same way, we may write for the second derivative of the frequency with respect to q_x

$$\Omega_{xx}(q\hat{x}) = M_s \, d \, v_R \, P , \tag{3.11)}$$

where P is dimensionless and only depends on the second order elastic moduli of the substrate. Numerical values of P and Q for several substances are given in Table 1.

Table 1. Numerical values for the quantities governing the parameters in (4.2) and (4.3) in the case of sagittal polarization

	Q	$-P$	$\Omega_{yy} \, q/v_R$	$v_R[10^3 \frac{m}{s}]$	α_r
NaF	0.079	0.337	0.889	3.045	real
Cu	0.387	0.388	1.761	2.015	complex
KCl	0.014	0.148	0.787	1.752	real
Si	0.023	0.579	1.189	4.917	complex

For a non-linear wave with a fundamental of shear horizontal polarization, we define the dimensionless quantities

$$R = \frac{v \, q^3}{\omega^{(2)}} \tag{3.12}$$

and

$$S = \Omega_{yy}(q\hat{x})q/v\ , \tag{3.13}$$

where $v = \sqrt{c_{44}/\rho}$ and Ω_{yy} is the second derivative of the frequency with respect to q_y, and note, that

$$\Omega_{xx} = -\,qv\,M_{\mathrm{h}}^2\,d^2 < 0, \tag{3.14}$$

where M_{h} is the shear horizontal mismatch parameter. In the case of an isotropic substrate, the following simple formula for $\omega^{(2)}$ is obtained:

$$\omega^{(2)} = \frac{q^3}{\sqrt{\rho\, c_{44}}}\left\{\frac{1}{8}(c_{4444} + 6c_{166} + 3c_{11}) - \frac{1}{4c_{11}}(c_{144} + c_{12})^2 - \right.$$

$$\left. - \frac{1}{4(c_{11} + c_{12})}(c_{166} + c_{11})^2\right\} + O(d)\ . \tag{3.15}$$

Here, all three contributions are of the same order of magnitude. The numerical values in Table 2 show that strong compensations occur between the parts involving the second harmonic and the static solution on the one hand and the quartic anharmonicity on the other. For the substances we have analysed, the cubic anharmonicity outweighs the quartic.

Table 2. Numerical values for the quantities governing the parameters in (4.3) in the case of predominantly shear horizontal polarization.

			δR^{-1}		
	R	Fourth order	Static	Second harmonic	S
NaCl	-0.56	1.23	-0.65	-2.36	1.538
KCl	-0.55	1.21	-0.79	-2.23	2.223
RbCl	-1.83	1.78	-0.70	-1.63	4.067
CsI	-0.10	3.16	-3.86	-9.10	5.212
Al	-0.16	3.33	-0.12	-9.61	0.115

4 Instabilities of the weakly non-linear waves

A non-linear dispersion relation of the form (3.1) and a continuity equation for the energy of the approximate form

$$\left\{\frac{\partial}{\partial t}+\frac{\partial}{\partial x_\alpha}V_\alpha(\mathbf{q})\right\}|A|^2=0\,, \tag{4.1}$$

which also holds for surface acoustic waves, are the ingredients into the formalism of Karpman and Krushkal' [4, 3] for the derivation of a governing equation for slow variations of the complex amplitude A. For modulations of the carrier wave along the propagation direction, the non-linear Schrödinger equation

$$i\frac{\partial}{\partial t}A+\frac{1}{2}\Omega_{xx}(\mathbf{Q})\frac{\partial}{\partial X^2}A-\omega^{(2)}|A|^2A=0 \tag{4.2}$$

is obtained, where $X=x-V_x(\mathbf{Q})t$, $\mathbf{Q}$ is the wave vector and $\mathbf{V}(\mathbf{Q})$ the group velocity of the carrier wave. Slow variations of the profile of a stationary beam in the y-direction can be described approximately by the non-linear Schrödinger equation in the form

$$i\,V_x(\mathbf{Q})\frac{\partial}{\partial x}A+\frac{1}{2}\Omega_{yy}(\mathbf{Q})\frac{\partial^2}{\partial y^2}A-\omega^{(2)}|A|^2A=0\,. \tag{4.3}$$

As is well known [11], the non-linear Schrödinger equation possesses an instability, which is associated with self-modulation of an initial pulse in the case of eq. (4.2) and self-channelling of an initial beam profile in the case of eq. (4.3). This instability occurs, if the coefficients $\Omega_{\alpha\alpha}(\mathbf{Q})$ and $\omega^{(2)}$ have different signs. From the data in Tables 1, 2 we may therefore conclude, that for non-linear surface waves of purely sagittal polarization, the self-modulation instability should occur, since the product of Ω_{xx} and $\omega^{(2)}$ is independent of the sagittal mismatch factor. For surface waves of predominantly shear horizontal polarization, the self-channelling instability can be expected. Whether the latter also occurs in the sagittal case partly depends on the sign of the mismatch factor.

5 Amplitude dependent attenuation

The elastic moduli of the substances in Table 2 satisfy the condition $c_{11}(c_{11}-c_{44})>(c_{44}+c_{12})^2$. If this condition is not fulfilled, the second harmonic of the shear horizontal surface wave (3.3) (i.e., the Love wave of the lowest branch) contains partial waves, which are of bulk character. In this case, the non-linear wave is no longer a pure surface wave, but becomes leaky by radiating energy into the substrate via the second harmonic. We attribute a damping coefficient Γ to the surface wave due to this attenuation mechanism. It is calculated by dividing the time averaged energy current density vertical to the surface, J_z, by the energy of the fundamental component of the wave per unit surface area, W:

$$\Gamma(q) = -\frac{1}{2} J_z(q)/W(q)\,, \tag{5.1}$$

where to leading order in qd

$$\Gamma(q) = \gamma\, q^2 \,|A|^2\,,$$

$$\frac{\gamma}{\omega_0} = \frac{1}{4}\left[\frac{c_{11}(c_{11}-c_{44}) - (c_{44}+c_{12})^2}{c_{44}\,(c_{44}-c_{11})}\right]^{1/2} \times$$

$$\times \left[\frac{(c_{44}-c_{11})\,(2c_{456}+c_{144}+2c_{44}+c_{12}) + (c_{12}+c_{44})\,(c_{166}+c_{11})}{(c_{11}+c_{12})\,(c_{12}-c_{11}+2c_{44})}\right]^2 . \tag{5.2}$$

$$(\omega_0^2 = q^2\, c_{44}/\rho).$$

To lowest order in qd, the ratio γ/ω_0 depends only on the elastic properties of the substrate.

The phenomenon of the coupling of a surface acoustic wave to a leaky second harmonic, which is generated by anisotropy in the case of Love waves, also appears in the case of surface waves propagating on a periodic grating. Here, the linear surface waves are of the Bloch form [2], i.e.

$$\mathbf{u}^{(1)}(\mathbf{x};\, t) = e^{-i\omega t} \sum_n \sum_r \mathbf{w}^{(r)}\,(q + nG,\, \omega \,|\, \mathbf{x})\, A_r(n) + \text{c.c.} \tag{5.3}$$

in the case of sagittal polarization, and we choose $A = A_{r=1}(0)$ as the amplitude of this wave. The second harmonic contains a part of the form

$$\mathbf{u}^{(2,h)}(\mathbf{x};\, t) = e^{-2i\omega t} \sum_n \sum_r E_r(n)\, \mathbf{w}^{(r)}(2q + nG,\, 2\omega \,|\, \mathbf{x})\, A^2 + \text{c.c.} \tag{5.4}$$

For an isotropic elastic medium, the decay constant in $\mathbf{w}^{(\ell,t)}$ is given by

$$\alpha_{\ell,t}(K,\omega) = \left[K^2 - \frac{\omega^2}{v_{\ell,t}^2}\right]^{1/2} \tag{5.5}$$

with v_t and v_ℓ the transverse and longitudinal sound velocity, and thus $\alpha_t(2q - G,\, 2\omega)$ will be purely imaginary for $q > q_t$, where

$$\omega(q_t) = v_t \cdot (\frac{1}{2}G - q_t) . \qquad (5.6)$$

For weak corrugation, the solution of (5.6) is approximately

$$q_t \approx \frac{1}{2}G \frac{v_t}{v_t + v_R} . \qquad (5.7)$$

For $q > q_\ell$ with q_ℓ determined by (5.6) with t substituted by ℓ, also the component of the second harmonic involving $\alpha_\ell (2q - G, 2\omega)$ becomes leaky. The attenuation coefficient can for $q << G/2$ be written to leading order in ζ_0 for the corrugation profile (2.6) as

$$\Gamma(q) = 2 | A / \zeta_0 |^2 \ q \, v_R \ \gamma(q/G) . \qquad (5.8)$$

The behaviour of the function $\gamma(q/G)$ is shown in Figure 1 for BaF_2 and tungsten using the isotropic approximation for the linear Rayleigh waves. For $q \approx G/2$, the above expression is no longer valid, and the attenuation coefficient is of zeroth order in ζ_0.

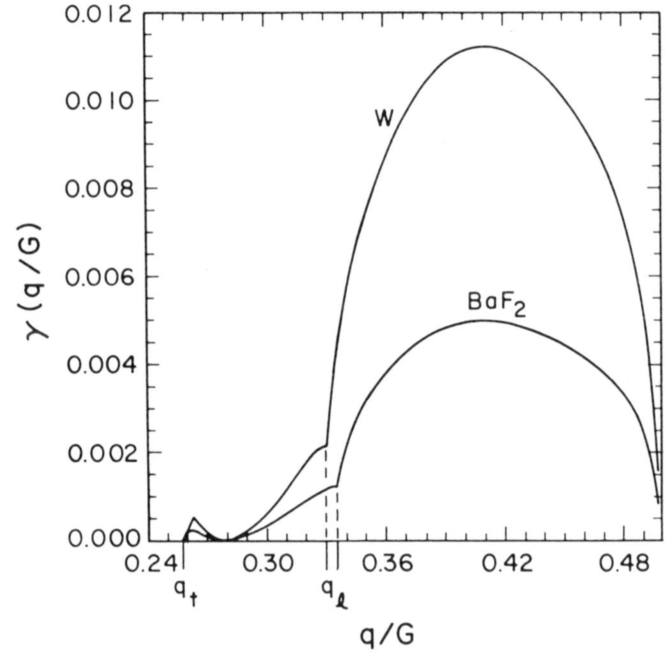

Figure 1. The function γ governing the amplitude dependent attenuation of a Rayleigh wave propagating on a shallow sinusoidal grating as function of the wave vector.

ACKNOWLEDGEMENTS

Financial support from the NSF (Grant No. DMR 88-15866) and the Deutsche Forschungsgemeinschaft is gratefully acknowledged.

References

[1] Gazis, D.C., Herman, R., and Wallis, R.F., Surface elastic waves in cubic crystals, *Phys. Rev.* **119**, 533 (1960).

[2] Glass, N.E., Loudon, R., and Maradudin, A.A., Propagation of Rayleigh surface waves across a large-amplitude grating, *Phys. Rev.* **B24**, 6843 (1981).

[3] Karpman, V.I., *Nonlinear waves in dispersive media*, (Pergamon, Oxford, 1974, russ. ed.: Nauka, Moscow, 1973).

[4] Karpman, V.I. and Krushkal', E.M., Modulated waves in nonlinear dispersive media, *Zh. Eksp. Teor. Fiz.* **55**, 530 (1968) [Soviet Physics - *JETP* **28**, 277 (1969)].

[5] Leibfried, G. and Ludwig, W., Gleichgewichtsbedingungen in der Gittertheorie, *Z. Phys.* **160**, 80 (1960).

[6] Maradudin, A.A., Surface acoustic waves on real surfaces, in: *Physics of Phonons*, ed. T. Paszkiewicz (Springer-Verlag, Berlin, 1987).

[7] Normandin, R., Fukui, M., and Stegeman, G.I., Analysis of parametric mixing and harmonic generation of surface acoustic waves, *J. Appl. Phys.* **50**, 81 (1979).

[8] Royer, D. and Dieulesaint, E., Rayleigh wave velocity and displacement in orthorhombic, tetragonal, hexagonal, and cubic crystals, *J. Acoust. Soc.* Am. **76** 1438 (1984).

[9] Tiersten, H.F., Elastic surface waves guided by thin films, *J. Appl. Phys.* **40**, 770 (1969).

[10] Tiersten, H.F. and Baumhauer, J.C., An analysis of second harmonic generation of surface waves in piezoelectric solids, *J. Appl. Phys.* **58**, 1867 (1985).

[11] Zakharov, V.E. and Shabat, A.B., Exact theory of two-dimensional self-focusing and one-dimensional self-modulation of waves in nonlinear media, *Zh. Eksp. Teor. Fiz.* **61**, 118 (1971) [Soviet Physics - *JETP* **34**, 62 (1972)].

Mayer A.P.
Department of Physics
University of California
Irvine, CA 92717
U.S.A.

Maradudin A.A.
Department of Physics and
Institute for Surface and Interface Science
University of California
Irvine, CA 92717
U.S.A.

D.F. PARKER

Pulse collisions for coupled NLS equations

Abstract

In a number of continuum and discrete systems, the equations governing modulations of harmonic wave trains have recently been shown to be a pair of coupled non-linear Schrödinger equations. Examples arise from analysis of optical fibres and of certain bimodal one-dimensional lattices. In some circumstances, the equations possess sech-envelope solutions having the same form as the solitons of the single NLS equation. However, even in these cases it is exceptional that a collision between two such pulses should be a true soliton collision, producing two emergent pulses identical in form to the incident pulses.

Some predictions of pulse collisions for the symmetric pair of NLS equations arising from the study of axisymmetric optical fibres are given. These show that it is normal that pulses governed by this pair of equations collide to give two pulses having permanent form. However, the emergent pulses are not usually the sech-profile pulses associated with a single NLS equation. For fibres which are not axisymmetric, it is shown that the governing equations are a more general pair of coupled NLS equations. Some preliminary predictions of their properties are given.

1 Introduction

Coupled equations of non-linear Schrödinger type have arisen recently from both continuum and discrete models of wave propagation in media which, according to linear theory, allow two independent modes, often corresponding to two lateral degrees of freedom. Remoissenet [8] discusses a *bi-lattice*, which is a scalar discrete system allowing two distinct wave modes. Using a semi-discrete approximation he shows that in some circumstances the complex envelopes A, B of the carrier waves satisfy coupled equations of the form

$$\begin{aligned} i\,A_\tau &= p\,A_{ss} + q\,|A|^2 A + r\,|B|^2 A\,, \\ i\,B_\tau &= p'\,B_{ss} + q'\,|B|^2 B + r'\,|A|^2 B\,, \end{aligned} \tag{1}$$

for some constants p, q, r, p', q', r'.

Clearly, if $B \equiv 0$, the remaining amplitude A satisfies the non-linear Schrödinger (NLS) equation which, in non-dimensional form, may be written as

$$i\,A_\tau = A_{ss} + |A|^2 A\,. \tag{2}$$

This equation is known to be completely integrable, possessing an inverse scattering transform and possessing both soliton and multi-soliton solutions. These latter show that two initially well-separated pulses having sech-envelope profiles may travel at distinct speeds until they collide and interact, but that after the collision the pulses regain their original profiles. This phenomenon is, of course, the reason why the pulses of permanent form have the right to be called solitons.

For the system (1) the only cases known to be completely integrable are (except for the trivial case $r = 0$, $r' = 0$ in which the equations uncouple) the cases $q = q'$, $r = r'$ with $p = \pm p'$, $q = \pm r$ (Zakharov and Schulman [9]). Nevertheless, computations [7] for interactions between the various types of sech-envelope pulse which may travel along an axially symmetric optical fibre suggest special properties even when the coefficients in (1) are not so severely restricted. The theory in this case gives $p = p'$, $q = q'$, $r = r'$ but $r \neq q$. It allows sech-envelope pulses of either left-handed or right-handed circular polarization and of linear polarization at any orientation to the fibre axis. Computations indicate that, after two of these pulses collide, the interaction separates into two pulses travelling at the original speeds but typically with some change in the pulse envelope. For example, a linearly polarized pulse emerges as a linearly polarized pulse ony if it collides with a linearly polarized pulse of identical or orthogonal orientation. The similarity of this result with predictions by Cadet [3], from both computations and parameter expansions for interactions between pulses describing transverse vibrations of an atomic chain, is the stimulus for the present investigation.

The analysis concentrates on the fibre-optic equations derived from Maxwell's equations using a continuum treatment. It describes the pulse interactions predicted for axisymmetric fibres and relates the input and output pulses to certain self-similar solutions of the equations. It also allows for non-axisymmetric fibres. Recent treatments [2, 5] of "birefringent fibres" (allowing for the slight differences in phase speeds and group speeds for linearized modes in distinct "basis" orientations) have led to equations generalizing (1). The present derivation, emphasizing the consequences of various symmetries in the fibre geometry, shows how a number of special parameter choices in (1) arise naturally. It also shows that the coupled NLS system is often more general than the system (1).

2 Waveguide modes in optical fibres

The material of optical fibres is usually treated as isotropic and non-magnetic, but dielectric with permittivity which is quadratic in the electric field intensity $\mathbf{E}$ (Kerr effect non-linearity). Then, if the fibre is longitudinally uniform, with cross-section $\mathcal{D}$ and material composition independent of the axial coordinate z, the electric displacement $\mathbf{D}$ is related to $\mathbf{E}$ by

$$\mathbf{D} = \{\varepsilon(x, y) + N(x, y)\mathbf{E}\cdot\mathbf{E}\}\,\mathbf{E}\,. \tag{3}$$

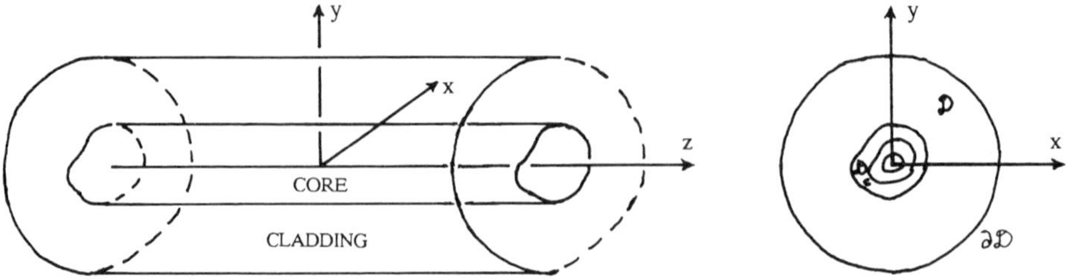

Figure 1. A typical optical fibre, showing contours of (e.g.) permittivity in the core.

Here, the (linear) permittivity ε and the non-linear coefficient N can vary with x and y within the *core* - a region $(x, y) \in D_c$, near the z-axis as shown in Figure 1. In the *cladding* $\mathcal{D} - \mathcal{D}_c$, it is usual to treat ε and N as constant. The equations connecting **E**, **D** and the magnetic field intensity **H** within the fibre are (3), together with Maxwell's equations in the absence of electric current:

$$\nabla\cdot\mathbf{D} = 0\,, \qquad \mu_0\nabla\cdot\mathbf{H} = 0\,, \tag{4}$$

$$\nabla\times\mathbf{E} = -\mu_0\frac{\partial\,\mathbf{H}}{\partial\,t}, \qquad \nabla\times\mathbf{H} = \frac{\partial\,\mathbf{D}}{\partial\,t}. \tag{5}$$

Since $\varepsilon(x, y)$ is slightly higher ($\sim$ 1%) in the core than in the cladding, linear theory predicts the existence of *guided modes*

$$\mathbf{E} = \tilde{\mathbf{E}}(x, y)\, e^{i(kz-\omega t)} + \text{c.c.}\,, \quad \mathbf{H} = \tilde{\mathbf{H}}(x, y)\, e^{i(kz-\omega t)} + \text{c.c.} \tag{6}$$

in which the *phase speed* ω/k exceeds the minimum speed of light within the core but is less than the speed of light within the cladding. The *mode shapes* defined by $\tilde{\mathbf{E}}(x, y)$ and $\tilde{\mathbf{H}}(x, y)$ satisfy an eigenvalue problem with boundary conditions imposed on the fibre surface $(x, y) \in \partial\mathcal{D}$ and with the wave number k and (angular) *frequency* ω appearing as parameters. The resulting eigenvalue condition connecting ω and k is the *dispersion relation*

$$D(\omega, k) = 0\,. \tag{7}$$

For axially-symmetric (circular) fibres it is natural to use cylindrical polar coordinates (r, θ, z) and corresponding base vectors $(\mathbf{e}_r, \mathbf{e}_\theta, \mathbf{e}_z)$. The circularly polarized modes are then found as

$$\tilde{\mathbf{E}} = \{ i\tilde{E}_1(r)\mathbf{e}_r \pm \tilde{E}_2(r)\mathbf{e}_\theta + \tilde{E}_3(r)\mathbf{e}_z\}\, \mathrm{e}^{\pm i\ell\theta} = \mathbf{E}^{\pm}(r, \theta)\mathrm{e}^{\pm i\ell\theta} ,$$
$$\tilde{\mathbf{H}} = \{ \pm\tilde{H}_1(r)\mathbf{e}_r + i\,\tilde{H}_2(r)\mathbf{e}_\theta \pm i\,\tilde{H}_3(r)\mathbf{e}_z \}\, \mathrm{e}^{\pm i\ell\theta} = \mathbf{H}^{\pm}(r, \theta)\mathrm{e}^{\pm i\ell\theta} , \tag{8}$$

where ℓ is an integer. Here, the imaginary factors i have been inserted so that all components $\tilde{E}_j, \tilde{H}_j$ $(j = 1,2,3)$ are real. The upper and lower signs $\pm$ correspond to modal patterns which rotate about the z-axis as in a left- or right-handed corkscrew, respectively. Both travel at the same phase speed ω/k, and have fields $\mathbf{E}$, of (4), which differ only by reflection in the Oxz plane. The six independent ordinary differential equations derived from (3)-(5) and connecting $\tilde{E}_j$ and $\tilde{H}_j$ may be found in [7]. Solutions satisfying the boundary conditions and which are finite at $r = 0$ do not exist for all combinations of ω and k. For each integer ℓ, there is typically a lowest frequency ω_c (the *cut-off frequency*) at which solutions exist. Above this frequency, there is a real wave number k corresponding to each ω, so defining a curve $k = k(\omega)$ in the ω, k plane, labelled by the integer ℓ. However, except for the case $\ell = 0$ describing axisymmetric fields, there are two distinct modes corresponding to each root of the dispersion relation (7).

For long-distance communications, fibres are usually designed as "mono-mode" fibres. This means that only one root of (7) has cut-off frequency ω_c below the operating frequency ω of the (laser) source. However, the root with lowest cut-off frequency is *not* the axisymmetric mode ($\ell = 0$). It is usually the root with $\ell = 1$, as in the step-index fibre [4] (essentially because plane electromagnetic waves are *transverse*). Consequently, in axisymmetric mono-mode fibres, the general propagating electric field at frequency ω is given by linear theory as

$$\mathbf{E} = \{ A^+\, \mathbf{E}^+(r, \theta)\mathrm{e}^{i\theta} + A^-\, \mathbf{E}^-(r, \theta)\mathrm{e}^{-i\theta}\}\, \mathrm{e}^{i(kz-\omega t)} + \mathrm{c.c.} , \tag{9}$$

where c.c. denotes a complex conjugate. It involves two complex amplitudes A^+, A^- (magnitude and phase) of the left- and right-handed modes (8).

When non-linear effects are retained in eqs. (3)-(5), equations governing the gradual variation of A^+ and A^- over the scale of many wavelengths have been rigorously derived in [7] as a pair of coupled NLS equations (1). Some salient steps and predictions of this procedure are first summarized, before the procedure is generalized to describe equations for non-axisymmetric fibres.

3 Coupled NLS equations for axisymmetric fibres

If the amplitudes A^+ and A^- in (9) vary over length scales much greater than the wavelength $2\pi/k$, it is well known that to first approximation they travel unattenuated at the *group speed* $c_g = [k'(\omega)]^{-1}$ associated with frequency ω. Thus, if a group variable χ is introduced by

$$\chi \equiv \nu\omega(t - sz), \quad s \equiv k'(\omega) = c_g^{-1}, \tag{10}$$

where ν $(<< 1)$ is a small parameter, it is found that the approximations $A^{\pm} \simeq A^{\pm}(\chi)$ hold over ranges of z as long as $O(\nu^{-1})$. In order that non-linear effects should be significant over scales $z = O(\nu^{-2})$ on which linear theory predicts dispersion effects, it is appropriate to introduce into (9) an amplitude factor ν and to define a "stretched" axial coordinate $Z = \nu^2 z$. Then, it is readily checked that eqs. (3)-(5) are satisfied to $O(\nu^2)$ by

$$\mathbf{E} = \nu\mathbf{E}^{(1)} + \nu^2\mathbf{E}^{(2)}, \; \mathbf{H} = \nu\mathbf{H}^{(1)} + \nu^2\mathbf{H}^{(2)}, \; \mathbf{D} = \varepsilon\mathbf{E}, \tag{11}$$

provided that $\mathbf{E}^{(1)}$ and $\mathbf{E}^{(2)}$ have the forms

$$\begin{aligned} \mathbf{E}^{(1)} &= A^+(\chi, Z)\,\mathbf{E}^+\,e^{i(\theta+\psi)} + A^-(\chi, Z)\,\mathbf{E}^-\,e^{i(-\theta+\psi)} + \text{c.c.}, \\ \mathbf{E}^{(2)} &= i\omega\left(\frac{\partial A^+}{\partial\chi}\frac{d\mathbf{E}^+}{d\omega}e^{i\theta} + \frac{\partial A^-}{\partial\chi}\frac{d\mathbf{E}^-}{d\omega}e^{-i\theta}\right)e^{i\psi} + \text{c.c.}, \end{aligned} \tag{12}$$

where $\psi \equiv kz - \omega t$ is the *phase*, where the derivatives $d\mathbf{E}^{\pm}/d\omega$ of the modal fields (8) give corrections to the field distributions associated with the amplitude modulations $\partial A^{\pm}/\partial\chi$ and where $\mathbf{H}^{(1)}, \mathbf{H}^{(2)}$ are given by expressions analogous to (12).

By seeking solutions in the form

$$\mathbf{E} = \nu\mathbf{E}^{(1)} + \nu^2\mathbf{E}^{(2)} + \nu^3\bar{\mathbf{E}}, \text{ etc.},$$

it is found by comparison of the $O(\nu^3)$ terms in (3) that

$$\bar{\mathbf{D}} = \varepsilon\bar{\mathbf{E}} + N\,|\,\mathbf{E}^{(1)}\,|^2\,\mathbf{E}^{(1)},$$

whilst eqs. (4) and (5) give a linear inhomogeneous system connecting $\bar{\mathbf{E}}, \bar{\mathbf{H}}$ and $\bar{\mathbf{D}}$. Without solving explicitly for $\bar{\mathbf{E}}$ and $\bar{\mathbf{H}}$, but merely by insisting that this system should possess solutions which are 2π-periodic in ψ (and hence free of unbounded "secular" terms) it is possible to derive compatibility conditions involving A^+, A^- and their derivatives [7]. These conditions reduce to the symmetric pair of coupled NLS equations

$$if_1 \frac{\partial A^+}{\partial Z} = \omega^2 g \frac{\partial^2 A^+}{\partial \chi^2} + (f_2 |A^+|^2 + f_3 |A^-|^2)A^+ ,$$

$$if_1 \frac{\partial A^-}{\partial Z} = \omega^2 g \frac{\partial^2 A^-}{\partial \chi^2} + (f_2 |A^-|^2 + f_3 |A^+|^2)A^- , \tag{13}$$

in which the coefficients are found to be

$$f_1 \equiv \iint \mathbf{e}_3 \cdot (\mathbf{H}^{+*} \times \mathbf{E}^+ + \mathbf{H}^+ \times \mathbf{E}^{+*})\, dS , \quad g \equiv \tfrac{1}{2} f_1 \frac{ds}{\partial \omega},$$

$$f_2 \equiv \omega \iint N(2|\mathbf{E}^+|^4 + |\mathbf{E}^+ \cdot \mathbf{E}^+|^2)\, dS , \tag{14}$$

$$f_3 \equiv 2\omega \iint N(|\mathbf{E}^+ \cdot \mathbf{E}^{-*}|^2 + |\mathbf{E}^+ \cdot \mathbf{E}^-|^2 + |\mathbf{E}^+|^2 |\mathbf{E}^-|^2)\, dS .$$

The coefficient f_1 may be identified as the Poynting vector of either circularly polarized mode, while f_2 and f_3 may be regarded as non-linear susceptibilities, since f_2/f_1 is the only coefficient of non-linearity when either mode exists in isolation. The ratio $h \equiv f_3/f_2$ is the *coupling parameter*. Study of (14) and the definitions (8) of $\mathbf{E}^+$ and $\mathbf{E}^-$, together with the fact that the electric fields are predominantly transverse, shows that typical values of h are close to 2.

A rescaling of coordinates according to

$$\tau = \frac{f_2}{f_1} Z , \quad x = \omega^{-1} \sqrt{\frac{f_2}{g}}\, \chi , \tag{15}$$

shows that eqs. (13) may be written concisely, using subscripts to denote differentiation, as

$$iA^{\pm}_{\tau} = A^{\pm}_{xx} + (|A^{\pm}|^2 + h|A^{\mp}|^2)A^{\pm} . \tag{16}$$

4 Pulse collisions

A number of types of exact solution to eqs. (16) are readily found.

(i) *Uniform (Periodic) Wave trains*. These have

$$A^+ = a_1\, e^{i(c_1x + d_1\tau)}\,,\ A^- = a_2\, e^{i(c_2x + d_2\tau)}\,, \tag{17}$$

in which the real parameters c_1, c_2, d_1, d_2 and (complex) amplitudes a_1, a_2 satisfy

$$d_1 = c_1{}^2 - |a_1|^2 - h|a_2|^2,\ d_2 = c_2{}^2 - |a_2|^2 - h|a_1|^2\,. \tag{18}$$

Equations (10) and (15) show that A^+ and A^- describe uniform wave trains having frequencies and wave numbers

$$\omega_j = \omega - \nu c_j(f_2/g)^{\frac{1}{2}},\ k_j = k - s\nu c_j(f_2/g)^{\frac{1}{2}} + \nu^2 d_j f_2/f_1\,,\quad (j = 1,2)$$

respectively. When the non-linear terms in (16) are neglected so that (18) reduces to $d_j = c_j^2$ $(j = 1,2)$, this relationship between k_j and ω_j follows immediately from Taylor series expansion of $k(\omega_j)$. More generally, eqs. (18) show how both the amplitudes affect each of the phase speeds ω_j/k_j.

For a single NLS equation (equivalently, for $h = 0$), uniform wave trains are unstable to long waves [1]. In the monochromatic case $c_1 = c_2$, solutions (17), (18) are also unstable to long waves, for all h [6]. It is therefore conjectured that uniform wave trains are always unstable.

(ii) *Circular Polarization* (CP). The assumption $A^- \equiv 0$ reduces (16) to the single NLS equation (cf. equation (2))

$$iA^+_\tau = A^+_{xx} + |A^+|^2A^+\,. \tag{19}$$

Consequently, if a signal (12) has purely left-handed modes, all modulations are governed by the NLS equation (19). The most important exact solution, the soliton, may be written as

$$A^+ = \gamma\sqrt{2}\, e^{-i\varphi}\, \text{sech}\, \gamma(x - 2\,\beta\tau)\,,\quad \varphi = \beta x + (\gamma^2 - \beta^2)\tau + 2\mu\,. \tag{20}$$

Here, γ is an arbitrary (real) amplitude, while β gives a frequency shift which determines the speed of the pulse envelope. The field pattern associated with this CP soliton is typified by the radial electric field

$$\mathbf{E}^{(1)}\cdot\mathbf{e}_r = -\,2\sqrt{2}\gamma\, \text{sech}\, \gamma(\tilde{k}z - \tilde{\omega}t)\, \tilde{E}_1(r)\sin(\theta + \bar{k}z - \bar{\omega}t - 2\mu)\,, \tag{21}$$

showing the left-handed helical twist with pitch $2\pi/\bar{k}$ and with pulsewidth $2(\gamma\tilde{\omega})^{-1}$ inversely proportional to the amplitude. Here

$$\tilde{\omega} = \nu(f_2/g)^{\frac{1}{2}}\,, \quad \tilde{k} = s\tilde{\omega} + \beta\tilde{\omega}^2 s'(\omega)\,,$$
$$\bar{\omega} = \omega + \beta\tilde{\omega}\,, \quad \bar{k} = k + s\beta\tilde{\omega} + \tfrac{1}{2}\beta^2\tilde{\omega}^2 s'(\omega) - \nu^2\gamma^2 f_2/f_1\,. \tag{22}$$

Similar solutions for A^-, with $A^+ \equiv 0$, describe right-handed solitons.

(iii) *Linear Polarization* (LP). Whenever $A^- = A^+ e^{2i\alpha}$ (α = const.) both equations (16) reduce to

$$iA^+_\tau = A^+_{xx} + (h+1)\,|A^+|^2 A^+\,. \tag{23}$$

This is a simple rescaling of (19) and so has LP soliton solutions in which the *only* change to (20) is that the amplitude is divided by $(h+1)^{\frac{1}{2}}$. The corresponding radial field, analogous to (21), is

$$\mathbf{E}^{(1)}\cdot\mathbf{e}_r = -4\gamma\left(\frac{2}{h+1}\right)^{\frac{1}{2}} \operatorname{sech}\gamma(\tilde{k}z - \tilde{\omega}t)\,\tilde{E}_1(r)\cos(\theta - \alpha)\sin(\alpha + \bar{k}z - \bar{\omega}t - 2\mu)\,.$$

This demonstrates that the signal has a fixed polarization angle $\theta = \alpha$.

Various collisions between LP and CP solitons have been computed. For two LP solitons at initial locations x_1, x_2, with peak amplitudes γ_1, γ_2, pulse speeds $2\beta_1, 2\beta_2$ and orientations α_1, α_2, appropriate initial conditions were chosen to agree at $\tau = 0$ with

$$A^+ = \left(\frac{2}{h+1}\right)^{\frac{1}{2}} \sum_{j=1,2} \gamma_j \exp -i\{\beta_j(x-x_j) - (\beta_j^2 - \gamma_j^2)\tau\} \operatorname{sech}\gamma_j(x-x_j-2\beta_j\tau)\,,$$
$$A^- = \left(\frac{2}{h+1}\right)^{\frac{1}{2}} \sum_{j=1,2} \gamma_j \exp -i\{\beta_j(x-x_j) - (\beta_j^2 - \gamma_j^2)\tau - 2\alpha_j\} \operatorname{sech}\gamma_j(x-x_j-2\beta_j\tau). \tag{24}$$

As anticipated, the linear superposition (24) of two distinct solitons well describes the solution until the pulse envelopes start to overlap. Then, after an interaction region in which the envelopes $|A^+|$ and $|A^-|$ differ substantially, it is found in all cases that pulses having the original speeds $2\beta_1, 2\beta_2$ emerge. Figure 2 shows the typical behaviour. Although the pulse with initial speed $2\beta_1 = 8$ emerges as a pulse with unchanged speed, the emergent pulse has $|A^-| \neq |A^+|$. It has symmetric profiles for both $|A^+|$ and $|A^-|$, with peak amplitudes in the ratio 1.3413:1. Similarly, the pulse with speed $2\beta_2 = -8$ emerges as a symmetric pulse with $|A^-|_{max}/|A^+|_{max} = 1.3413$.

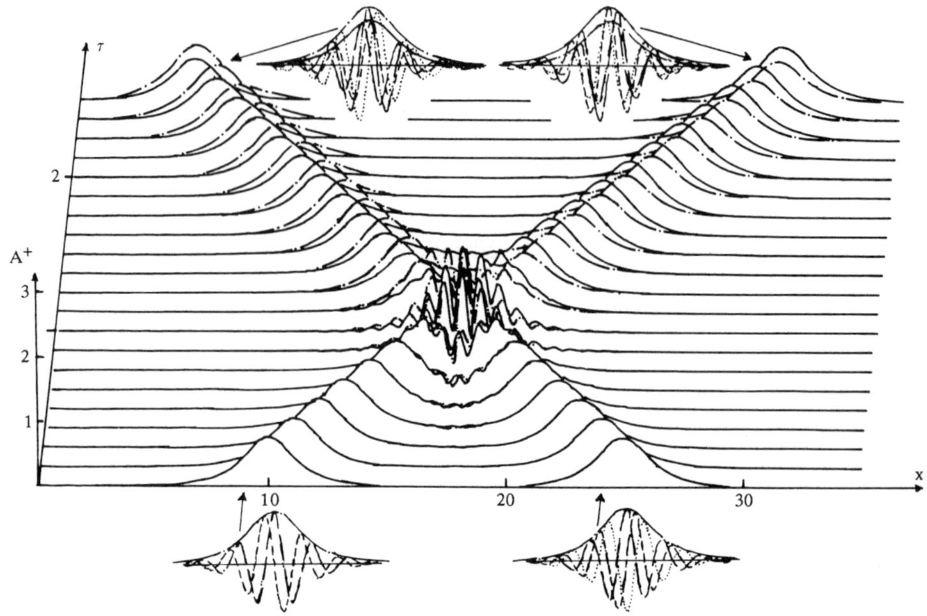

Figure 2. Colliding pulses with initial conditions (24). $\gamma_1 = \gamma_2 = 1$, $\beta_1 = -\beta_2 = 4$, $x_1 = 10$, $x_2 = 25$, $\alpha_1 = 0$, $\alpha_2 = \frac{1}{4}\pi$, $h = 2.5$. For A^+; | |——, real part— —. imaginary part– –: For A^-; | |—·—, real part—··—, imaginary part ······ .

These non-distorting pulses are similarity solutions of (16) having the forms

$$A^+ = F_1(\sigma) \exp -i(h_1\tau + \beta\sigma + \delta_1)\ , \ A^- = F_2(\sigma) \exp -i(h_2\tau + \beta\sigma + \delta_2)\ ,$$

where $\sigma \equiv x - 2\beta\tau$ and where $F_1(\sigma)$, $F_2(\sigma)$ satisfy the coupled ordinary differential equations

$$F_1''(\sigma) + (F_1^2 + hF_2^2 + \beta^2 - h_1)F_1 = 0,\ F_2''(\sigma) + (F_2^2 + hF_1^2 + \beta^2 - h_2)F_2 = 0\ .$$

Solutions with $F_1(\sigma) \equiv F_2(\sigma)$ (and $h_1 \equiv h_2$) are sech-envelopes. However, as described in [7], other choices of h_1 and h_2 allow solutions having $F_1(\sigma)$ and $F_2(\sigma)$ symmetric about $\sigma = 0$. The choice allowing $F_1(0)/F_2(0) = 1.3413$ describes the pulse emerging from the collision with speed $2\beta_1 = 8$.

Collisions between LP and CP solitons, between left- and right-handed CP solitons as well as between LP solitons with $\gamma_2 \neq \gamma_1$ and with $\beta_2 \neq -\beta_1$ have been computed. Each case has shown the emergence of two distinct pulses. Except in the cases when eqs. (16) reduce to either (19) or (23), the only cases in which the emergent pulses are indistinguishable from the incident pulses are collisions between left- and right-handed CP solitons and collisions between two LP solitons having orthogonal orientations, $|\alpha_1 - \alpha_2| = \frac{1}{2}\pi$. These cases are analogous to those in which Cadet [3] found no tendency towards "elliptic polarization".

5 Non-axisymmetric fibres

When ε, N or the cross-section $\mathcal{D}$ are not axisymmetric, the modes (6) of linear theory must be left in Cartesian form. Then, in expression (11) the electric fields have the form

$$\mathbf{E}^{(1)} = \hat{A}(\chi, Z)\, \hat{\mathbf{E}}(x, y)\, e^{i\hat{\psi}} + \check{A}(\chi, Z)\, \check{\mathbf{E}}(x, y)\, e^{i\check{\psi}} + \text{c.c.}\,, \tag{25}$$

where $\hat{\psi} = \hat{k}z - \omega t$, $\check{\psi} = \check{k}z - \omega t$ and $\hat{k}$, $\check{k}$ denote two real roots of the dispersion relation (7).

When $\hat{k} \neq \check{k}$, the $\hat{\mathbf{E}}$ and $\check{\mathbf{E}}$ modes interact significantly only if $\hat{k}'(\omega) - \check{k}'(\omega) = O(\nu)$, so that linear theory predicts pulse overlap over ranges at least as long as $O(\nu^{-2})$. Then, definitions (10) hold for χ, provided that $\hat{k}' = s + \nu\delta(\omega)$, $\check{k}' = s - \nu\delta(\omega)$ so that s is the *mean group slowness*. Various cases now arise. If the differences $|\omega/\hat{k} - \omega/\check{k}|$ and $2\nu\delta$ in the phase speeds and the group slownesses are both $O(\nu)$ quantities, there is no phase-matching over the interaction range. The modulation equations are found to be

$$\begin{aligned} i\hat{f}_1(\hat{A}_Z + \delta\hat{A}_\chi) &= \omega^2\hat{g}\hat{A}_{\chi\chi} + (\hat{f}_2|\hat{A}|^2 + f_3|\check{A}|^2)\hat{A}\,, \\ i\check{f}_1(\check{A}_Z - \delta\check{A}_\chi) &= \omega^2\check{g}\check{A}_{\chi\chi} + (\check{f}_2|\check{A}|^2 + f_3|\hat{A}|^2)\check{A}\,, \end{aligned} \tag{26}$$

where the coefficients $\hat{f}_1, \check{f}_1, \hat{g}, \check{g}, \hat{f}_2, \check{f}_2$ and f_3 are defined analogously to those in (14), except that the fields $\mathbf{E}^+$ and $\mathbf{E}^-$ are replaced by $\hat{\mathbf{E}}$ and $\check{\mathbf{E}}$, respectively.

A further assumption, that the deviation from axial symmetry is small, implies that $\hat{f}_1 - \check{f}_1 = O(\nu)$, etc. Then, neglecting these differences and defining x, τ analogously to (15) replaces eqs. (26) by

$$i(\hat{A}_\tau + 2\Delta\hat{A}_x) = \hat{A}_{xx} + (|\hat{A}|^2 + h|\check{A}|^2)\hat{A}\,,\quad i(\check{A}_\tau - 2\Delta\check{A}_x) = \check{A}_{xx} + (|\check{A}|^2 + h|\hat{A}|^2)\check{A}\,,$$

where $2\Delta = f_1\delta\omega^{-1}(f_2 g)^{-\frac{1}{2}}$. A further substitution $\hat{B} = \hat{A}\exp i(\Delta^2\tau - \Delta x)$, $\check{B} = \check{A}\exp i(\Delta^2\tau + \Delta x)$ shows that $\hat{B}$ and $\check{B}$ satisfy the pair of eqs. (16). Thus, birefringence characterized by δ merely adjusts the interpretation of the terms in (16). The pulse collision phenomena are unchanged.

If the difference between the pulse speeds, $\omega/\hat{k}$, $\omega/\check{k}$ is $O(\nu^2)$, the substitutions $\hat{k} = k + \nu^2\kappa$, $\check{k} = k - \nu^2\kappa$ show that (25) may be written as

$$\mathbf{E}^{(1)} = \{\hat{C}(\chi, Z)\,\hat{\mathbf{E}}(x, y) + \check{C}(\chi, Z)\,\check{\mathbf{E}}(x, y)\}e^{i\psi} + \text{c.c.} \;, \tag{27}$$

where $\hat{C} = \hat{A}e^{i\kappa Z}$, $\check{C} = \check{A}e^{-i\kappa Z}$. For $\mathbf{E}^{(2)}$, an expression analogous to (12) is also found. Pursuing the technique which led to eqs. (13) leads to modulation equations

$$\begin{aligned} i\hat{f}_1(\hat{C}_Z - i\kappa\hat{C} + \delta\hat{C}_\chi) &= \omega^2\hat{g}\hat{C}_{\chi\chi} + \hat{N}\,, \\ i\check{f}_1(\check{C}_Z + i\kappa\check{C} - \delta\check{C}_\chi) &= \omega^2\check{g}\check{C}_{\chi\chi} + \check{N}\,, \end{aligned} \tag{28}$$

in which $\hat{N}, \check{N}$ represent expressions of third degree in $\hat{C}, \check{C}, \hat{C}^*$ and $\check{C}^*$. When the functions $\varepsilon(x, y)$, $N(x, y)$ and cross-section $\mathcal{D}$ are arbitrary, $\hat{N}$ and $\check{N}$ each contain six different terms. However, if ε, N and $\mathcal{D}$ all have reflectional symmetry in one, or both, of the Ox or Oy axes, the expressions for $\hat{N}$ and $\check{N}$ reduce to

$$\hat{N} = \hat{f}_2|\hat{C}|^2\hat{C} + f_3|\check{C}|^2\hat{C} + f_4\check{C}^2\hat{C}^*\,,\ \hat{N} = \check{f}_2|\check{C}|^2\check{C} + f_3|\hat{C}|^2\check{C} + f_4\hat{C}^2\check{C}^*\,,$$

with $\hat{f}_2, \check{f}_2$ and f_3 as in (26). Although this allows eqs. (28) to be reduced to the system

$$\begin{aligned} i\hat{B}_Z &= \tilde{p}\hat{B}_{\chi\chi} + \hat{f}_1^{-1}\{\hat{f}_2|\hat{B}|^2\hat{B} + f_3|\check{B}|\check{B}^2\hat{B} + f_4{}^2\hat{B}^*\,e^{-2i(\tilde{\theta}-\check{\theta})}\}\,, \\ i\check{B}_Z &= \check{p}\check{B}_{\chi\chi} + \check{f}_1^{-1}\{\check{f}_2|\check{B}|^2\check{B} + f_3|\hat{B}|^2\check{B} + f_4\hat{B}^2\check{B}^*e^{2i(\tilde{\theta}-\theta)}\}\,, \end{aligned} \tag{29}$$

by the sequence of substitutions $\hat{B} = \hat{A}\exp -i(\hat{\beta}\chi - \hat{p}\hat{\beta}^2 Z) = \hat{C}e^{-i\tilde{\theta}}$, $\check{B} = \check{A}\exp i(\check{\beta}\chi + \check{p}\check{\beta}^2 Z) = \check{C}e^{-i\check{\theta}}$, where

$$\hat{\beta} = \tfrac{1}{2}\omega\delta/\hat{p},\ \check{\beta} = \tfrac{1}{2}\omega\delta/\check{p},\ \hat{p} = \omega^2\hat{g}/\hat{f}_1,\ \check{p} = \omega^2\check{g}/\check{f}_1\,,$$

the resulting system (29) is usually more general than system (1). When the difference

$2\nu\delta$ in the slownesses is (like $2\nu^2\kappa$) an $O(\nu^2)$ quantity, the terms $\delta, \hat{\beta}, \check{\beta}$ may be neglected, so that phase $2(\hat{\theta} - \check{\theta})$ reduces to $4\kappa Z$. This applies, for example, when deviations from axial symmetry are small. Since, in this case, differences $\hat{f}_1 - \check{f}_1$, etc., may also be neglected, eqs. (29) may then be non-dimensionalized as

$$i\hat{B}_\tau = \hat{B}_{xx} + (h+1)\,|\hat{B}\,|^2\hat{B} + 2\,|\check{B}\,|^2\,\hat{B} + (H-1)\check{B}^2\hat{B}^* e^{-4i\kappa Z}\,,$$

$$i\check{B}_\tau = \check{B}_{xx} + (h+1)\,|\check{B}\,|^2\,\check{B} + 2\,|\hat{B}\,|^2\,\check{B} + (H-1)\hat{B}^2\check{B}^* e^{4i\kappa Z}\,. \qquad (30)$$

This has the form considered by Blow, Doran and Wood (eq. (1) of [2]) for birefringent fibres, when $H = h$ (i.e., $f_2 = f_3 + f_4$).

A separate simplification of eqs. (28) arises when ε, N and $\mathcal{D}$ have the *symmetry of the square* (reflectional symmetry in each of the lines $x = 0$, $y = 0$ and $y = \pm x$). In this case, $k = \check{k}$, $\hat{f}_1 = \check{f}_1$, etc., so that $\kappa \equiv 0$, $\delta \equiv 0$. Without any assumption that deviations from axial symmetry are small, the modulation equations may be reduced to form (30) with $\kappa \equiv 0$. In the case $H \equiv h$, this new system is equivalent to eqs. (16) rewritten in terms of the amplitudes $\hat{B} = \frac{1}{2}i(A^+ + A^-)$, $\check{B} = \frac{1}{2}(A^- - A^+)$ of linearly polarized modes in an axisymmetric fibre. Consequently, for $H = h$ (again $f_2 = f_3 + f_4$), pulse collisions behave exactly as for the axisymmetric fibre. More generally, the system possesses sech-envelope solutions only in the cases

(i) $\check{B} = 0$, or $\hat{B} = 0$: orientation along Ox, or Oy, axes;

(ii) $\check{B} = \pm\hat{B}$: orientation along 45° lines, $y = \pm x$;

(iii) $\check{B} = \pm i\hat{B}$: left- or right-handed rotation.

Since the parameter $H - h$ measures the effective deviation from axial symmetry, it would be of interest to investigate how the behaviour of collisions between solitons of the three types (i)-(iii) depends on $H - h$.

References

[1] Bespalov, V.I. and Talanov, V.I., *JETP Lett.* **3**, 307 (1966).
[2] Blow, K.J., Doran, N.J. and Wood, D., *Optics Lett.* **12**, 202 (1987).
[3] Cadet, S., *J. de Physique* **C3**, 21 (1989).
[4] Marcuse, D., *Light Transmission Optics* (Van Nostrand, 1972).
[5] Menyuk, C.R., *IEEE Jnl. Quantum Elect.* **23**, 174 (1987).

[6] Parker, D.F., in preparation.
[7] Parker, D.F. and Newboult, G.K., *J. de Physique* **C3**, 137 (1989).
[8] Remoissenet, M., contribution to workshop 'Integrable systems and applications', Ile d'Oléron (1988).
[9] Zakharov, V.E. and Schulman, E.I., *Physica* **4D**, 270 (1982).

Parker D.F.
Department of Mathematics,
University of Edinburgh,
Edinburgh EH9 3JZ
U.K.

Former Address:
Department of Theoretical Mechanics,
University of Nottingham,
Nottingham, NG7 2RD
U.K.

M. PEYRARD, T. DAUXOIS AND A.R. BISHOP

Dynamics of the thermal denaturation of DNA

1 Introduction

Although the *static* structure of DNA has been known for a long time, the idea that the *dynamics* of the double helix is also essential in the basic mechanisms of life is more recent. Modern biology has shown that the molecule is permanently agitated by large amplitude motions which are of physiological importance [3]. Fast local fluctuations can be large enough to allow small molecules to intercalate between the bases that connect the two strands of the double helix. This phenomenon has been proposed as a possible cause for cancer. It is also well established that DNA transcription, i.e., the reading of the genetic code during the life of a cell, involves a local denaturation of the molecule during which the two strands separate locally. The denaturation is activated by an enzyme but the origin of the energy required to break the hydrogen bonds connecting the base pairs is not yet understood. At physiological temperature, simple thermal fluctuations would be far from being sufficient.

The study of thermal denaturation of DNA is a preliminary step for understanding the more complex transcription. In the last few years the idea that non-linear excitations could play a role in the process has become increasingly popular. Englander et al. [2] suggested first a theory of soliton excitations as an explanation of the open states of DNA and then various non-linear models were proposed [17, 16, 18].

However, although all these models exhibit solitary-wave solutions, the *formation* of these non-linear excitations had not been investigated. In this paper, we examine this question in the framework of a model proposed originally to explain the hydrogen bond melting suggested by the normal mode analysis of infrared or Raman experiments [13, 5, 4, 7].

We treat intrinsically the non-linearities associated with the large motions involved in the denaturation. A transfer integral technique is used to investigate the statistical mechanics of the model and we determine the inter-strand separation in the double helix as a function of temperature. The results are then discussed in terms of non-linear excitations and we show that a spontaneous localization of energy is possible in the model. This would explain the rather low denaturation temperature. The results are tested with numerical simulations.

2 The model and its statistical mechanics

For each base pair, the model includes two degrees of freedom u_n and v_n which correspond to the displacements of the bases from their equilibrium positions along the direction of the hydrogen bonds that connect the two strands. Following the previous investigations on DNA [5, 4], the potential for the hydrogen bonds is approximated by a Morse potential. An harmonic coupling due to the stacking is assumed between neighbouring bases so that the Hamiltonian of the model is

$$H = \sum_n \frac{1}{2} m \left(\dot{u}_n^2 + \dot{v}_n^2\right) + \frac{1}{2} k \left[(u_n - u_{n-1})^2 + (v_n - v_{n-1})^2\right] + V(u_n - v_n) \qquad (1)$$

with

$$V(u_n - v_n) = D\left\{\exp\left[-a(u_n - v_n)\right] - 1\right\}^2 . \qquad (2)$$

In a first approach, the inhomogeneities due to the base sequence and the asymmetry of the two strands are neglected so that we use a common mass m for the bases and the same coupling constant k along the two strands. The Morse potential V is an average potential representing the 2 or 3 bonds which connect the two bases in a pair. It can be estimated from the parameters obtained for the individual bonds by the analysis of the vibrational modes of DNA [4].

With these assumptions, the motions of the two strands can be described in terms of the variables

$$x_n = \frac{1}{\sqrt{2}}(u_n + v_n), \text{ and } y_n = \frac{1}{\sqrt{2}}(u_n - v_n) \qquad (3)$$

which represent the in-phase and out-of-phase motions respectively. Only the out-of-phase displacements y_n stretch the hydrogen bonds. The Hamiltonian (1) becomes

$$H = H(x) + H(y) = \sum_n \left\{\frac{p_n^2}{2m} + \frac{1}{2} k (x_n - x_{n-1})^2\right\}$$

$$+ \sum_n \left\{\frac{q_n^2}{2m} + \frac{1}{2} k (y_n - y_{n-1})^2 + D\left[\exp(-a\sqrt{2} y_n) - 1\right]^2\right\}, \qquad (4)$$

where $p_n = m\dot{x}_n$ and $q_n = m\dot{y}_n$. The classical partition function of a chain containing N base pairs

$$Z = \int_{-\infty}^{+\infty} \prod_{n=1}^{N} \left(dx_n dy_n dp_n dq_n e^{-\beta H(p_n, x_n, q_n, y_n)}\right)$$

factors into different parts

$$Z = Z_p\, Z_x\, Z_q\, Z_y\,. \tag{6}$$

The two momentum parts are readily integrated and, since the motions in x and y are decoupled, Z_x is simply the contribution of the potential energy in an harmonic chain. As the coupling involves only nearest neighbours interactions, Z_y can be expressed in the form

$$Z_y = \int_{-\infty}^{+\infty} \prod_{n=1}^{N} dy_n e^{-\beta f(y_n, y_{n-1})}\,, \tag{7}$$

where f denotes the potential energy part in $H(y)$. This integral can be evaluated exactly in the limit of large systems using the eigenfunctions and eigenvalues of a transfer integral operator [15, 8]

$$\int dy_{n-1} e^{-\beta f(y_n, y_{n-1})}\, \varphi_i(y_{n-1}) = e^{-\beta \varepsilon_i}\, \varphi_i(y_n). \tag{8}$$

The calculation is similar to the one performed by Krumhansl and Schrieffer [8] for the statistical mechanics of the Φ^4 field. It yields

$$Z_y = e^{-N\beta\varepsilon_0}, \tag{9}$$

where ε_0 is the lowest eigenvalue of a Schrödinger-like equation which determines the eigenfunctions of the transfer integral operator (8)

$$-\frac{1}{2\beta^2 k}\frac{\partial^2 \varphi_i}{\partial y^2} + D\left(e^{-4ay} - 2e^{-2ay}\right)\varphi_i(y) = (\varepsilon_i - s_0 - D)\,\varphi_i(y)\,, \tag{10}$$

with $s_0 = (1/2\beta)\ln(\beta k/2\pi)$.

Equation (10) is formally identical to the Schrödinger equation for a particle in a Morse potential, so that it can be solved exactly [10, 1]. This equation has a discrete spectrum when $d = (\beta/a)\sqrt{k\,D} > 1/2$ and the eigenvalue and the normalized eigenfunction for the ground state are then

$$\varepsilon_0 = \frac{1}{2\beta}\ln\left(\frac{\beta k}{2\pi}\right) + \frac{\alpha}{\beta}\sqrt{\frac{D}{k}} - \frac{a^2}{4\beta^2 k} \tag{11}$$

$$\varphi_0(y) = (\sqrt{2}a)^{1/2}\,\frac{(2d)^{d-1/2}}{\sqrt{\Gamma(2d-1)}}\exp\left(-\,de^{-\sqrt{2}ay}\right)\exp\left[-\,(d-1/2)\sqrt{2}ay\right]. \tag{12}$$

These results can be used to compute the mean stretching $\bar{y}_m$ of the hydrogen bonds. It is given by

$$\bar{y}_m = \frac{1}{Z} \int \prod_{n=1}^{N} y_m e^{-\beta H} \, dx_n dy_n dp_n dq_n \, ,$$

but, since H separates into a sum over the x, y, p, q variables, $\bar{y}_m$ reduces to

$$\bar{y}_m = \frac{1}{Z_y} \int \prod_{n=1}^{N} y_m e^{-\beta f(y_n, y_{n-1})} dy_n \, . \tag{13}$$

As the model is assumed to be homogeneous, the result does not depend on the particular site considered. The integral can again be calculated with the transfer integral method [15] and yields

$$\bar{y} = \frac{\sum_{i=1}^{N} \langle \varphi_i(y) \mid y \mid \varphi_i(y) \rangle e^{-N \beta \varepsilon_i}}{\sum_{i=1}^{N} \langle \varphi_i(y) \mid \varphi_i(y) \rangle e^{-N \beta \varepsilon_i}} . \tag{14}$$

In the limit of large N, the result is again dominated by the lowest eigenvalue ε_0 so that $\bar{y}$ is given by

$$\bar{y} = \langle \varphi_0(y) \mid y \mid \varphi_0(y) \rangle = \int \varphi_0^2(y) y \, dy \tag{15}$$

for a normalized eigenfunction $\varphi_0(y)$. This integral has been evaluated numerically with the expression (12) of $\varphi_0(y)$. The results are shown in Figure 1 for three values of the coupling constant k. In the calculation, we have used $D = 0.33$ eV and $a = 1.8 = \text{Å}^{-1}$ which corresponds to mean values for the $N - H \cdots N$ and $N - H \cdots O$ bonds in the $A - T$ and $G - C$ base pairs [4].

This figure shows a rapid increase of $\bar{y}$ around a particular temperature which is a characteristic of DNA denaturation as observed for instance by measuring its absorbance of ultraviolet light at 260 nm [3]. The denaturation temperature is indeed sensitive to the parameters of the hydrogen bonds which bind the two strands, but Figure 1 shows that it is also very sensitive to the intra-strand interaction constant, a parameter which is not so well determined by experimental results. As k increases the denaturation temperature increases. This is similar to the results obtained with the plane base-rotator model and consistent with the increase observed experimentally in the presence of reagents that increase the hydrophobic interactions [3]. Our results indicate that, in the absence of inhomogeneities, k must be of the order of 3.0×10^{-3} eV/Å^2

to get a reasonable denaturation temperature. This value characterizes a weak coupling between the transverse base motions in DNA which means that discreteness effects have to be taken into account in the models describing DNA denaturation [2, 17, 16, 18].

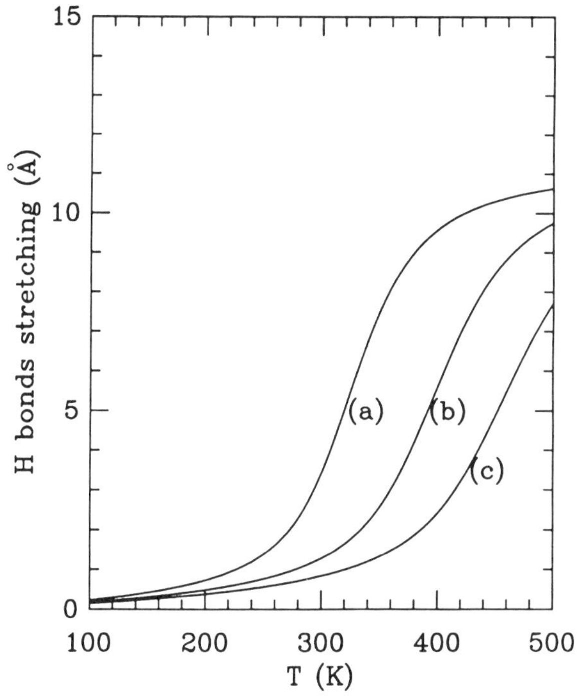

Figure 1. Variation of $\bar{y}$ as a function of temperature for three values of the coupling constant k : (a) $k = 2.0\ 10^{-3}$ eV/Å^2, (b) $k = 3.0\ 10^{-3}$ eV/Å^2 and (c) $k = 4.0\ 10^{-3}$ eV/Å^2.

3 A possible mechanism for denaturation

Although it shows that denaturation can occur at temperatures well below the temperature corresponding to the breaking of a single hydrogen bond, the calculation of $\bar{y}$ does not indicate *how* it occurs. To explore this aspect, it is interesting to relate our model to those involving non-linear excitations. Since the non-linearities appear in terms of the variable y, let us consider the equations of motion which derive from $H(y)$ (Eq. (4))

$$m \frac{\partial^2 y_n}{\partial t^2} = k(y_{n+1} + y_{n-1} - 2y_n) - 2\sqrt{2}Dae^{-\sqrt{2}ay_n}\left(e^{-\sqrt{2}ay_n} - 1\right) = 0\,. \qquad (16)$$

The phenomena precursor to denaturation can be investigated by expanding eq. (16) for small y as

$$m \frac{\partial^2 y_n}{\partial t^2} = k(y_{n+1} + y_{n-1} - 2y_n) + 4Da^2 y_n - 6\sqrt{2}Da^3 y_n^2 + \frac{28}{3} Da^4 y_n^3 = 0 . \quad (17)$$

A solution of this equation can be obtained with a multiple-scale expansion [6] under the form

$$y_n = F_1(X_n, \tau)e^{i\theta_n} + \text{c.c.} + \varepsilon[F_0(X_n, \tau) + F_2(X_n, \tau)e^{2i\theta_n} + \text{c.c.}] \quad (18)$$

with $X_n = \varepsilon x = \varepsilon nl$ and $\tau = \varepsilon t$, l being the lattice spacing i.e. the mean distance between adjacent bases. In this expression y is written as a modulated wave in which the carrier wave $\exp(i\theta_n) = \exp[i(qnl - \omega t)]$, with ω and q related by the dispersion relation of the lattice, includes the discreteness effect while the modulation factor is treated in the continuum limit (the so called *semi-discrete* approximation [12]). However, for a qualitative understanding of the denaturation, we can restrict ourselves to a continuum approximation for the carrier wave as well, which simplifies significantly the calculations. The d.c. and first harmonic terms in (18), F_0 and F_2, are necessary because eq. (17) contains an even power of y. The multiple-scale expansion yields a non-linear Schrödinger (NLS) equation for $F_1(X, \tau)$ [14]

$$iF_{1s} + PF_{1ZZ} + Q|F_1|^2 F_1 = 0 \quad (19)$$

with $Z = X - V_g T$, $s = \varepsilon\tau$ and

$$V_g = \frac{d\omega}{dq} = \frac{q}{\omega}\frac{k^2 l^2}{m}, \quad P = \frac{(k^2 l^2/m) - V_g^2}{2\omega}, \quad Q = \frac{4}{3}\frac{Da^4}{\omega}. \quad (20)$$

The solitary waves in y that result from this equation are the breathing modes suggested by the infrared and Raman experiments [13].

The statistical mechanics of the NLS equation has been investigated recently by Lebowitz, Rose and Speer [9]. They show that the system can develop singularities in a finite time. Such singularities which correspond to self-focusing phenomena in plasmas may well be responsible for DNA denaturation because they occur when the L^2 norm of the field

$$N(F_1) = \int_0^L |F_1|^2 \, dx ,$$

(where L denotes the size of the system) is such that $N\beta$ exceeds some threshold. In our calculation, $N(F_1)$ is given at temperature T by $\overline{y^2}$ which can be calculated similarly to $\bar{y}$ and is given by an expression analogous to eq. (15):

$$\overline{y^2} = \langle \varphi_0(y) \mid y^2 \mid \varphi_0(y) \rangle = \int \varphi_0^2(y) y^2 dy \,. \tag{21}$$

The calculation shows that $\overline{y^2}$ rises by several orders of magnitude in a small temperature range around DNA denaturation so that the threshold for energy localization could be reached. This suggests that the denaturation "bubble" which is observed experimentally at the beginning of the denaturation process could be created by *energy localization due to non-linear effects*. It is however difficult to provide a more quantitative analysis of this phenomenon within the framework of the present model because the coefficients of the NLS derived in the limit of small y depend on the frequency of the carrier wave introduced in the expansion. Although this frequency must lie within the frequencies of the phononmodes in the lattice, it is not well defined for the thermal fluctuations that we consider here.

Thus in order to check *the possible existence of an energy localization mechanism in DNA* we have used numerical simulations of the equations of motions which derive from the Hamiltonian (1). In a typical calculation the molecule contains 128 or 256 cells with periodic boundary conditions. The equations are solved with a fourth-order Runge Kutta scheme with a time step chosen to preserve energy to an accuracy better than 0.1% during the whole calculation. In the initial state the particles are in their equilibrium positions with a Gaussian distribution of velocities. The initial condition contains only kinetic energy and, assuming that equipartition of energy is roughly valid to define the temperature in the system, the variance of the distribution is set to $\sqrt{2k_BT/m}$. We start from a rather low temperature (≈ 100 K) and the simulation is run until the system reaches an equilibrium. Then the system is "heated" for a few time steps by multiplying the velocities by a factor slightly larger than unity and we wait until the new equilibrium is reached. The same process is repeated up to a complete denaturation of the molecule. The mean hydrogen bond stretching $\bar{y}$ and the displacements correlation function are monitored during the computation. In addition to the fast increase of $\bar{y}$ when the denaturation temperature is reached, the results show also an increase of the correlation length in the system. More important for our purpose, *these calculations confirm the existence of an energy localization mechanism* before denaturation. Instead of being roughly uniform in the system, the denaturation starts on some specific sites characterized by larger thermal fluctuations. Then it grows around these sites while the remaining of the molecules shows low hydrogen bonds stretching. This creates the "bubbles" that are observed experimentally when DNA is denaturated. A typical result close to denaturation temperature is shown in Figure 2.

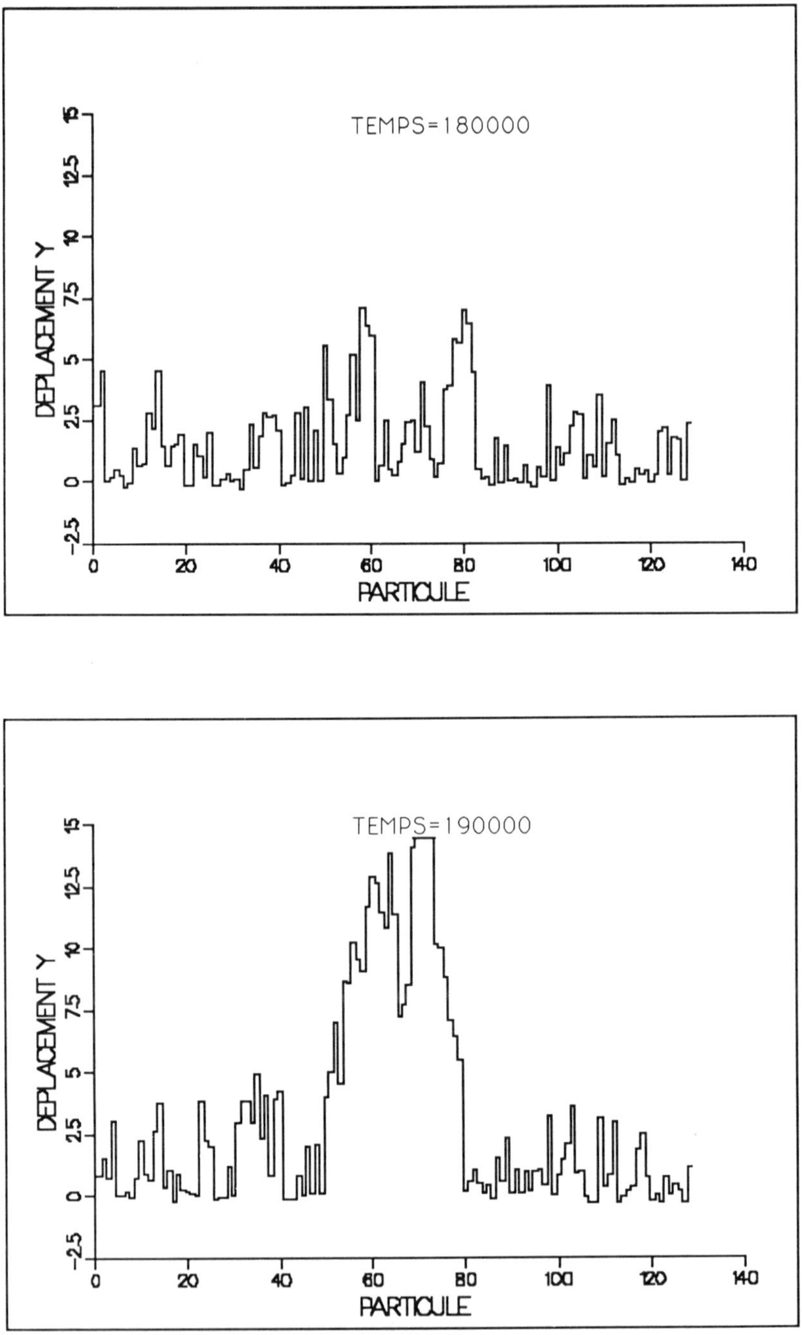

Figure 2. Hydrogen bond stretching along the DNA chain close to denaturation as shown by the numerical simulation of the dynamics of the molecule. Note the formation of denaturation "bubbles" while other regions exhibit only a small stretching.

Then, as the temperature is increased further, the "bubbles" grow and finally the whole molecule is denaturated. This process, shown by the simulations, is in agreement with the experimental observations.

Besides the role of energy localization, another important feature that emerges from our results is the *role of discreteness* in the dynamics of DNA. The model gives a reasonable denaturation temperature only if the coupling between adjacent bases is weak so that the energy can localize itself in regions containing only one or two base pairs to initiate the opening.

In spite of its extreme simplicity compared to the real structure of DNA, the non-linear model that we have investigated exhibits properties which are close to the experimental observations. However, it must certainly be complicated to provide more inhomogeneities which are an essential feature of DNA because they contain the genetic code and must not be ignored. They can act as extrinsic nucleation centres for the energy localization mechanism and significantly lower the denaturation temperature. This effect, which has been observed in the numerical simulations of a different model [11] will be investigated in future works.

Acknowledgements

One of us (M.P.) would like to thank D.K. Campbell for his hospitality in the Center for Nonlinear Studies of the Los Alamos National Laboratory, where part of this work has been performed. We would like to thank H.A. Rose, A. Garcia, M. Remoissenet and C. Reiss for helpful discussions.

References

[1] Dana Brabson, G., *J. of Chemical Education* **50**, 397 (1973).

[2] Englander, S.W., Kallenbach, N.R., Heeger, A.J., Krumhansl, J.A. and Litwin, S., *Proc. Nat. Acad. Sci. U.S.A* **777**, 7222 (1980).

[3] Freifelder, D., *Molecular Biology*, Jones and Bartlett, Boston 1987.

[4] Gao, Y., Devi-Prasad, K.V. and Prohofsky, E.W., *J. Chem. Phys.* **8O**, 6291 (1984).

[5] Gao, Y. and Prohofsky, E.W., *J. Chem. Phys.* **80**, 2242 (1984).

[6] Kaup, D.J. and Newell, A.C., *Phys. Rev.* **B18**, 5162 (1978).

[7] Kim, Y., Devi-Prasad, K.V. and Prohofsky, E.W., *Phys. Rev.* **B32**, 5185 (1985).

[8] Krumhansl, J.A. and Schrieffer, J.R., *Phys. Rev.* **B11**, 3535 (1975).

[9] Lebowitz, J.L., Rose, H.A. and Speer, E.R., *J. Stat. Phys.* **50**, 657 (1988).

[10] Morse, P.M., *Phys. Rev* **1**, 57 (1929).

[11] Peyrard, M. and Bishop, A.R., in Proceedings of the 6th Interdisiplinary Workshop on Nonlinear Coherent Structures in Physics, Mechanics and Biological Systems (1989), M. Bartes, J. Leon eds, Springer (1990).

[12] Pnevmatikos, St., Flytzanis, N. and Remoissenet, M., *Phys. Rev.* **33**, 2308 (1986).

[13] Prohofsky, E.W., Lu, K.C., Van Zandt, L.L. and Putnam, B.E., *Physics Letters* **70A**, 492 (1979).

[14] Remoissenet, M., *Phys. Rev.* **B33**, 2386 (1986).

[15] Scalapino, D.J., Sears, M. and Ferrel, R.A., *Phys. Rev.* **B6**, 3409 (1972).

[16] Takeno, S. and Homma, S., *Prog. Theor. Phys.* **77**, 548 (1987).

[17] Yomosa, S., *Phys. Rev.* **A27**, 2120 (1983) and **A30**, 474 (1984).

[18] Zhang, Chun-Ting, *Phys. Rev.* **A35**, 886 (1987).

Peyrard M. and Dauxois Th.
Physique non linéaire: Ondes
et Structures Cohérents (OSC)
Université de Bourgogne
6 Boulevard Gabriel
F-21000 Dijon
FRANCE

Bishop A.R.
Theoretical Division and Center
for Nonlinear Studies
Los Alamos National Laboratory
Los Alamos, New Mexico 87545
U.S.A.

M. PLANAT

Observation of acoustic envelope solitary waves generated by metallic interdigital transducers on a quartz crystal

1 Introduction

Instabilities resulting from the combined action of non-linearity and dispersion in wave propagation is a subject of growing interest in mathematical physics and engineering [13, 12]. Amplitude and frequency modulational instability, pulse compression and splitting, Fermi Pasta Ulam (FPU) recurrence and solitons have already been observed in hydrodynamics [2], electrical transmission lines [7], plasma physics [5], and optical waveguides [8]. In addition to longitudinal instabilities (in one-dimensional systems), transverse instabilities (in two or three-dimensional systems) have also been studied in plasma and laser physics [14]. No such behaviour has yet been seen in solid state acoustics at room temperature.

In this work we show that the bulk acoustic wave radiation from interdigital transducers deposited on a quartz crystal plate manifests longitudinal and transverse instabilities characteristic of non-linear dispersive waves. Pulse envelope modulation, envelope splitting and compression are observed when the incident pulse energy is increased by varying frequency, input pulse length or amplitude. This behaviour is in rough agreement with a theoretical model based on non-linear geometrical optics [6].

2 First remarks on the experimental setup

A schematic of the experimental setup is shown in Figure 1. We used a Y-cut quartz substrate (10×9 mm^2 and thickness 1 mm) with two interdigital transducers deposited on the top normal to the Z-axis. Each transducer had 100 periods, a 4.5 mm width, a spatial periodicity $\lambda_0/2 = 17.2\,\mu$m and 800 Å thick aluminium. In the linear regime such a cut supports only shear horizontal motion [16, 9]. Since the excitation transducer is very long, propagation takes place with the wave number k along the direction of constructive interference θ_c given by the relation

$$k_1 = 2\pi/\lambda_0 = k \cos \theta_c \,. \tag{1}$$

In other words the propagation angle θ_c can be scanned by the applied frequency f according to the relation

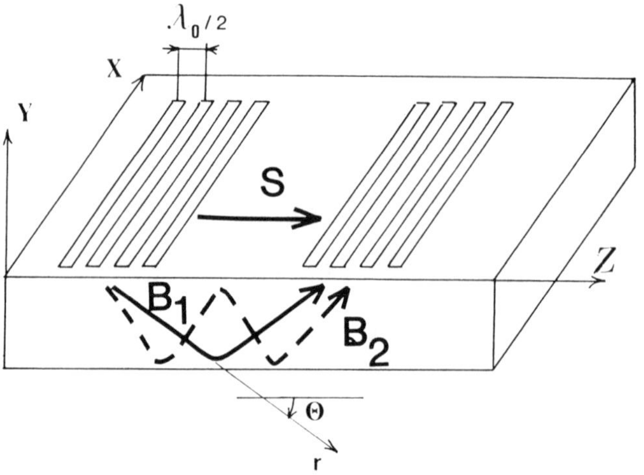

Figure 1. Schematic for the experimental setup.

$$V_c = f\lambda_0 \cos\theta_c . \tag{2}$$

For a monochromatic wave, the angle θ_0 for the propagation of energy is normal to the slowness curve and consequently there will be a cut-off in the device response when $\theta_0 = 0$. This angle corresponds to the propagation of the so-called surface skimming bulk wave (SSBW). In our device this happens for $\theta_c = 24.2°$ and the corresponding cut-off frequency is $f_0 = 126$ MHz. Radiation from finite length transducers can be studied asymptotically at large distances using the linear geometrical optics approximation [16]. First experiments were performed in a continuous wave excitation regime [9]. Then pulse experiments have revealed a well-resolved structure (see Section 7) and will be introduced in the next paragraph.

3 Non-linear dispersion relation

The slowness curve for transverse waves has the form

$$\rho\,\omega^2 = C_{55}\,k_1^2 + C_{44}\,k_2^2 + 2C_{45}\,k_1\,k_2 . \tag{3}$$

For a Y-Z cut of quartz, $C_{44} = 4 \times 10^{10}$ N/m², $C_{55} = 5.8 \times 10^{10}$ N/m² and $C_{45} = -1.8 \times 10^{10}$ N/m². If we assume a very long transducer, the horizontal wave number k_1 can be considered constant; introducing the wave number k along the propagation direction the corresponding dispersion relation $\omega = W(k)$ takes the form

$$\rho\,\omega^2 = C_{55}\,k_1^2 + C_{44}(k^2 - k_1^2) + 2C_{45}\,k_1(k^2 - k_1^2)^{1/2} . \tag{4}$$

It should be observed that dispersion arises due to the anisotropy. For an isotropic substrate there is no dispersion.

An estimate of non-linear propagation effects has been obtained from amplitude frequency measurements in resonators. For a travelling plane wave of amplitude a, the frequency shift is found to be [10]

$$\Delta\omega = k_0\,\Delta V = -\,q\,a^2 \ \text{ where } \ q = \frac{1}{16}\omega_0\,k_0^2\,\frac{C^{(4)}}{C^{(2)}}, \tag{5}$$

where $C^{(2)}$ and $C^{(4)}$ are the effective second-order and fourth-order elastic constants. In our device the wave propagates in singly rotated orientations (rotation around X-axis) and third-order constants vanish by symmetry. The non-linear coefficient $C = C^{(4)}/16\,C^{(2)}$ is known to have an order of magnitude $C \simeq 10$ and is < 0 for AT-cuts and > 0 for BT-cuts [3].

4 Propagation equation for the pulse envelope

From (3) and (5) we obtain the non-linear dispersion relation in Stokes form

$$\omega = W(k) - q\,a^2\,. \tag{6}$$

For an input pulse propagating along the r direction and centred on the wave number k_0 corresponding to the frequency $\omega_0 = W(k_0)$, the wave function has the form

$$\Psi(\mathbf{x}, t) = \Phi(\mathbf{x}, t)\,\exp\{i(k_0 r - \omega_0 t)\}\,, \tag{7}$$

where $\mathbf{x}$ is an arbitrary spatial vector.

If the complex amplitude $\Phi(\mathbf{x}, t)$ varies fairly slowly, the wave remains quasi-monochromatic; then the wave function can be expanded to first order around the carrier and the wave envelope can be shown to obey the eq. [6]

$$i\left\{\frac{\partial\Phi}{\partial t} + \omega_0'\,\frac{\partial\Phi}{\partial r}\right\} + \frac{1}{2}\,\omega_0''\,\frac{\partial^2\Phi}{\partial r^2} + \frac{\omega_0'}{2k_0}\,\frac{\partial^2\Phi}{\partial s^2} + q\,|\Phi|^2\,\Phi = 0\,, \tag{8}$$

where ω'_0 and ω''_0 are respectively the linear group velocity and the concavity of dispersion curve evaluated for the carrier wave number k_0 and s is the transverse coordinate. The first term inside brackets represents the wave propagation in the main direction r with the group velocity ω'_0. The last three terms represent, respectively, effects of dispersion, diffraction and non-linearity. If we neglect the non-linear term and the dispersion term, we obtain the so-called parabolic equation of the approximate diffraction theory.

Equation (8) without the longitudinal r derivatives (respectively without s derivative) is in the form of the non-linear Schrödinger equation. If $q\omega'_0 > 0$, transverse modulations are unstable and the medium is self-focusing (self-channelling); in the other case it is defocusing. If $q\omega''_0 > 0$, longitudinal modulations become unstable and the so-called self-modulation (modulational instability) takes place; in the other case envelope modulations remain stable. In the general case where both dispersion and diffraction effects have the same order of magnitude both phenomena can play a role and the character of propagation can be quite complicated. These effects will be examined now.

5 Automodulation effects

We first examine the case of longitudinal modulations [9]. They can be studied by assuming a solution with a uniform spatial state of amplitude a plus an infinitesimal propagating perturbation of frequency Ω and wave number K [1]. Substituting this form for the solution into (8) and neglecting transverse effects yields the following dispersion relation

$$\Omega = K \left| \frac{1}{4} \omega''^2_0 K^2 - q\, \omega''_0 a^2 \right|^{1/2}. \tag{9}$$

For $q\omega''_0 a^2 > \omega''^2_0 K^2/4 > 0$ the frequency will become imaginary and the perturbation in amplitude will grow exponentially in time.

Self-modulation of waves arises in this case. The maximum growth rate occurs when $K = K_{max} = [2qa^2/\omega''_0]^{1/2}$ with the value $\Omega = \Omega_{max} = qa^2$. Assuming that the wave packet propagates approximately with the linear group velocity ω'_0 this allows us to introduce the distance $x_{n\ell}$ for maximum instability and the corresponding modulating frequency f_m at which the growth rate is maximum [1]

$$x_{n\ell} = \omega'_0/qa^2 \text{ and } f_m = \frac{\omega'_0 K_{max}}{2\pi}. \tag{10}$$

A second glance to modulational instabiity can be given by observing that (8) (without transverse effects) admits a family of localized solutions of the form [13, 4]

$$\Phi(r, t) = \Phi_1 \operatorname{sech} \left(\frac{t - t_0 - x/v_g}{\tau_0} \right) \exp i(r_0 x - s_0 t)\,, \tag{11}$$

where τ_0 is the pulse half width, r_0 and s_0 are wave number and frequency shifts, $v_g = \omega'_0 + \omega''_0 r$ is the pulse velocity and Φ_1 is the pulse amplitude given by

$$\Phi_1^2 = \omega''_0/\tau_0^2 v_g^2 q\,. \tag{12}$$

For large values of the input pulse amplitude and length, higher-order effects are to be expected including modulational instabiity, narrowing, splitting and spatial recurrence of wave packets [8, 15, 11].

6 Autofocusing effects

Analysis of transverse instabilities is performed by using eq. (8) without r derivatives. Self-focusing effects arise when $\omega'_0 q > 0$. Equation (8) has a solution in the form of a stationary waveguide

$$\Phi(s, t) = U_2 \operatorname{sech} \left\{ U_2 \sqrt{\nu}\, | s - s_0 | \right\} \exp \left\{ i\nu\, U_2^2 \frac{\omega'_0}{2k_0} (t - t_0) \right\}, \qquad (13)$$

where $\nu = C\,\omega_0\, k_0^3/\omega'_0$. The half-width of this plane beam is $1/U_2\sqrt{\nu}$. Since the transducer of length L, radiating in the direction θ_c such that $k_1 = k \cos \theta_c$ (see eq. 1) approximately determines this width Δs, focusing will appear in the form of a soliton [4] when the amplitude U_2 is such that $\Delta s = 1/U_2\sqrt{\nu}$. We obtain

$$U_2 = \frac{2}{L} \sqrt{\omega'_0 / C\, \omega_0 k_0 (k_0^2 - k_1^2)}\,. \qquad (14)$$

For higher amplitudes, we expect that the wave will decay into several plane-parallel channels. This self-channelling process is analogous to the self-modulation process described earlier.

7 Experimental instabilities and solitons for the Y-Z cut substrate

The slowness curve for transverse waves in the Y-Z cut substrate was plotted in Ref. [9]. The dispersion curve is given in Figure 2. The critical distance $x_{n\ell}$ and critical amplitudes U_1 and U_2 for longitudinal and transverse instabilities are given in Figure 3.

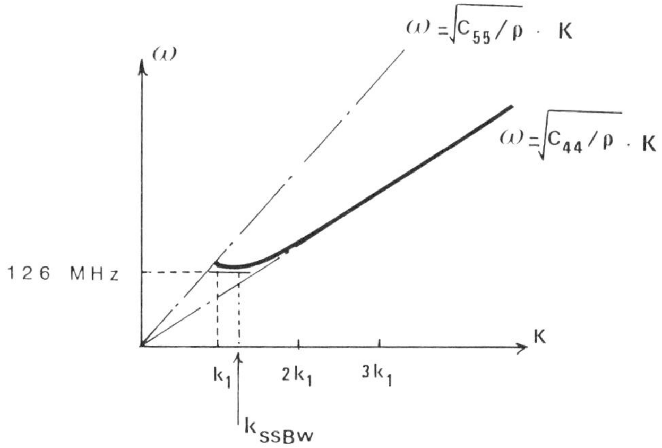

Figure 2. Dispersion curves for waves under a long transducer on a Y-Z cut quartz substrate.

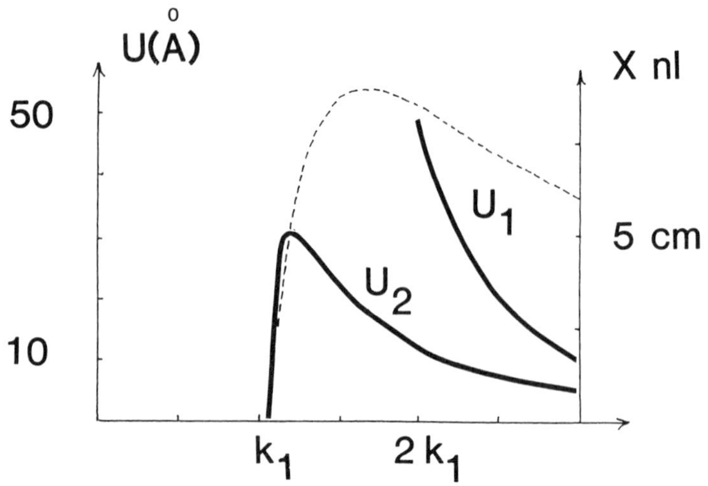

Figure 3. Distance $x_{n\ell}$ of maximum longitudinal instability amplitude U_2 of the transverse soliton U_1 for L = 4.5 mm and of the longitudinal soliton for a = 100 Å and τ = 100 ns versus carrier wave number.

By varying the applied wave number from $k_{SSBW} \simeq 1.1\ k_1$ to $3k_1$, corresponding to carrier frequencies between the cut-off frequency 126 MHz to about 350 MHz the wave propagates in a range of angles θ_c corresponding to the so-called BT cuts. It is then reasonable to expect that q wil be positive in this region [3]. Since $\omega'_0 q > 0$ *and* $\omega''_0 q > 0$ the medium is both autofocusing and automodulating.

The strongest effect is to be expected close to the cut-off frequency. In this region group velocity ω'_0 is zero and according to eqs. (11) and (15) maximum self-focusing and self-modulation occurs. By increasing slightly the applied frequency the wave should demodulate and defocus quickly (Figure 3). At about 130 MHz the threshold U_2 for the fundamental transverse soliton reaches its maximum value (30 Å) while the corresponding value for the longitudinal soliton is considerably higher. At still higher frequencies self-focusing again becomes significant and longitudinal instabilities are also expected at high amplitudes and input lengths.

To check these expectations, incident pulses with 70 ns length and 10 V amplitude were fed into the transducer (Figure 4). Considerable instabilities are obtained near the cut-off frequency 126 MHz. At 137 MHz the wave has defocused enough to present a main part in the response; at the same time envelope (longitudinal) instability gradually diminishes. At 150 MHz the response shows two main parts in time corresponding to the idea of two bounces (Figure 1). Modulational instability has become very slight (Figure 3). At 170 MHz self-focusing increases again and four bounces are visible in the response. Each of them has a stable envelope. At 270 MHz, more and more bounces are seen and each of them again becomes unstable due to longitudinal modulational instability.

In a second set of experiments incident pulses with fixed frequency 270 MHz and variable length τ (Figure 5) and amplitude v (Figure 6) were used. Plots were taken

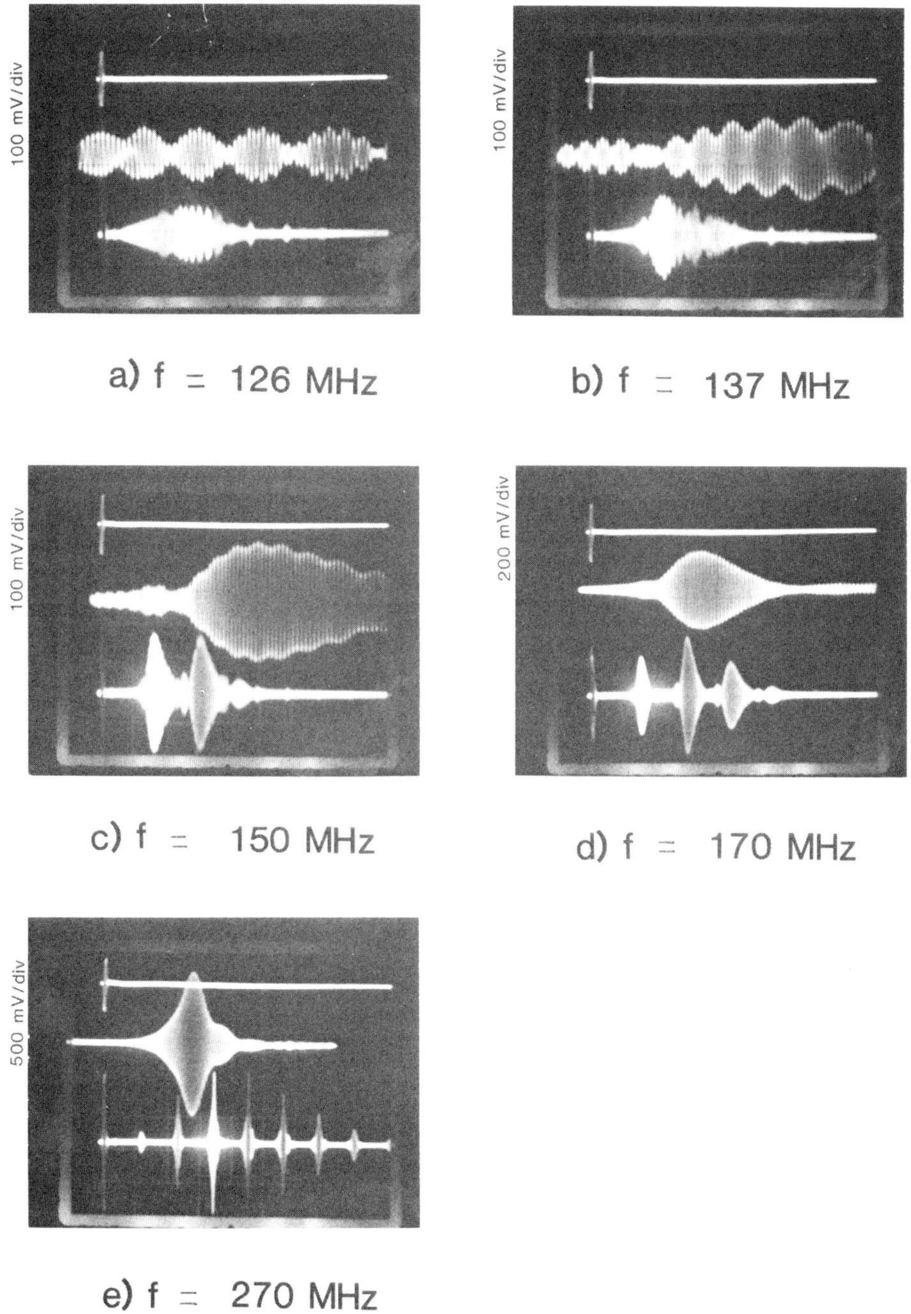

Figure 4. Instabilities in the radiation from an interdigital transducer on a Y–Z cut substrate: input pulse length: τ = 70 ns; input pulse amplitude: τ = 10 V; response amplification: 45 dB; main time base: 500 ns/div; delayed time base: 50 ns/div.

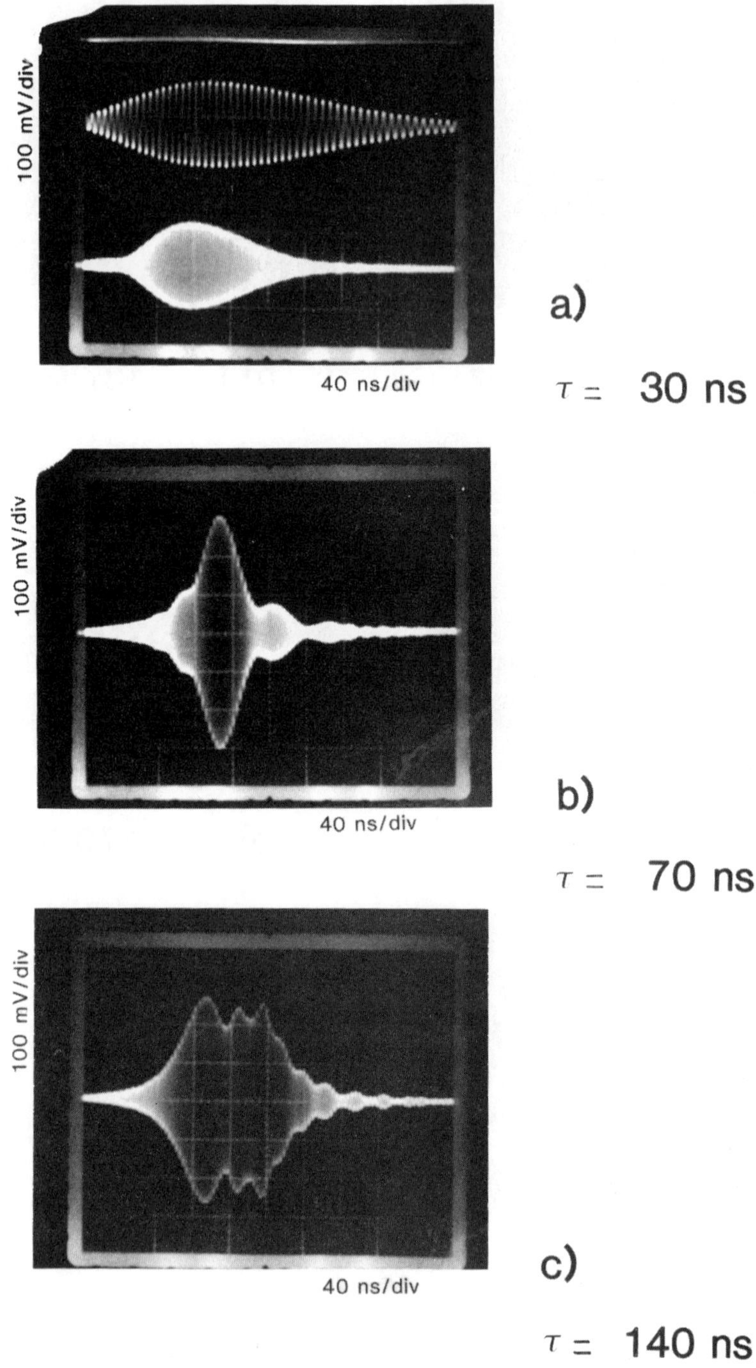

Figure 5. Dependence of input length on the shape of the pulse: input frequency: 270 MHz; input amplitude: 20 V.

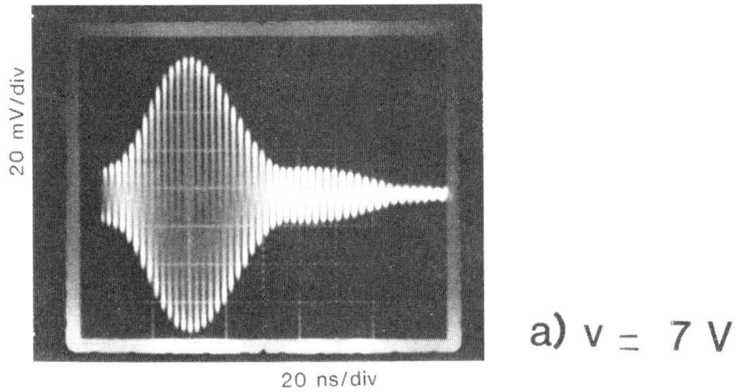

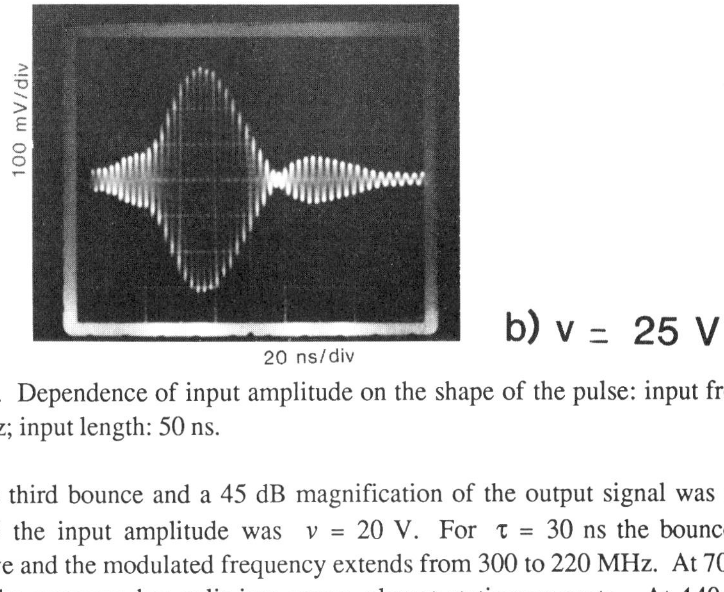

Figure 6. Dependence of input amplitude on the shape of the pulse: input frequency: 270 MHz; input length: 50 ns.

from the third bounce and a 45 dB magnification of the output signal was used. In Figure 5 the input amplitude was v = 20 V. For τ = 30 ns the bounce is very dispersive and the modulated frequency extends from 300 to 220 MHz. At 70 ns input length, the response has split into many, almost stationary parts. At 140 ns input length, strong modulational instability is seen. In Figure 6 the input pulse length was τ = 50 ns. At 7 V input amplitude the response pulse is extended over a broad time window, the main part has an 60 ns half-width. The shape of the envelope mainly results from the autofocusing effect. At 25 V input amplitude the response pulse has split into two stationary parts at frequencies 260 MHz and 240 MHz and the main pulse has been slightly narrowed to about 50 ns. Both autofocusing and automodulation effects contribute.

In a third set of experiments an incident pulse with a 10 V amplitude, a 70 ns input length and a 210 MHz carrier frequency was used. The spectrum of time response is given in Figure 7. As observed, a mode spectrum has been stimulated. By increasing frequency, modes get thinner and progressively disappear.

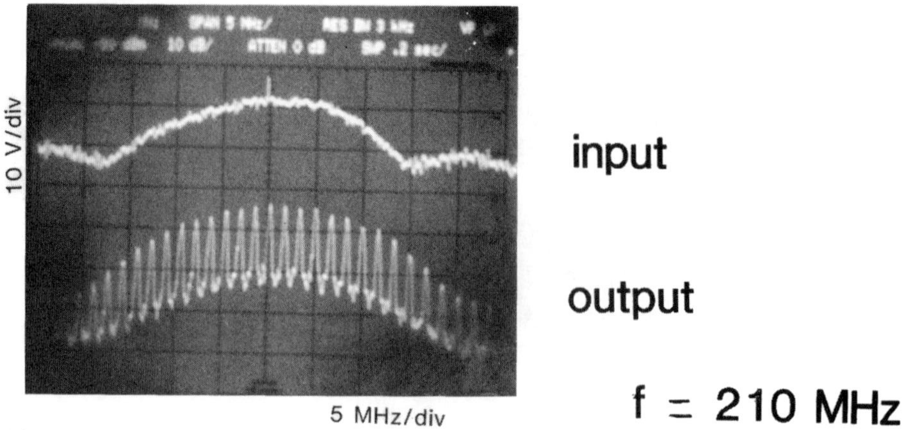

Figure 7. Stimulated mode spectrum in pulse experiments: input pulse length: 70 ns; input pulse amplitude: 10 V.

8 Conclusion and perspectives

Longitudinal and transverse acoustic instabilities in the transverse wave radiation from finite length transducers on a quartz crystal have been studied. This was shown to result from the combined action of anisotropy induced dispersion, diffraction and non-linearity.

Two important aspects were neglected. First, theory was performed for incident waves while experiments also include reflected waves. Experiments will be performed later with a detection transducer at the bottom of the plate.

Second, the time response exhibits two parts. The acoustic part propagates at the velocity of sound and was studied in this paper; the electromagnetic part was also observed to have an interesting non-linear structure. A useful concept to study for both aspects is that of effective permittivity function and of polariton. Due to the lack of space such aspects will be considered in a future paper.

References

[1] Ewen, J.F., Gunshor, R.L. and Weston, V.H., *J. Appl. Phys.* **53**, 5682 (1982).

[2] Feir, J.E., *Proc. Roy. Soc. London*, **A299**, 54 (1967).

[3] Gagnepain, J.J. and Besson, R., Nonlinear effects in piezoelectric quartz crystals, in *Physical Acoustics*, vol. XI, Academic Press, New York, 1975.

[4] Hasegawa, A. and Tappert, F., *Appl. Phys. Lett.* **23**, 142 (1973).

[5] Ikezi, H., Wojtowicz, S.S., Woltz, R.E., de Cramie, J.S. and Baker, D.R., *J. Appl. Phys.* **64**, 3277 (1988).

[6] Karpman, V.I., in *Nonlinear Waves in Dispersive Media*, Pergamon Press, Oxford, 1975.

[7] Lonngren, K. and Scott, A., eds. *Solitons in Action*, (Academic Press, New York, 1978).

[8] Mollenauer, L.F., Stolen, R.H. and Gordon, J.P., *Phys. Rev. Lett.* **45**, 1095 (1980).

[9] Planat, M. and Hoummady, M., *Appl. Phys. Lett.* **55**, 103 (1989).

[10] Planat, M., Théobald, G. and Gagnepain, J.J., *L'Onde Electrique* **60** (8-9) and **60** (11), (1980).

[11] Satsuma, J. and Yajima, N., *Suppl. Prog. Theor. Phys.* **55**, 284 (1974).

[12] Scott, A.C., Chu, F.Y.F. and McLaughlin, D.W., *Proc. IEEE* **61**, 1443 (1973).

[13] Witham, G.B., *Linear and Nonlinear Waves*, John Wiley, New York, 1974.

[14] Zakharov, V.E. and Rubenchik, A.M. *Sov. Phys. JETP*, **38**, 494 (1974).

[15] Zakharov, V.E. and Shabat, A.B., *Sov. Phys. JETP* **34**, 62 (1972).

[16] Zhang, Y. and Planat, M., *Elect. Lett.* **32**, 68 (1987).

Planat M.
L.P.M.O.-C.N.R.S.,
32 avenue de l'Observatoire
F-2500 Besançon
FRANCE

H. ZORSKI

One-dimensional chain of dipoles

We propose to examine in this paper a classical problem of one-dimensional chain of dipoles with nearest-neighbour interaction. It turns out that this chain has a number of interesting properties and for certain interaction energies there are soliton solutions.

1 General equations

The dipole will here be regarded as a pair of charged material points (Figure 1) with the constraint

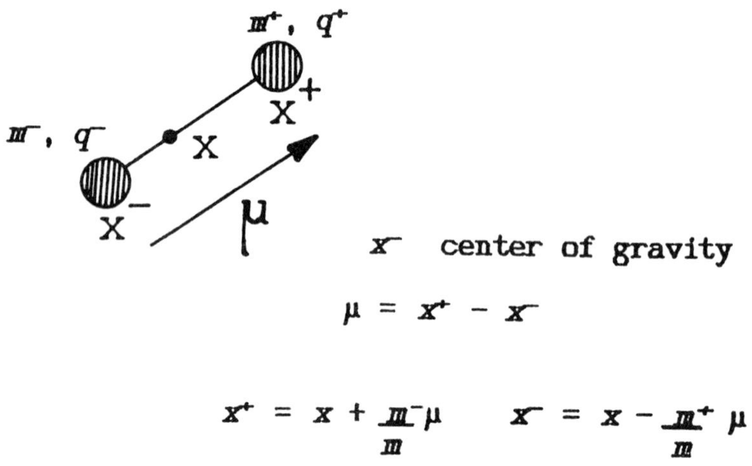

Figure 1.

$$|x^+ - x^-| = |\mu| = l(t) \tag{1.1}$$

and in this paper $l(t) = \text{const.} = l_0$. The equations of motion have the form

$$\begin{aligned} m^+ \ddot{x}^+ &= F(x^+) + \lambda(x^+ - x^-), \ F(x^+) = q^+E(x^+), \\ m^- \ddot{x}^- &= F(x^-) + \lambda(x^+ - x^-), \ F(x^-) = q^-E(x^-), \end{aligned} \tag{1.2}$$

where $E(x)$ is the external electric field and λ is the Lagrangian multiplier; these equations are equivalent to the system of translational and rotational equations, derived by eliminating λ:

$$
\begin{gathered}
m\ddot{x} = F(x^+) + F(x^-), \\
\mu \times \left[\ddot{\mu} - \left(\frac{F(x^+)}{m^+} - \frac{F(x^-)}{m^-}\right)\right] = 0.
\end{gathered}
\tag{1.3}
$$

Consider the force due to a dipole μ_y at y, acting on a particle q^+ at x^+, with the potential φ assumed later to be central

$$
\begin{aligned}
F(\mu_y \to x^+) &= -(q^+)^2\, \nabla_{x^+} \left[\varphi(y^+ - x^+) + \frac{q^-}{q^+}\,\varphi(y^- - x^+)\right] \\
&= -\frac{(q^+)^2 m}{m^-}\nabla_{\mu_x}\left[\varphi(y^+ - x^+) + \frac{q^-}{q^+}\,\varphi(y^- - x^+)\right]
\end{aligned}
\tag{1.4'}
$$

and similarly

$$
\begin{aligned}
F(\mu_y \to x^-) &= \frac{q^+ q^- m}{m^+}\nabla_{\mu_x}\left[\varphi(y^+ - x^-) + \frac{q^-}{q^+}\,\varphi(y^- - x^-)\right] = \\
&= -q^+ q^- \nabla_{x^-}\left[\varphi(y^+ - x^-) + \frac{q^-}{q^+}\,\varphi(y^- - x^-)\right].
\end{aligned}
\tag{1.4''}
$$

Thus, denoting by $I = \dfrac{m^- m^+}{m}$ the moment of inertia of the dipole divided by l_0^2 we obtain the equations of motion of the dipole at x in the form

$$
\begin{gathered}
m\ddot{x} = -\nabla_x \sum_y \varphi^{\mathrm{T}}_{y,x}, \\
\mu_x \times \left(\ddot{\mu}_x + \nabla_{\mu_x}\frac{1}{I}\sum_y \varphi^{\mathrm{T}}_{y,x}\right) = 0,
\end{gathered}
\tag{1.5}
$$

where

$$\varphi^{\mathrm{T}}_{y,x} = (q^{+})^2 \left\{ \varphi(y^{+} - x^{+}) + \frac{\overset{-}{q}}{\overset{+}{q}} \left[\varphi(y^{-} - x^{+}) + \varphi(y^{+} - x^{-}) \right] + \right.$$

$$\left. + \left(\frac{\overset{-}{q}}{\overset{+}{q}}\right)^2 \varphi(y^{-} - x^{-}) \right\} \tag{1.6}$$

The above equations contain (in the argument of φ) x^{+} and x^{-}; we recall that

$$x^{+} = x + \frac{\overset{-}{m}}{m}\mu_x \,, \quad x^{-} = x - \frac{\overset{+}{m}}{m}\mu_x$$

where x is the position of the centre of gravity and we introduce the displacement u_x and u_y, i.e., $y - x = R + (\mu_y - \mu_x)$ where $R = Y - X$; hence

$$y^{+} - x^{+} = R + (u_y - u_x) + \frac{\overset{-}{m}}{m}(\mu_y - \mu_x) \,,$$

$$y^{+} - x^{-} = R + (u_y - u_x) + \left(\frac{\overset{-}{m}}{m}\mu_y + \frac{\overset{+}{m}}{m}\mu_x \right),$$

$$y^{-} - x^{+} = R + (u_y - u_x) - \left(\frac{\overset{+}{m}}{m}\mu_y + \frac{\overset{-}{m}}{m}\mu_x \right),$$

$$y^{-} - x^{-} = R + (u_y - u_x) - \frac{\overset{+}{m}}{m}(\mu_y - \mu_x) \,.$$

Denoting $\Delta u = u_y - u_x$ we expand $\varphi(y^{+} - x^{+})$, etc., into Taylor series neglecting terms of the order higher than quadratic in both $|\Delta u|$ and l_0 and we set $\nabla_{\mathrm{R}}\varphi(R) = 0$ anticipating elimination of these terms in equilibrium:

$$\varphi(y^{+} - x^{+}) = \varphi(R) + \frac{1}{2}\left[\Delta u \Delta u + 2\,\frac{\overset{-}{m}}{m}\Delta u(\mu_y - \mu_x) + \left(\frac{\overset{-}{m}}{m}\right)^2 (\mu_y - \mu_x)^2 \right] : \nabla_{\mathrm{R}}\nabla_{\mathrm{R}}\varphi,$$

$$\varphi(y^{+} - x^{-}) = \varphi(R) + \frac{1}{2}\left[\Delta u \Delta u + 2\Delta u\left(\frac{\overset{-}{m}}{m}\,\mu_y + \frac{\overset{+}{m}}{m}\mu_x\right) + \left(\frac{\overset{-}{m}}{m}\mu_y + \frac{\overset{+}{m}}{m}\mu_x\right)^2\right] : \nabla_{\mathrm{R}}\nabla_{\mathrm{R}}\varphi,$$

$$\varphi(y^{-} - x^{+}) = \varphi(R) + \frac{1}{2}\left[\Delta u \Delta u - 2\Delta u\left(\frac{\overset{+}{m}}{m}\mu_y + \frac{\overset{-}{m}}{m}\mu_x\right) + \left(\frac{\overset{+}{m}}{m}\,\mu_y + \frac{\overset{-}{m}}{m}\mu_x\right)^2\right] : \nabla_{\mathrm{R}}\nabla_{\mathrm{R}}\varphi,$$

$$\varphi(y^- - x^-) = \varphi(R) + \frac{1}{2}\Big[\Delta u \Delta u - 2\Delta u \frac{\overset{+}{m}}{m}(\mu_y - \mu_x) + \Big(\frac{\overset{+}{m}}{m}\Big)^2 (\mu_y - \mu_x)^2\Big] : \nabla_R \nabla_R \varphi \,.$$

Thus, the interaction energy between the dipoles μ_x and μ_y takes the form (we omit the constant term)

$$\varphi^{\mathrm{T}}_{y,x} = (q^+)^2 \Big\{ K_0^2 \frac{1}{2} \Delta u \Delta u + K_0 K_1 \frac{\overset{-}{m}}{m} \Delta u(\mu_y - \mu_x)$$

$$+ \frac{1}{2}\Big(\frac{\overset{-}{m}}{m}\Big)^2 \big[K_0 K_2(\mu_x\mu_x + \mu_y\mu_y) - K_1^2(\mu_x\mu_y + \mu_y\mu_{xy})\big]\Big\} : \nabla_R \nabla_R \varphi \,, \qquad (1.7)$$

where

$$K_0 = 1 + \frac{\overset{-}{q}}{\overset{+}{q}}, \quad K_1 = 1 - \frac{\overset{-}{q}}{\overset{+}{q}} \frac{\overset{+}{m}}{\overset{-}{m}}, \quad K_2 = 1 + \frac{\overset{-}{q}}{\overset{+}{q}} \Big(\frac{\overset{+}{m}}{\overset{-}{m}}\Big)^2 .$$

We note that for the neutral dipoles $K_0 = 0$ and

$$\varphi^{\mathrm{T}}_{y,x} = -(q^+)^2 \, \mu_x \mu_y : \nabla_R \nabla_R \varphi \,. \qquad (1.8)$$

We now proceed to examine a one-dimensional chain of dipoles with nearest-neighbour interaction; setting $x = x_n$, $y = x_{n+1}$ and $y = x_{n-1}$ we have

$$\sum_y \varphi^{\mathrm{T}}_{y,x} = \varphi^{\mathrm{T}}_{n+1,n} + \varphi^{\mathrm{T}}_{n,\,n-1}$$

and the equations of motion take the form

$$m\,\ddot{x}_n + \nabla_{x_n}\big(\varphi^{\mathrm{T}}_{n+1,n} + \varphi^{\mathrm{T}}_{n,\,n-1}\big) = 0, \text{ or } m\,\ddot{u}_n + \nabla_{u_n}\big(\varphi^{\mathrm{T}}_{n+1,n} + \varphi^{\mathrm{T}}_{n,\,n-1}\big) = 0\,,$$

$$\mu_n \times \big[I\ddot{\mu}_n + \nabla_{\mu_n}\big(\varphi^{\mathrm{T}}_{n+1,n} + \varphi^{\mathrm{T}}_{n,\,n-1}\big) = 0 \,. \qquad (1.9)$$

Substituting from (1.7) we obtain

$$\varphi^{\mathrm{T}}_{n+1,n} + \varphi^{\mathrm{T}}_{n,\,n-1} = \varphi^{(u)}_n + \varphi^{(\mu)}_n + \varphi^{(\mu,\,u)}_n \,,$$

where

$$\varphi_n^{(u)} = \frac{1}{2}(q^+)^2 K_0 \left[(u_{n+1} - u_n)^2 + (u_n - u_{n+1})^2\right] : \nabla_R \nabla_R \varphi ,$$

$$\varphi_n^{(\mu)} = \frac{1}{2}(q^+)^2 \left(\frac{\bar{m}}{m}\right)^2 \Big[K_0 K_2(\mu_{n+1}\mu_{n+1} + \mu_{n-1}\mu_{n-1} + 2\mu_n)$$

$$- 2K_1^2 \mu_n(\mu_{n+1} + \mu_{n-1})\Big] : \nabla_R \nabla_R \varphi ,$$

$$\varphi_n^{(\mu, u)} = (q^+)^2 \frac{\bar{m}}{m} K_0 K_1 \Big[(u_{n+1} - u_n)(\mu_{n+1} - \mu_n)$$

$$+ (u_n - u_{n-1})(\mu_n - \mu_{n-1})\Big] : \nabla_R \nabla_R \varphi ,$$

are the translation, rotation and mixed energies, respectively. Thus

$$\ddot{u}_n - \omega_{(1)}^2 C\cdot(\Delta^2 u_n + \alpha \Delta^2 \mu_n) = 0 ,$$

$$\mu_n \times \left[\ddot{\mu}_n - \frac{1}{\alpha}\omega_{(2)}^2 C\cdot\cdot(\Delta^2 u_n + \alpha\Delta^2\mu_n + \beta\mu_n)\right] = 0, \ |\mu_n| = l_0 , \tag{1.10}$$

where $\Delta^2 u_n = u_{n+1} + u_{n-1} - 2u_n$ is the second finite difference,

$$\omega_{(1)}^2 = \frac{(q^+)^2 K_0^2}{m} , \ \omega_{(2)} = \left(\frac{\bar{m}}{m}\right)^2 \frac{K_1^2 (q)^{+2}}{I} , \ \alpha = \frac{\bar{m}}{m}\frac{K_1}{K} ,$$

$$\beta = 2\left(1 - \frac{K_0 K_2}{K_1}\right)\alpha .$$

$$C = \nabla_R \nabla_R \varphi = \frac{\varphi'(|R|)}{|R|}\left(\delta_{ij} - \frac{R_i R_j}{|R|^2}\right) + \frac{R_i R_j}{|R|^2}\varphi''(|R|) ,$$

(for the considered central potential); in our case, assuming that the chain is situated on the x_1-axis we have

$$C_{11} = \varphi''(|R|), \ C_{22} = \frac{\varphi'(|R|)}{|R|} = C_{33}$$

and the remaining components of the above tensor vanish. For the three-dimensional Coulomb potential

$$\varphi(|R|) = \frac{\varphi_0}{|R|},\ C_{11} = \frac{2\varphi_0}{|R|^3},\ C_{22} = C_{33} = -\frac{\varphi_0}{|R|^3}$$

i.e., $C_{11} > 0,\ C_{22} = C_{33} < 0$, while for the two-dimensional potential

$$\varphi(|R|) = \varphi_0\,(|R|)\ln\frac{|R_0|}{|R|},\ C_{11} = \frac{\varphi_0}{|R|^2},\ C_{22} = -\frac{\varphi_0}{|R|^2}.$$

We confine ourselves hereafter to the two-dimensional displacement $u_n = (u_n, v_n)$ and the two-dimensional dipole $(\mu_n{}^1, \mu_n{}^2) = l_0\,(\cos\theta_n, \sin\theta_n)$ but generally the potential can be either three- or two-dimensional; in the second case we imagine an infinite row of chains in the third direction. The equations of motion in the two-dimensional case have the form

$$\ddot{u}_n - \omega_{(1)}^2 C_{11}\,\Delta^2(u_n + \alpha\mu_n^1) = 0\,,$$

$$\ddot{v}_n - \omega_{(2)}^2 C_{22}\,\Delta^2(v_n + \alpha\mu_n^2) = 0\,,$$

$$(\mu_n^1\mu_n^2 - \mu_n^1\mu_n^2) - \frac{\beta}{\alpha}\,\omega_{(2)}^2\,(C_{22} - C_{11})\mu_n^1\,\mu_n^2 -$$

$$-\frac{1}{\alpha}\left(\mu_n^1\,\ddot{v}_n - \frac{\omega_{(2)}^2}{\omega_{(2)}^2}\mu_n^2\,\ddot{\mu}_n\right) = 0\,. \tag{1.11}$$

A more detailed investigation of the above system will be presented elsewhere; now we note the following facts: (i) since for the typical potential $\varphi(|R|)$, $C_{22} < 0$ the second eq. (1.11) in the non-polar case ($l_0 = 0$) reduces in the continuum limit to the Laplace rather than the wave equation. This is due to the fact that there are no reactions from above or below; (ii) The static solution is determined by the condition $\mu_n^1\mu_n^2 = 0$ (the displacements are then found from the first two equations), i.e., at every point the dipole is either vertical or horizontal; (iii) There is an exact solution of the system with $\ddot{\theta} = 0$:

$$\mu_n^1 = l_0\cos(kn - \Omega t),\ \mu_n^2 = l_0\sin(kn - \Omega t)\,,$$

$$u_n = a\cos(kn - \Omega t),\ v_n = b\ \sin(kn - \Omega t)\,, \tag{1.12}$$

with a dispersion equation quadratic in Ω^2. There are (as well as for the solutions of the linearized system) stop bands in the relation $\Omega(\kappa)$, i.e., regions for which Ω is imaginary, for some realistic values of the parameters in (1.11).

2 The one-dimensional continuum

The transition to the continuum in (1.11) is carried out by replacing Δ^2 by $a^2 \dfrac{\partial^2}{\partial x^2}$.

Consider first neutral dipoles; then $K_0 = 0$, $\ddot{u} = \ddot{v} = 0$ and the third equation (for $C_{12} = C_{21} = 0$) takes the form

$$(\mu_1 \ddot{\mu}_2 - \ddot{\mu}_1 \mu_2) - c_{(2)}^2 \Big[(C_{22} \mu_1 \mu_2'' - C_{11} \mu_1'' \mu_2) + \\ + \frac{2}{a^2}(C_{22} - C_{11})\mu_1 \mu_2\Big] = 0, \qquad (2.1)$$

where $c_{(2)}^2 = a^2 \omega_{(2)}^2$. Setting $\mu_1 = l_0 \cos\theta$, $\mu_2 = l_0 \sin\theta$ we arrive at an equation for θ:

$$\ddot{\theta} - \frac{1}{2} c_{(2)}^2 \Big\{ \Big[(C_{11} + C_{22}) + (C_{22} - C_{11}) \cos 2\theta \Big] \theta'' + \\ + \frac{2}{a^2}(C_{22} - C_{11})\,(1 - \frac{a^2}{2}\theta'^2) \sin 2\theta \Big\} = 0. \qquad (2.2)$$

For

$$\varphi = \frac{\varphi_0}{|R|} = \frac{\varphi_0}{a},$$

$$\ddot{\theta} + \frac{3\varphi_0 c_{(2)}^2}{2a^3} \Big[\Big(\cos 2\theta - \frac{1}{3}\Big)\theta'' + \frac{2}{a^2}\Big(1 - \frac{a^2}{2}\theta'^2\Big) \sin 2\theta \Big] = 0$$

while for $\varphi = \varphi_0 \, ln \dfrac{|R_0|}{|R|}$,

$$\ddot{\theta} + \frac{\varphi_0 c_{(2)}^2}{a^3} \Big[\cos 2\theta \, \theta'' + \frac{2}{a^2}\Big(1 - \frac{a^2}{2}\theta'^2\Big) \sin 2\theta \Big] = 0.$$

We note that both equations change the type, the results however are essentially different for the Lennard-Jones potential $\varphi = \varphi_0\left[\left(\frac{\sigma}{|R|}\right)^{12} - \left(\frac{\sigma}{|R|}\right)^{6}\right]$. Now (we recall that all calculations are performed at the equilibrium point $\nabla_R\varphi = 0$)

$$C_{11} + C_{22} = \frac{33\varphi_0}{|R|}, \quad C_{22} - C_{11} = -\frac{18\varphi_0}{|R|},$$

and therefore the equation is hyperbolic. Introducing the new angle $\beta = 2\theta$ and the variables $\bar{x} = \frac{2}{a}x$, $\bar{t} = \left(\frac{6\,\varphi_0 c^2_{(2)}}{a^5}\right)^{1/2} t$ we can write eq. (2.3) in the form

$$\ddot{\beta} + \left(\cos\beta - \frac{1}{3}\right)\beta'' + \left(1 - \frac{1}{2}\beta'^2\right)\sin\beta = 0, \tag{2.4}$$

where now the differentation takes place with respect to $\bar{x}$ and $\bar{t}$. Multiplying this equation in turn by $\dot{\beta}$ and β' we arrive at the conservation laws for the energy and field momentum

$$\dot{H} + R' = 0, \quad \dot{H}_* + R'_* = 0,$$

where

$$H = \frac{1}{2}\dot{\beta}^2 + (1 - \cos\beta) - \frac{1}{2}\beta'^2\left(\cos\beta - \frac{1}{3}\right), \quad R = \dot{\beta}\beta'\left(\cos\beta - \frac{1}{3}\right),$$

$$H_* = \dot{\beta}\beta', \quad R_* = \frac{1}{2}\beta'^2\left(\cos\beta - \frac{1}{3}\right) + \left(1 - \cos\beta\right) - \frac{1}{2}\dot{\beta}^2.$$

The Hamiltonian H differs from that for a pendulum by the term

$$-\frac{1}{2}\beta'^2\left(\cos\beta - \frac{1}{3}\right).$$

There are two constant solutions of interest, namely $\beta_0 = 0$ and $\beta_0 = \pi$; the linearized equation has the form $(\beta = \beta_0 + \varphi)$

$$\ddot{\varphi} + \left(\cos\beta_0 - \frac{1}{3}\right)\varphi'' + \cos\beta_0\varphi = 0 \tag{2.5}$$

and is hyperbolic for $\beta_0 = \pi$. The dispersion equations are the following: for $\beta_0 = 0$, $\omega^2 = 1 - \frac{2}{3}k^2$ whereas for $\beta_0 = \pi$, $\omega^2 = -1 + \frac{4}{3}k^2$.

In the coupled case, for $C_{12} = C_{21} = 0$ we have the system

$$\ddot{u} - c_{(1)}^2 C_{11}\left[u'' + \alpha l_0(\cos\theta)''\right] = 0,$$

$$\ddot{v} - c_{(1)}^2 C_{22}\left[v'' + \alpha l_0(\sin\theta)''\right] = 0,$$

$$\ddot{\theta} - \frac{1}{2}c_{(2)}^2 \Big\{\left[(C_{22}+C_{11}) + (C_{22}-C_{11})\cos 2\theta\right]\theta'' + \tag{2.6}$$

$$+\frac{\beta}{\alpha a^2}(C_{22}-C_{11})\left(1 - \frac{\alpha a^2}{\beta}\theta'^2\right)\sin 2\theta + \frac{2}{\alpha l_0}\left(C_{22}\cos\theta\, v'' - C_{11}\sin\theta\, u''\right)\Big\} = 0,$$

and for $\varphi = \frac{\varphi_0}{|\beta|}$, $\beta = 2\theta$

$$\ddot{u} - \gamma_{(1)}^2\left[u'' + \alpha l_0\left(\cos\frac{\beta}{2}\right)''\right] = 0,$$

$$\ddot{v} + \frac{1}{2}\gamma_{(1)}^2\left[v'' + \alpha l_0\left(\sin\frac{\beta}{2}\right)''\right] = 0,$$

$$\ddot{\beta} + \gamma^2\left[\left(\cos\beta - \frac{1}{3}\right)\beta'' + \left(\eta - \frac{1}{2}\beta'^2\right)\sin\beta\right. \tag{2.7}$$

$$\left. + \rho\left(\cos\frac{\beta}{2}v'' + 2\sin\frac{\beta}{2}u''\right)\right] = 0,$$

where

$$\gamma_{(1)}^2 = \frac{2\varphi_0 c_{(1)}^2}{a^3},\ \eta = -\frac{4}{a^2}\left(\frac{q^-}{q^+}\right)\left(\frac{m}{m^-}\right)^2 K_1^{-2},\ \rho = \frac{4}{3\alpha l_0}(\eta > 0, \rho < 0).$$

We note that the coupling terms contain the second derivatives only. There are static solutions of the above system in the form

$$\theta = \pm\frac{n\pi}{2},\ n = 0,1,2,...,$$

$$u = -\alpha l_0 \cos\theta + (a_1 + b_1 x),\ v = -\alpha l_0 \sin\theta + (a_2 + b_2 x).$$

To derive solutions depending on one variable $\xi = x - vt$ only, we elimiante u'' and v'' from the first two equations and for β we have

$$(A + B\cos\beta)B'' + (C - \frac{1}{2}B\beta'')\sin\beta = 0,$$

where

$$A(v) = v^2 - \frac{1}{3}\gamma_{(1)}^2\left[1 + \frac{1}{2}\gamma_{(1)}^2\frac{\gamma_{(1)}^2 + 5v^2}{(v^2 + \frac{1}{2}\gamma_{(1)}^2)(v^2 - \gamma_{(1)}^2}\right],\ A(0) = 0,$$

$$B(v) = \frac{\gamma^2 v^4}{(v^2 + \frac{1}{2}\gamma_{(1)}^2)(v^2 - \gamma_{(1)}^2}, \quad C = \gamma^2\eta > 0,$$

with the solution

$$\pm\sqrt{2}\left(\xi - \xi_0\right) = \int_{\beta_0}^{\beta}\sqrt{\frac{A + B\cos\beta}{(E - C) + C\cos\beta}}\,d\beta, \qquad (2.9)$$

where E is the energy constant $E = \frac{1}{2}\beta_0'^2(A + B\cos\beta_0) + (1 - \cos\beta_0)C$. The right-hand side of (2.9) can be written in terms of the elliptic integral of the third kind. We denote here a solution in the particular case $A = -B, B < 0,\ E = 2C$ (i.e., $v^2 < \gamma_{(1)}^2$; this occurs for $\gamma^2 > \gamma_{(1)}^2$

$$\cos\frac{\beta}{2} = \cos\frac{\beta_0}{2}\exp\left[-\sqrt{\frac{C}{-2\beta}}\,|\xi - \xi_0|\right]$$

with the property $\cos\frac{\beta}{2} \to 0$ as $|\xi - \xi_0| \to \infty$; there is discontinuity of the first derivative at $\xi = \xi_0$ and the solution is monotonic for $\xi > \xi_0$ and (symmetrically) for

$\xi < \xi_0$. The condition $A(v) = -B(v)$ yields the following solution for v^2 :

$$v^2 = \frac{1}{2}\gamma^2_{(1)}\left[\left(\frac{5}{6} - \frac{\gamma^2}{\gamma^2_{(1)}}\right) + \sqrt{\left(\frac{5}{6} - \frac{\gamma^2}{\gamma^2_{(1)}}\right) + \frac{14}{3}}\right].$$

3 Interaction energy leading to soliton solutions

We recall that the interaction energy between dipoles at y and x has the form

$$\varphi^+_{y,x} = \frac{1}{2}(q^+)^2 K_0{}^2\, C : (u_y - u_x)^2 + \frac{1}{4}\,(c - 2e) : (\mu_x^2 + \mu_y^2) + \tag{3.1}$$

$$+ e : \mu_x\mu_y + g{:}(u_y - u_x)\,(\mu_y - \mu_x),$$

where

$$C = \nabla_{\rm R}\nabla_{\rm R}\varphi,\ \frac{1}{4}\,(c - 2e) = \frac{1}{2}(q^+)^2\left(\frac{\bar{m}}{m}\right)^2 K_0\, K_2\, \nabla_{\rm R}\nabla_{\rm R}\varphi\,,$$

$$e = -(q^+)^2\left(\frac{\bar{m}}{m}\right)^2 K_1{}^2\, \nabla_{\rm R}\nabla_{\rm R}\varphi\,,\ g = (q^+)^2\frac{\bar{m}}{m}\ K_0\, K_1\, \nabla_{\rm R}\nabla_{\rm R}\varphi\,.$$

It was derived by expansion in terms of the dipole length and the displacement difference.

Now let the above matrices be arbitrary constant matrices but subject to the following conditions: $C, c,\ e$ are symmetric (if e is symmetric, then c is also symmetric). We have, replacing $\varphi^{\rm T}_{y,x}$ by $\varphi^{\rm T}_{n+1,\,n} + \varphi^{\rm T}_{n,\,n-1}$ (since only the nearest-neighbour interaction is consdered)

$$\frac{\partial}{\partial u^i_n}(\varphi^{\rm T}_{n+1,\,n} + \varphi^{\rm T}_{n,\,n-1}) = -\,(q^+)^2 K_0{}^2\, C_{ip}\,\Delta^2 u^p_n - g_{ip}\,\Delta^2 \mu^p_n,$$

$$\frac{\partial}{\partial \mu^i_n}(\varphi^{\rm T}_{n+1,\,n} + \varphi^{\rm T}_{n,\,n-1}) = C_{ip}\,\mu^p_n + e_{ip}\,\Delta^2 \mu^p_n - g_{p\,i}\,\Delta^2 u^p_n,$$

Assuming, furthermore, that $g_{12} = g_{21} = 0$ and passing to the continuum we arrive at the equations of motion in the form

$$\ddot{u} - c^2_{(1)} C_{11} (u'' + \alpha_1 \mu_1'') = 0 ,$$

$$\ddot{v} - c^2_{(1)} C_{22} (v'' + \alpha_1 \mu_2'') = 0 , \tag{3.2}$$

$$(\mu_1 \ddot{\mu}_2 - \ddot{\mu}_1 \mu_2) + \frac{a^2}{I}(e_{22} \mu_1 \mu_2'' - e_{11} \mu_1'' \mu_2)$$

$$+ \frac{1}{I}(c_{22} - c_{11})\mu_1\mu_2 - \frac{a^2}{I} (g_{22}\mu_1 v'' - g_{11}\mu_2 u'') = 0,$$

where

$$c^2_{(1)} = \omega^2_{(1)} a^2, \ \alpha_1 = \frac{a^2 g_{11}}{m} \frac{1}{c^2_{(1)} C_{11}}, \ \alpha_2 = \frac{a^2 g_{22}}{m} \frac{1}{c^2_{(1)} C_{22}} .$$

Writing $\mu_1 = l_0 \cos \theta$, $\mu_2 = l_0 \sin \theta$ we have for the last equation

$$\ddot{\theta} + \frac{a^2}{I}\left[(e_{22} \cos^2 \theta + e_{11} \sin^2 \theta)\theta'' - (e_{22} - e_{11}) \sin \theta \cos \theta \, \theta'^2\right]$$

$$+ \frac{1}{I}(c_{22} - c_{11}) \sin \theta \cos \theta - \frac{a^2}{l_0 I}(g_{22} \cos \theta v'' - g_{11} \sin \theta u'') = 0. \tag{3.3}$$

We observe that in order to obtain for the uncoupled problem ($g_{11} = g_{22} = 0$) the sine-Gordon equation, we must assume that $e_{11} = e_{22}$; then the term $e : \mu_x \mu_y$ in (3.1) becomes the scalar product $e_{11} \mu_x \cdot \mu_y$ and the term containing θ'^2 vanishes. We set also $e_{11} = e_{22} = -e, e > 0$ whence

$$\theta'' - \frac{ea^2}{I}\theta'' + \frac{1}{2I}(c_{22} - c_{11}) \sin 2\theta - \frac{a^2}{l_0 I}(g_{22} \cos \theta v'' - g_{11} \sin \theta u'') = 0 .$$

To simplify the notation we set

$$c^2_{(1)} C_{11} = \gamma^2_{(1)} , \ c^2_{(1)} C_{22} = - \gamma^2_{(2)} , \ \frac{ea^2}{I} = \gamma^2 ,$$

$$\frac{1}{2I}(c_{22} - c_{11}) = \beta,\ \frac{a^2}{l_0 I}\, g_{22} = g_2\,,\ \frac{a^2}{l_0 I}\, g_{11} = g_1\,,$$

and assume for convenience $\beta > 0$. Thus, our system takes the final form

$$\begin{aligned} &\ddot{u} - \gamma_{(1)}^2 (u'' + \alpha_1 \mu_1'') = 0\,, \\ &\ddot{v} + \gamma_{(1)}^2 (v'' + \alpha_2 \mu_2'') = 0\,, \\ &\ddot{\theta} - \gamma^2 \theta'' + \beta \sin 2\theta - (g_2 \cos \theta v'' - g_1 \sin \theta u'') = 0\,. \end{aligned} \tag{3.4}$$

We observe that the above system has the exact solution $\theta = kx - \omega t$

$$\mu_1 = l_0 \cos(kx - \omega t)\,,\ \mu_2 = l_0 \sin(kx - \omega t)\,,$$

$$u = a \cos(kx - \omega t),\ v = b \sin(kx - \omega t)\,,$$

for which $\mu_1 \ddot{\mu}_2 - \ddot{\mu}_1 \mu_2 = 0,\ \mu_1 \mu_2'' - \mu_1'' \mu_2 = 0$ with the dispersion equation

$$(\omega^2 - \gamma_{(1)}^2 k^2)(\omega^2 + \gamma_{(1)}^2 k^2) -$$

$$- \frac{l_0 k^4}{2\beta}\left[g_1 \gamma_{(1)}^2 \alpha_1 (\omega^2 + \gamma_{(2)}^2 k^2) + g_2 \gamma_{(2)}^2 \alpha_2 (\omega^2 - \gamma_{(1)}^2 k^2) \right] = 0\,.$$

We now seek a solution of the system (3.4) depending on one variable $\xi = x - vt$; thus

$$u'' = \frac{\alpha_1 \gamma_{(1)}^2}{v^2 - \gamma_{(1)}^2} \mu_1'',\ v'' = - \frac{\alpha_1 \gamma_{(2)}^2}{v^2 - \gamma_{(2)}^2} \mu_2''\,,$$

$$A_{(2)} \mu_1 \mu_2'' + A_{(1)} \mu_1'' \mu_2 + \mu_1 \mu_2 = 0,$$

where

$$A_{(1)} = - \frac{1}{2\beta}\left[(v^2 - \gamma^2) + l_0 g_1 \frac{\alpha_1 \gamma_{(1)}^2}{\gamma_{(1)}^2 - v^2} \right].$$

$$A_{(2)} = -\frac{1}{2\beta}\Big[\,(v^2 - \gamma^2) + \log g_1 \frac{\alpha_2\,\gamma_{(2)}^2}{\gamma_{(2)}^2 - v^2}\Big].$$

In order to obtain the sine–Gordon equation we require that

$$A_{(2)} = -A_{(1)}$$

whence

$$v^2 = c_{(1)}^2\,C_{11}\,\frac{\frac{1}{2}g_1^2 + g_2^2}{g_2^2 - g_1^2}$$

Thus we must assume that the coupling constants satisfy the condition $g_2 > g_1$. Our equation now takes the form

$$\mu_1\mu_2'' - \mu_1''\,\mu_2 + \frac{1}{A_{(2)}}\mu_1\mu_2 = 0$$

i.e.,

$$\theta'' + \frac{1}{2A_{(2)}}\sin 2\theta = 0$$

or for $\varphi = 2\theta$,

$$\varphi'' + \frac{1}{A_{(2)}}\sin\varphi = 0\,.$$

We have arrived at the pendulum equation. Substituting for v^2 we have

$$A_{(2)} = \frac{c_{(1)}^2\,C_{11}}{2\beta}\Big[\frac{\frac{1}{2}g_1^2 + g_2^2}{g_2^2 - g_1^2} - \frac{ea^2}{I}\Big(1 + \frac{2l_0 I}{3me\,c_{(1)}^4}\,\frac{g_2^2 - g_1^2}{C_{11}^2}\Big)\Big].$$

Thus in this case, as well known, there are soliton solutions of the coupled problem (3.4).

Acknowledgements

This paper was written during the author's stay in the University of Stuttgart. I am grateful to this University for hospitality and to the Deutsche Forschungsgemeinschaft for the financial support.

References

[1] L. Brillouin, *Wave Propagation in Periodic Structures*. McGraw-Hill, New York, 1946.

[2] I.G. Kaplan, *Introduction to the Theory of Intermolecular Interactions* (in Russian). Moscow, Nauka, 1982.

[3] J. Pouget and G.A. Maugin, *Phys. Rev.* **B30**, 5306 (1984) and *Phys. Rev.* **B31**, 4633 (1985).

Zorski H.
I.P.P.T.-P.A.N.,
Department of Fluid Mechanics
Swietokrzyska 21
PL-00-049 Warsaw
POLAND